Polyphenols in Plants: Isolation, Purification and Extract Preparation

Polyphenols in Plants: Isolation, Purification and Extract Preparation

Edited by

Ronald Ross Watson
College of Public Health,
University of Arizona,
Tucson, AZ, USA

AMSTERDAM • BOSTON • HEIDELBERG • LONDON
NEW YORK • OXFORD • PARIS • SAN DIEGO
SAN FRANCISCO • SINGAPORE • SYDNEY • TOKYO

Academic Press is an imprint of Elsevier

Academic Press is an imprint of Elsevier
32 Jamestown Road, London NW1 7BY, UK
225 Wyman Street, Waltham, MA 02451, USA
525 B Street, Suite 1800, San Diego, CA 92101-4495, USA

First edition 2014

British Library Cataloguing-in-Publication Data
A catalogue record for this book is available from the British Library

Library of Congress Cataloging-in-Publication Data
A catalog record for this book is available from the Library of Congress

ISBN: 978-0-12-397934-6

For information on all Academic Press publications
visit our website at elsevierdirect.com

Typeset by TNQ Book and Journals
www.tnq.co.in

Contents

PART 1 MODIFICATION BY PLANT GROWTH AND ENVIRONMENT

SECTION 1A   STRESS AND POLYPHENOLS IN PLANTS

SECTION 1B   PLANT SYSTEMS OF POLYPHENOL
                MODIFICATION

PART 2   ISOLATION AND ANALYSIS OF POLYPHENOL STRUCTURE

SECTION 2A   ANALYSIS TECHNIQUES FOR POLYPHENOLS

Preface

Polyphenols are a unique group of phytochemicals present in fruits, vegetables and other plants. Their activity is based upon functional groups capable of accepting a free radical's negative charge. Polyphenols are being studied extensively for their health promoting and disease treating activities. This leads to substantial interest in environmental factors that change their concentration in plant sources as well as methods to isolate, describe and identify them. The latter is the primary focus of this book.

Two chapters review external factors that also produce different amounts in the plant. One describes phenolic compounds and saponins whose levels and relative composition change under different irrigation regimens and thus soil salt levels. Lichens, a very old type of plant that is used to investigate changes in the environment is described for such studies. Events in the plants' growth, stage of maturity and cultivar type change their relative amounts as reviewed in the book. Such factors are important for reproducible production of bioactive polyphenols that are consumed as part of the plant or isolated for use as a dietary supplement. Other writers describe plant polyphenol profiles as a method to trace and support biodiversity through characterization of these key compounds and their plant sources. The role of these compounds in controlling various activities of plants are reviewed, as their primary function is to benefit the producing plants, distinct from human's interest in their biomedical actions.

Plant systems modify and change polyphenols structure and concentration. Classification is based upon their chemical structure which contains multiple linked phenol groups. Such diverse chemical structures are the background and cause of polyphenols' many functions. The compounds' molecular weight ranges show small compounds in the molecular weight range of 3000 to very large compounds of 20,000. The latter are susceptible to biomodification by human as well as microbial enzymes providing challenges to identify bioactive compounds.

This knowledge is vital for other studies of biomedical benefits, to be able to identify various biologically important polyphenols. A major identification tool is gas chromatography-mass spectrometry and recent advances are reviewed. Improvements in this approach are described along with tools and methods for rapid, non-destructive measurements in fruits and vegetables. Finally novel techniques for identification are reviewed.

Methods to recover, purify, fractionate and finally isolate desired polyphenols are vital including specific examples of purification procedures to obtain biological activity. Of course not all polyphenols are isolatable from plants. Those that are extractable are the primary focus of researchers. The more novel and less well studied non-extractable polyphenols in foods need new methods of isolation, analysis and composition which are reviewed for future biomedical research.

Because of their structure polyphenols lend themselves to absorption to various resins and ion exchange chromatography with subsequent removal to purify and fractionate them. Chapters by those doing such studies, combined with a specific example of characterization of isolated polyphenols from *Hibiscus* flowers, are provided. Finally, in the commercial production of vegetables the effects of water combined with heat is described in hydrothermal processing of polyphenols.

Examples of polyphenol characterization techniques as well as occurrence in various plant materials are reviewed. Improved characterization of polyphenols using liquid chromatography with specific chapters using fruit polyphenols is included in the book as methods and examples. Different plant parts have distinct needs for varying amounts and quantities of polyphenols, thus offering researchers special raw material sources. The book ends with a determination of polyphenols and their related flavonoids' antioxidant activities in seeds whereas most chapters have focused on polyphenol composition in the leaf and fruit parts of plants.

Ronald Ross Watson

Acknowledgement

The editor acknowledges the efforts and time of the book's editorial assistant, Bethany L. Stevens. Her work made this book possible. She worked long hours reviewing and aiding the authors to get their chapters done in a timely fashion using the style and format required. The support of staff at Elsevier were extremely important in supporting Ms. Stevens interactions with the authors and getting the excellent reviews converted into text and published. A grant from Elsevier that helped support Ms. Stevens work is appreciated.

Contributors

Hanno Bährs
Humboldt-Universität zu Berlin, Laboratory of Freshwater & Stress Ecology, Berlin, Germany

Barbara Borczak
Department of Human Nutrition, Faculty of Food Technology, Agricultural University in Kraków, Kraków, Poland

Maria Fiorenza Caboni
Inter-Departmental Centre for Agri-Food Industrial Research (CIRI Agroalimentare), University of Bologna, Italy, Department of Agro-Food Sciences and Technologies, Alma Mater Studiorum Università di Bologna, Bologna, Italy

Natalia Campillo
Department of Analytical Chemistry, Faculty of Chemistry, University of Murcia, Murcia, Spain

Reinhold Carle
Hohenheim University, Institute of Food Science and Biotechnology, Stuttgart, Germany

Shumon Chakrabarti
Humboldt-Universität zu Berlin, Laboratory of Freshwater & Stress Ecology, Berlin, Germany

Dennis Dannehl
Faculty of Agriculture and Horticulture, Biosystems Engineering Division, Humboldt-Universität zu Berlin, Berlin, Germany

M Elena Díaz-Rubio
Department of Metabolism and Nutrition, Institute of Food Science, Technology and Nutrition, Madrid, Spain

Alberto Fernández-Gutiérrez
Department of Analytical Chemistry, University of Granada, Granada, Spain, Functional Food Research and Development Centre (CIDAF), Armilla (Granada), Spain

Shmuel Galili
Agricultural Research Organization, The Volcani Center, Bet Dagan, Israel

Ana Maria Gómez-Caravaca
Department of Analytical Chemistry, University of Granada, Granada, Spain, Functional Food Research and Development Centre (CIDAF), Armilla (Granada), Spain

Ran Hovav
Agricultural Research Organization, The Volcani Center, Bet Dagan, Israel

Olga Jáuregui
Scientific and Technical Services, University of Barcelona, Barcelona, Spain

Melanie Josuttis
Institut für Produktqualität, Berlin, Germany

Judith Kammerer
Hohenheim University, Institute of Food Science and Biotechnology, Stuttgart, Germany

Dietmar Rolf Kammerer
Hohenheim University, Institute of Food Science and Biotechnology, Stuttgart, Germany

Vassiliki G. Kontogianni
Section of Organic Chemistry and Biochemistry, Department of Chemistry, University of Ioannina, Ioannina, Greece

Rosa María Lamuela-Raventós
CIBEROBN Fisiopatología de la Obesidad y la Nutrición and RETIC, Instituto de Salud Carlos III, Spain, Nutrition and Food Science Department, Pharmacy School, University of Barcelona, Barcelona, Spain

Pauline Laue
Humboldt-Universität zu Berlin, Laboratory of Freshwater & Stress Ecology, Berlin, Germany, Lausitz University of Applied Sciences, Senftenberg, Germany

Matías Libuy
Molecular and Clinical Pharmacology Program, Institute of Biomedical Sciences, Faculty of Medicine, University of Chile, Santiago, Chile

Miriam Martínez-Huélamo
CIBEROBN Fisiopatología de la Obesidad y la Nutrición and RETIC, Instituto de Salud Carlos III, Spain, Nutrition and Food Science Department, Pharmacy School, University of Barcelona, Barcelona, Spain

Jara Pérez-Jiménez
Department of Metabolism and Nutrition, Institute of Food Science, Technology and Nutrition, Madrid, Spain

Paola Quifer-Rada
CIBEROBN Fisiopatología de la Obesidad y la Nutrición and RETIC, Instituto de Salud Carlos III, Spain, Nutrition and Food Science Department, Pharmacy School, University of Barcelona, Barcelona, Spain

Ramón Rodrigo
Molecular and Clinical Pharmacology Program, Institute of Biomedical Sciences, Faculty of Medicine, University of Chile, Santiago, Chile

Giuseppe Ruberto
Istituto del CNR di Chimica Biomolecolare, Catania, Italy

Leonardo Sabatino
University of Catania, Central Inspectorate Department of Protection and Prevention of Fraud quality of food products (ICQRF), Catania, Italy

Josline Y. Salib
Department of Chemistry of Tanning Materials, National Research Center, Dokki, Egypt

Fulgencio Saura-Calixto
Department of Metabolism and Nutrition, Institute of Food Science, Technology and Nutrition, Madrid, Spain

Andreas Schieber
Institute of Nutritional and Food Sciences, University of Bonn, Bonn, Germany

Nadine Schulze-Kaysers
Institute of Nutritional and Food Sciences, University of Bonn, Germany

Monica Scordino
University of Catania, Central Inspectorate Department of Protection and Prevention of Fraud quality of food products (ICQRF), Catania, Italy

Antonio Segura-Carretero
Department of Analytical Chemistry, University of Granada, Granada, Spain, Functional Food Research and Development Centre (CIDAF), Armilla (Granada), Spain

Gajendra Shrestha
Department of Biology and the M.L. Bean Life Science Museum, Brigham Young University, Provo, UT, USA

Elżbieta Sikora
Department of Human Nutrition, Faculty of Food Technology, Agricultural University in Kraków, Kraków, Poland

Laura Siracusa
Istituto del CNR di Chimica Biomolecolare, Catania, Italy

Larry L. St Clair
Department of Biology and the M.L. Bean Life Science Museum, Brigham Young University, Provo, UT, USA

Christian E.W. Steinberg
Humboldt-Universität zu Berlin, Laboratory of Freshwater & Stress Ecology, Berlin, Germany

Anna Vallverdú-Queralt
CIBEROBN Fisiopatología de la Obesidad y la Nutrición and RETIC, Instituto de Salud Carlos III, Spain, Nutrition and Food Science Department, Pharmacy School, University of Barcelona, Barcelona, Spain

Vito Verardo
Inter-Departmental Centre for Agri-Food Industrial Research (CIRI Agroalimentare), University of Bologna, Italy

Pilar Viñas
Department of Analytical Chemistry, Faculty of Chemistry, University of Murcia, Murcia, Spain

Fabian Weber
Institute of Nutritional and Food Sciences, University of Bonn, Bonn, Germany

Modification by Plant Growth and Environment

Cultivar and Production Effects on Bioactive Polyphenols

1

Dennis Dannehl*, Melanie Josuttis†

*Faculty of Agriculture and Horticulture, Biosystems Engineering Division, Humboldt-Universität zu Berlin, Berlin, Germany, †Institut für Produktqualität, Berlin, Germany

CHAPTER OUTLINE HEAD

1.1 Introduction

The demand for improved quality control of horticultural products is increasing among consumers. Several years ago, the external qualities of fruit and vegetables were the main factors that influenced buyer behavior. However, buyer behavior has recently been changing. The population's increased awareness of health contributes to the fact that, in addition to conventional quality assessment, the nutritional quality of fruit and vegetables is gaining in importance for consumers. One of the reasons for this increased health awareness in society is the apparent relationship between the intake of fruit and vegetables and numerous health benefits for consumers. The reasons for these positive effects are attributed to different phytochemicals with antioxidant properties especially phenolic compounds, which can detoxify reactive oxygen species in the human body (Bazzano *et al.*, 2002). Therefore, industrialized countries are encouraged to develop technical and control methods to guarantee the production of the quantity and quality of food and food products, in a sustainable way. To realize these goals, different methods to increase the levels of secondary metabolites in cell cultures and intact fruit, as well as vegetables, have been applied in the horticultural sector. In order to represent the diversity of these scientific approaches, the objective of this chapter is to review a selection of several studies.

Polyphenols in Plants. http://dx.doi.org/10.1016/B978-0-12-397934-6.00001-2

1.2 Effects of plant species, cultivar and breeding success

Phenolic compounds are classified into different groups and their occurrence in plants primarily depends on the plant species. High amounts of phenolic acids (hydroxycinnamic and hydroxybenzoic acids) are contained in kale (12 mg/100 g fresh weight [FW]) and wheat (50 mg/100 g FW) (Hermann, 2001; Watzl and Leitzmann, 2005). High contents of anthocyanins were found in blackcurrant (250 mg/100 g FW), in blueberry (50 mg/100 g FW), and cherry (275 mg/100 g FW) (Mazza and Miniati, 1993). As an example for flavonols, different levels of quercetin were analysed in the peel of apple (14 mg/100 g FW), in pear (2.8 mg/100 g FW), in lettuce (4.5 mg/100 g FW), and in tomatoes (0.8 mg/100 g FW), and the levels of kaempferol (0.3 mg/100 g FW, 1.2 mg/100 g FW, <0.1 mg/100 g FW, and 0.01 mg/100 g FW) also were determined in these fruit and vegetables, respectively (Herrmann, 1976; Hertog *et al.*, 1992; Krause and Galensa, 1992; Souci *et al.*, 2008). Furthermore, polymeric phenolics such as condensed or hydrolysed tannins occur in fruit and nuts, as reviewed by Serrano *et al.* (2009).

Besides the plant species, numerous investigations showed huge differences in secondary plant components even among cultivars of one species. Therefore, conventional breeding could be a potential approach to increase the accumulation of phenolic compounds. Comparing different studies, a 14-fold difference in total phenol content between 30 cultivars of tomatoes was established (Bahorun *et al.*, 2004; Frusciante *et al.*, 2007; Kaur and Kapoor, 2002; Marsic *et al.*, 2011). Il Park *et al.* (2011) analysed a 149-fold difference in total anthocyanin content in the skin of three cultivars of radish, where the color of the skin ranged between white and red. Atkinson *et al.* (2006) reported more than a five-fold difference in total ellagic acid between 44 strawberry cultivars and Wang and Lin (2000) observed a 1.6-fold difference in total phenolics of eight strawberry cultivars. In studies conducted on fruits of jujube, eggplant, blueberry, apple, and raspberry, 1.3-fold, 6-fold, 10-fold, 5-fold, and 1.8-fold differences in total phenol content were detected, respectively (Anttonen and Karjalainen, 2005; Ehlenfeldt and Prior, 2001; Lata *et al.*, 2005; Prohens *et al.*, 2007; San and Yildirim, 2010).

1.3 Light effects

The quantity and quality of light play a major role in terms of the accumulation of secondary plant compounds in fruit and vegetables. Ju *et al.* (1999) covered the orchard floor with different reflecting films to enhance the content of anthocyanin in apple skin. The light intensity inside the tree canopy increased from 30% daylight to 68% daylight using by crinkled aluminium foil. This higher light exposure resulted in a two-fold increase of anthocyanin in "Fuji" apple skin, whereas the total flavonoid content was not affected compared with the control fruit. Awad *et al.* (2001) found that the skin of "Jonagold" apples from the top of the canopy, which received up to 62% more light, contained a 27.5-fold higher content of cyanidin 3-galactoside and a two-fold higher accumulation of total flavonoids than those from the canopy interior. Similar results were detected by Merzlyak *et al.* (2002), who compared shaded and sunlit fruit, and found an increase in amounts of phenolics (by 100%) in sunlit skin of "Zhigulevskoye" apples. However, shaded and non-shaded strawberry plants showed no difference in total phenolic content in their fruit, but a small reduction in anthocyanins (by 9%) was found in the shaded fruit (Anttonen *et al.*, 2006). Furthermore, the accumulation of total phenolics in lettuce plants was approximately three-fold higher than in control plants after a one-day exposure to high photosynthetic active radiation at $800\,\mu mol/m^2 s$ (Myung-Min *et al.*, 2009). Wilkens *et al.* (1996) reported a four-fold increase in chlorogenic acid and rutin in tomatoes which were exposed to a 40% higher light intensity. These accumulations of secondary plant compounds may be attributed to photooxidative stress or to increased photosynthesis, providing an elevated carbon supply for biosynthesis of phytochemicals (Poiroux-Gonord *et al.*, 2010; Treutter, 2010).

Protected cultivation of crops is a common method, not only for out-of-season production, but for producing other advantages as well. Ordinary plastic films block UV-B radiation. Compared to those films, the positive effects were achieved using UV-B transparent plastic films during pre-harvest processes, which induced an accumulation of phenolics in lettuce and tomatoes but also a growth inhibition of the lettuce (Garcia-Macias *et al.*, 2007; Luthria *et al.*, 2006). The contents of caffeic acid, cyanidin, and quercetin for lettuce cultivated under the highest UV exposure plastic films increased up to 79%, 381%, and 349%, respectively. A similar trend was observed regarding the levels of caffeic acids, *p*-coumaric acid and ferulic acid in tomatoes, which were approximately 20% higher with the influence of an UV-B translucent cover material (Luthria *et al.*, 2006). Furthermore, Josuttis *et al.* (2010) have demonstrated that individual phenolics in strawberries were affected in the absence of UV radiation. In comparison to an UV-B blocking plastic film, the contents of cyanidin 3-glucoside, quercetin 3-glucoronide, and kaempferol 3-glucoside in strawberry fruit were increased up to 64%, 609%, and 59%, respectively, when they were grown in a tunnel covered with an UV-B transparent (70%) plastic film. However, the content of all other phenolics, total anthocyanins, and the antioxidant capacity were not affected by the different treatments. Comparable results were shown for raspberry, blueberry, and strawberry fruit by Ordidge *et al.* (2010), who compared three films differing in UV-B transparency.

1.4 **Effects of growing temperature**

Regarding different temperatures, alterations in levels of secondary metabolites can occur depending on the specific fruit or vegetable. A daily mean temperature of 27.5°C inhibited the synthesis of the anthocyanin content in chicory by over 90% compared to a mean temperature of 12.5°C (Boo *et al.*, 2006). Yamane *et al.* (2006) found that the anthocyanin accumulation in grape berry skins was significantly higher after two-week treatments at 20°C rather than at 30°C. In this context, only 50% of the anthocyanin content in grape skin was analysed at a temperature of 35°C compared with those matured at a temperature of 25°C (Mori *et al.*, 2007).

In contrast, Wang and Zheng (2001) demonstrated that the content of pelargonidin 3-glucoside in the fruit of strawberry cultivars "Kent" and "Earliglow" increased 2.7-fold with elevated temperatures ranging from (day/night) 18/12°C to 30/22°C. This temperature effect for strawberry phenolics was confirmed by Josuttis *et al.* (2011). Remberg *et al.* (2010) showed that anthocyanins in raspberry were also increased by elevated growing temperatures. However, this result was not valid for each anthocyanin. Cyanidin-3-*O*-sophoroside, one of the major anthocyanins in raspberry, was enhanced by lower temperature. In tomatoes an increase in the total phenol content by higher temperature was oberved during ripening processes between 18°C and 22°C (Toor *et al.*, 2006), as well as 20°C and 22°C (Dannehl *et al.*, 2012a). All these studies showed that high and low temperatures change the synthesis of secondary metabolites in fruit and vegetables in a species-specific way.

1.5 Impact of water management

Regarding the effects of water deficiency on phenolic compounds in various plants, there are conflicting results from different studies that give negative and positive effects. Servili *et al.* (2007) reported that wide differences in concentrations of total phenolics were analysed in virgin olive oil, when the olive trees were either fully irrigated or irrigated to a lesser extent. The latter treatment resulted in a pronounced increase by 134%, whereas the fruit fresh weight was reduced by 44%. Recently, Shinohara *et al.* (2011) found that heads of globe artichoke plants grown under drought stress produced high amounts of phenolic stress metabolites. The highest levels of total phenolics and chlorogenic acid in artichoke heads were achieved when the water supply was reduced by 50%, where these plant responses were accompanied by a yield reduction of up to 35%. Similar results were obtained concerning grape berries (Esteban *et al.*, 2001). The content of total phenolics and total anthocyanins were observed to be higher in the skin of grape berries from non-irrigated plants, whereas greater water availability led to a five-fold yield increase. Other results indicated that water stress at 50% field capacity resulted in a reduced accumulation of phenolic compounds (e.g., hydroxycinnamic acids and flavonoids) in tomatoes "Josefina" (Sanchez-Rodriguez *et al.*, 2011). However, the same investigations showed that a moderate water stress can increase PAL activity and synthesized phenolics in more tolerant tomato cultivars, e.g., "Zarina." From these observations, it is concluded that the accumulation of secondary metabolites as a result of water stress is not only species-specific but also cultivar-specific.

Because the levels of relative humidity (RH) can affect the stomata of plants and, thus, the photosynthetic activity as well as the plant transpiration, this climatic factor may cause a stress-induced accumulation of secondary metabolites. Nevertheless, there are only a small handful of studies regarding the effects of RH on phytochemicals in plants during plant production. It was reported that a lower level of RH led to abiotic stress, which triggered an antioxidant response based on higher concentrations of phenolic compounds in tomatoes (Rosales *et al.*, 2011).

1.6 **Effects of elevated carbon dioxide concentrations**

Generally, elevated atmospheric CO_2 can increase photosynthetic activity, which can promote the synthesis of carbohydrates in plants, e.g., in leaves and heads of broccoli plants (Krumbein *et al.*, 2010; Schmidt *et al.*, 2008). The resulting higher carbon supply can accelerate plant growth and can provide resources for carbon-based secondary plant compounds (Dannehl *et al.*, 2012a; Treutter, 2010). In this context, CO_2 levels maintained at 1000 ppm led to a pronounced increase in PAL activity and in the content of chlorogenic acid in the upper leaves of tobacco plants (Matros *et al.*, 2006). Wang *et al.* (2003) treated strawberry plants using CO_2 concentrations ranging from 350 to 950 ppm. The higher CO_2 growing conditions enhanced the content of quercetin 3-glucoside (by 300%), kaempferol 3-glucoside (by 61%), and cyanidin 3-glucoside (by 105%) in strawberries compared with those grown under 350 ppm. Bindi *et al.* (2001b) used three different free air CO_2 enrichment arrays (350, 550, and 700 ppm) to fumigate grapevine plants, where this fumigation was realized using polyethylene pipes positioned on the soil surface (Bindi *et al.*, 2001a). Grapes exposed to 550 ppm changed the concentrations of total flavonoids and anthocyanins in wine, which increased by 38% and 41%, respectively. However, no significant differences in secondary metabolites were detectable among grapes treated with 550 or 700 ppm. Furthermore, Dannehl *et al.* (2012a) found a higher level of phenolic compounds in tomatoes, which were increased by 16% when the fruit were treated with a maximum mean CO_2 difference of 177 ppm.

1.7 **Macro- and micronutrients and heavy metals induce changes in polyphenols**

Several studies showed that fertilization may affect secondary metabolites in plants. For example, calcium deficiency induced the accumulation of anthocyanin in red-cabbage seedlings and stimulated the formation of naringenin-7-glucoside in shoots of *Prunus avium* L., which were grown *in vitro* (Bassim and Pecket, 1975; Yuri *et al.*, 1990). However, Fanasca *et al.* (2006) found that the content of caffeic acid in tomatoes remained unaffected using the same fertilization. Durst (1976) demonstrated that manganese increased the PAL activity in artichoke tuber tissue and Fanasca *et al.* (2006) showed that an increased magnesium supply in the nutrient solution led to higher levels of caffeic acid in tomatoes. However, blueberry plants treated with an excess of boron revealed a reduced concentration of phenolic compounds in their fruit, probably because boron is able to inactivate polyphenols and favours the formation of pectins (Eichholz *et al.*, 2011). Furthermore, iron application can induce positive effects on the fruit yield of lemon trees but not on the bioactive phenolics in lemon juice (Mellisho *et al.*, 2011). An average decrease in the total phenol content by 33% was observed. Vallejo *et al.* (2003) investigated the effects of poor (15 kg/ha) and rich (150 kg/ha) sulfur fertilization levels on phenolic compounds in different broccoli cultivars, which were grown in the spring. In comparison with the poorly fertilized broccoli plants, it was assumed that rich sulfur fertilization induced stress, whereby the contents of total flavonoids and chlorogenic acid in broccoli were increased by up to 142% and 91%, respectively. However, a phosphate deficiency led to an enrichment of total flavonoids in tomatoes (Zornoza and Esteban, 1984). Zagoskina *et al.* (2007) treated root-, stem-, and leaf-derived callus cultures of tea plants with a high concentration of cadmium. The results showed that the fresh weight of stem and leaf calli was decreased by 50% compared with the control cultures, whereas these negative responses were accompanied by enhanced levels of phenolic compounds in root (by 50%) and stem (by 87%) calli. Another study demonstrated that the total phenolic content increased by 5%, 44%, 72%, or 102% in leaves of *Vallisneria natans*, which were exposed to lead concentrations of 10 micromole (μM), 25 μM, 50 μM, or 75 μM, respectively (Wang *et al.*, 2011).

Moreover, the effect of nitrogen (N) fertilization can generally be considered as negative regarding the accumulation of phenolic compounds in plants. In this context, Radi *et al.* (2003) found that apricot fruit from trees fertilized with 80 kg/N ha contained 60% more total phenolics and a reduced fresh weight, than those fertilized with 150 kg/N ha. Approximately the same application levels of N were used in experiments with "Elstar" apples, where a reduction in the contents of total flavonoids (by 18%) and chlorogenic acid (by 5%) in apple skin was observed at higher N fertilization (Awad and de Jager, 2002). Similar results were found in terms of contents of quercetin and kaempferol in tomato fruit skins, which were exposed to high concentrations of N and harvested in May (Stewart *et al.*, 2001). However, Benard *et al.* (2009) demonstrated that different N supplies did not significantly affect the content of chlorogenic acid in tomatoes. These results are supported by Gravel *et al.* (2010). Another study revealed no effects on total phenolics in red pepper, when treatments were increased from a low nitrate level (4 mol NO_3^-/m³) to a high nitrate level (20 mol NO_3^-/m³) (Flores *et al.*, 2004).

Finally, several studies refer to the effects of high salinity on secondary metabolites in different plants, with results ranging from positive to negative effects. As such, Keutgen *et al.* (2007) found that salt stress increased the contents of anthocyanins and total phenolics in the fruit of the strawberry cultivar "Korona" by approximately 22% compared with the control fruit. Similar results were analysed regarding phenolics in tomatoes, whereas the FW of these fruit was reduced dramatically (Krauss *et al.*, 2006). However, Anttonen *et al.* (2006) detected that the levels of quercetin, kaempferol, and FW in "Bounty" strawberries decreased with increasing values of the electrical conductivity (EC) in the nutrient solution. Salt treated lettuce plants responded in the same manner, resulting in reduced concentrations of chlorogenic and caffeic acid (Kim *et al.*, 2008).

1.8 The application of electricity as a new approach in the horticultural sector

Knowledge of the effects of electricity on secondary plant compounds is limited. The relevance of pulsed electric field (PEF) technology in food- and biotechnology has increased substantially during the last few years (Janositz and Knorr, 2010). PEF technology consists of the application of short, high-power electrical pulses to products placed in a treatment chamber, confined between electrodes (Soliva-Fortuny et al., 2009). In basic research it was found that low PEF treatments can induce changes on a plant cellular level followed by stress reactions, e.g., the accumulation of secondary metabolites (Janositz et al., 2010; Soliva-Fortuny et al., 2009). In comparison with the reference samples, Guderjan et al. (2005) demonstrated that the level of isoflavonoids in soybeans increased by 20.5% after a PEF treatment (0.6 kV/cm) in a range of reversible cell permeabilization. However, other experimental setups permit an intermittent-direct-electric-current (IDC) flow through whole plants, using electrodes fixed to plant components and their growth medium. Currently, in preharvest processes, electricity is incorporated in order to enhance secondary plant compounds in various plant species (Dannehl et al., 2009; Dannehl et al., 2012b). Compared with untreated plants, an enhancement of phenolic compounds (by 12.8%) and associated antioxidant activity (by 9.1%) was analysed in radish tubers treated with an IDC of 1000 mA for 1 h per day (Dannehl et al., 2009). In particular, the anthocyanin content was increased by 28% but only in plant segments that were positioned in the IDC flow. However, a further IDC increase of up to 1400 mA was sufficient to promote the formation of phenolic compounds and associated antioxidant activity in garden cress sprouts by 16% and 18%, respectively (Dannehl et al., 2012b). In this context, the scientists concluded that IDC elicited the biosynthesis of secondary metabolites in a broad range of plant species without adverse effects on the physiological functions of the cells or tissue.

References

Anttonen, M.J., Karjalainen, R.O., 2005. Environmental and genetic variation of phenolic compounds in red raspberry. J. Food Compos. Anal. 18, 759–769.

Anttonen, M.J., Hoppula, K.I., Nestby, R., Verheul, M.J., Karjalainen, R.O., 2006. Influence of fertilization, mulch color, early forcing, fruit order, planting date, shading, growing environment, and genotype on the contents of selected phenolics in strawberry (Fragaria x ananassa Duch.) fruits. J. Agric. Food Chem. 54, 2614–2620.

Atkinson, C.J., Dodds, P.A.A., Ford, Y.Y., Le Miere, J., Taylor, J.M., Blake, P.S., et al., 2006. Effects of cultivar, fruit number and reflected photosynthetically active radiation on Fragaria x ananassa productivity and fruit ellagic acid and ascorbic acid concentrations. Ann. Bot. 97, 429–441.

Awad, M.A., Wagenmakers, P.S., de Jager, A., 2001. Effects of light on flavonoid and chlorogenic acid levels in the skin of "Jonagold" apples. Sci. Hortic. Amsterdam. 88, 289–298.

Awad, M.A., de Jager, A., 2002. Relationships between fruit nutrients and concentrations of flavonoids and chlorogenic acid in "Elstar" apple skin. Sci. Hortic. Amsterdam. 92, 265–276.

Bahorun, T., Luximon-Ramma, A., Crozier, A., Aruoma, O.I., 2004. Total phenol, flavonoid, proanthocyanidin and vitamin C levels and antioxidant activities of Mauritian vegetables. J. Sci. Food Agric. 84, 1553–1561.

Bassim, T.A.H., Pecket, R.C., 1975. The effect of membrane stabilizers on phytochrome-controlled anthocyanin biosynthesis in Brassica oleraceae. Phytochemistry 14, 731–733.

Bazzano, L.A., He, J., Ogden, L.G., Loria, C.M., Vupputuri, S., Myers, L., et al., 2002. Fruit and vegetable intake and risk of cardiovascular disease in US adults: the first National Health and Nutrition Examination Survey Epidemiologic Follow-up Study. Am. J. Clin. Nutr. 76, 93–99.

Benard, C., Gautier, H., Bourgaud, F., Grasselly, D., Navez, B., Caris-Veyrat, C., et al., 2009. Effects of Low Nitrogen Supply on Tomato (*Solanum lycopersicum*) Fruit Yield and Quality with Special Emphasis on Sugars, Acids, Ascorbate, Carotenoids, and Phenolic Compounds. J. Agric. Food Chem. 57, 4112–4123.

Bindi, M., Fibbi, L., Lanini, M., Miglietta, F., 2001a. Free Air CO_2 Enrichment (FACE) of grapevine (*Vitis vinifera* L.): I. Development and testing of the system for CO_2 enrichment. Eur. J. Agron. 14, 135–143.

Bindi, M., Fibbi, L., Miglietta, F., 2001b. Free Air CO_2 Enrichment (FACE) of grapevine (*Vitis vinifera* L.): II. Growth and quality of grape and wine in response to elevated CO_2 concentrations. Eur. J. Agron. 14, 145–155.

Boo, H.O., Chon, S.U., Lee, S.Y., 2006. Effects of temperature and plant growth regulators on anthocyanin synthesis and phenylalanine ammonia-lyase activity in chicory (*Cichorium intybus* L.). J. Hort. Sci. Biotech. 81, 478–482.

Dannehl, D., Huyskens-Keil, S., Eichholz, I., Ulrichs, C., Schmidt, U., 2009. Effects of intermittent-direct-electric-current (IDC) on polyphenols and antioxidant activity in radish (*Raphanus sativus* L.) during growth. J. Appl. Bot. Angew. Bot. 83, 54–59.

Dannehl, D., Huber, C., Rocksch, T., Huyskens-Keil, S., Schmidt, U., 2012a. Interactions between changing climate conditions in a semi-closed greenhouse and plant development, fruit yield, and health-promoting plant compounds of tomatoes. Sci. Hortic. Amsterdam. 138, 235–243.

Dannehl, D., Huyskens-Keil, S., Wendorf, D., Ulrichs, C., Schmidt, U., 2012b. Influence of intermittent-direct-electric-current (IDC) on phytochemical compounds in garden cress during growth. Food Chem. 131, 239–246.

Durst, F., 1976. The correlation of phenylalanine ammonia-lyase and cinnamate 4-hydroxylase activity in Jerusalem artichoke tuber tissue. Planta 132, 221–227.

Ehlenfeldt, M.K., Prior, R.L., 2001. Oxygen radical absorbance capacity (ORAC) and phenolic and anthocyanin concentrations in fruit and leaf tissues of highbus blueberry. J. Agric. Food Chem. 49, 2222–2227.

Eichholz, I., Huyskens-Keil, S., Kroh, L.W., Rohn, S., 2011. Phenolic compounds, pectin and antioxidant activity in blueberries (*Vaccinium corymbosum* L.) influenced by boron and mulch cover. J. Appl. Bot. Angew. Bot. 84, 26–32.

Esteban, M.A., Villanueva, M.J., Lissarrague, J.R., 2001. Effect of irrigation on changes in the anthocyanin composition of the skin of cv Tempranillo (*Vitis vinifera* L) grape berries during ripening. J. Sci. Food Agric. 81, 409–420.

Fanasca, S., Colla, G., Maiani, G., Venneria, E., Rouphael, Y., Azzini, E., et al., 2006. Changes in antioxidant content of tomato fruits in response to cultivar and nutrient solution composition. J. Agric. Food Chem. 54, 4319–4325.

Flores, P., Navarro, J.M., Garrido, C., Rubio, J.S., Martinez, V., 2004. Influence of Ca^{2+}, K^+ and NO_3-fertilization on nutritional quality of pepper. J. Sci. Food Agric. 84, 569–574.

Frusciante, L., Carli, P., Ercolano, M.R., Pernice, R., Di Matteo, A., Fogliano, V., et al., 2007. Antioxidant nutritional quality of tomato. Mol. Nutr. Food Res. 51, 609–617.

Garcia-Macias, P., Ordidge, M., Vysini, E., Waroonphan, S., Battey, N.H., Gordon, M.H., et al., 2007. Changes in the flavonoid and phenolic acid contents and antioxidant activity of red leaf lettuce (Lollo Rosso) due to cultivation under plastic films varying in ultraviolet transparency. J. Agric. Food Chem. 55, 10168–10172.

Gravel, V., Blok, W., Hallmann, E., Carmona-Torres, C., Wang, H., Van de Peppel, A., et al., 2010. Differences in N uptake and fruit quality between organically and conventionally grown greenhouse tomatoes. Agron. Sustainable Dev. 30, 797–806.

Guderjan, M., Topfl, S., Angersbach, A., Knorr, D., 2005. Impact of pulsed electric field treatment on the recovery and quality of plant oils. J. Food Eng. 67, 281–287.

Herrmann, K., 1976. Contents and localization of phenolics in vegetables. Qualitas Plantarum-Plant Foods for Human Nutrition 25, 231–245.

Hermann, K., 2001. Inhaltsstoffe von Obst und Gemüse Verlag Eugen Ulmer. Germany, Stuttgart.

Hertog, M.G.L., Hollman, P.C.H., Katan, M.B., 1992. Content of potentially anticarcinogenic flavonoids of 28 vegetables and 9 fruits commonly consumed in the Netherlands. J. Agric. Food Chem. 40, 2379–2383.

Il Park, N., Xu, H., Li, X., Jang, I.H., Park, S., Ahn, G.H., et al., 2011. Anthocyanin Accumulation and Expression of Anthocyanin Biosynthetic Genes in Radish (*Raphanus sativus*). J. Agric. Food Chem. 59, 6034–6039.

Janositz, A., Knorr, D., 2010. Microscopic visualization of Pulsed Electric Field induced changes on plant cellular level. Innovative Food Sci. Emerg. Technol. 11, 592–597.

Josuttis, M., Dietrich, H., Treutter, D., Will, F., Linnemannstons, L., Kruger, E., 2010. Solar UVB Response of Bioactives in Strawberry (*Fragaria x ananassa* Duch. L.): A Comparison of Protected and Open-Field Cultivation. J. Agric. Food Chem. 58, 12692–12702.

Josuttis, M., Dietrich, H., Patz, C.D., Krüger, E., 2011. Effects of air and soil temperatures on the chemical composition of fruit and agronomic performance in strawberry (*Fragaria x ananassa* Duch.). J. Hort. Sci. Biotech. 86, 415–421.

Ju, Z., Duana, Y., Ju, Z., 1999. Effects of covering the orchard floor with reflecting films on pigment accumulation and fruit coloration in "Fuji" apples. Sci. Hortic. Amsterdam. 82, 47–56.

Kaur, C., Kapoor, H.C., 2002. Anti-oxidant activity and total phenolic content of some Asian vegetables. Int. J. Food Sci. Technol. 37, 153–161.

Keutgen, A.J., Pawelzik, E., 2007. Modifications of strawberry fruit antioxidant pools and fruit quality under NaCl stress. J. Agric. Food Chem. 55, 4066–4072.

Kim, H.-J., Fonseca, J.M., Choi, J.-H., Kubota, C., Kwon, D.Y., 2008. Salt in irrigation water affects the nutritional and visual properties of romaine lettuce (*Lactuca sativa* L.). J. Agric. Food Chem. 56, 3772–3776.

Krause, M., Galensa, R., 1992. Bestimmung von Naringenin und Naringenin-Chalkon in Tomatenschalen mit RP-HPLC nach Festphasenextraktion. Z. Lebensm. Unters. For. 194, 29–32.

Krauss, S., Schnitzler, W.H., Grassmann, J., Woitke, M., 2006. The influence of different electrical conductivity values in a simplified recirculating soilless system on inner and outer fruit quality characteristics of tomato. J. Agric. Food Chem. 54, 441–448.

Krumbein, A., Klaering, H.-P., Schonhof, I., Schreiner, M., 2010. Atmospheric Carbon Dioxide Changes Photochemical Activity, Soluble Sugars and Volatile Levels in Broccoli (*Brassica oleracea* var. italica). J. Agric. Food Chem. 58, 3747–3752.

Lata, B., Przeradzka, M., Binkowska, M., 2005. Great differences in antioxidant properties exist between 56 apple cultivars and vegetation seasons. J. Agric. Food Chem. 53, 8970–8978.

Luthria, D.L., Mukhopadhyay, S., Krizek, D.T., 2006. Content of total phenolics and phenolic acids in tomato (*Lycopersicon esculentum* Mill.) fruits as influenced by cultivar and solar UV radiation. J. Food Compos. Anal. 19, 771–777.

Marsic, N.K., Gasperlin, L., Abram, V., Budic, M., Vidrih, R., 2011. Quality parameters and total phenolic content in tomato fruits regarding cultivar and microclimatic conditions. Turk. J. Agric. For. 35, 185–194.

Matros, A., Amme, S., Kettig, B., Buck-Sorlin, G.H., Sonnewald, U., Mock, H.P., 2006. Growth at elevated CO_2 concentrations leads to modified profiles of secondary metabolites in tobacco cv. SamsunNN and to increased resistance against infection with potato virus Y. Plant Cell Environ. 29, 126–137.

Mazza, G., Miniati, E., 1993. Anthocyanins in fruits, vegetables, and grains. CRC, Boca Raton, USA.

Mellisho, C.D., Gonzalez-Barrio, R., Ferreres, F., Ortuno, M.F., Conejero, W., Torrecillas, A., et al., 2011. Iron deficiency enhances bioactive phenolics in lemon juice. J. Sci. Food Agric. 91, 2132–2139.

Merzlyak, M.N., Solovchenko, A.E., Chivkunova, O.B., 2002. Patterns of pigment changes in apple fruits during adaptation to high sunlight and sunscald development. Plant Physiol. Biochem. 40, 679–684.

Mori, K., Goto-Yamamoto, N., Kitayama, M., Hashizume, K., 2007. Loss of anthocyanins in red-wine grape under high temperature. J. Exp. Bot. 58, 1935–1945.

Myung-Min, O., Carey, E.E., Rajashekar, C.B., 2009. Environmental stresses induce health-promoting phytochemicals in lettuce. Plant Physiol. Biochem. 47, 578–583.

Ordidge, M., Garc¡a-Marc¡as, P., Battey, N.H., Gordon, M.H., Hadley, P., John, P., et al., 2010. Phenolic contents of lettuce, strawberry, raspberry, and blueberry crops cultivated under plastic films varying in ultraviolet transparency. Food Chem. 119, 1224–1227.

Poiroux-Gonord, F., Bidel, L.P.R., Fanciullino, A.-L., Gautier, H., Lauri-Lopez, F., Urban, L., 2010. Health Benefits of Vitamins and Secondary Metabolites of Fruits and Vegetables and Prospects To Increase Their Concentrations by Agronomic Approaches. J. Agric. Food Chem. 58, 12065–12082.

Prohens, J., Rodriguez-Burruezo, A., Raigon, M.D., Nuez, F., 2007. Total phenolic concentration and browning susceptibility in a collection of different varietal types and hybrids of eggplant: Implications for breeding for higher nutritional quality and reduced browning. J. Am. Soc. Hortic. Sci. 132, 638–646.

Radi, M., Mahrouza, M., Jaouad, A., Amiot, M.J., 2003. Influence of mineral fertilization (NPK) on the quality of apricot fruit (cv. Canino). The effect of the mode of nitrogen supply. Agronomie 23, 737–745.

Remberg, S.F., Sonsteby, A., Aaby, K., Heide, O.M., 2010. Influence of postflowering temperature on fruit size and chemical composition of glen ample raspberry (*Rubus idaeus* L.). J. Agric. Food Chem. 58, 9120–9128.

Rosales, M.A., Cervilla, L.M., Sanchez-Rodriguez, E., Rubio-Wilhelmi, M.D., Blasco, B., Rios, J.J., et al., 2011. The effect of environmental conditions on nutritional quality of cherry tomato fruits: evaluation of two experimental Mediterranean greenhouses. J. Sci. Food Agric. 91, 152–162.

San, B., Yildirim, A.N., 2010. Phenolic, alpha-tocopherol, beta-carotene and fatty acid composition of four promising jujube (*Ziziphus jujuba* Miller) selections. J. Food Compos. Anal. 23, 706–710.

Sanchez-Rodriguez, E., Moreno, D.A., Ferreres, F., del Mar Rubio-Wilhelmi, M., Manuel Ruiz, J., 2011. Differential responses of five cherry tomato varieties to water stress: Changes on phenolic metabolites and related enzymes. Phytochemistry 72, 723–729.

Schmidt, U., Huber, C., Rocksch, T., 2008. Evaluation of combined application of fog system and CO_2 enrichment in greenhouses by using phytomonitoring data. Acta Hort. 801, 1301–1308.

Serrano, J., Puupponen-Pimi, R., Dauer, A., Aura, A.M., Saura-Calixto, F., 2009. Tannins: Current knowledge of food sources, intake, bioavailability and biological effects. Mol. Nutr. Food Res. 53, 310–329.

Servili, M., Esposto, S., Lodolini, E., Selvaggini, R., Taticchi, A., Urbani, S., et al., 2007. Irrigation effects on quality, phenolic composition, and selected volatiles of virgin olive oils cv. Leccino. J. Agric. Food Chem. 55, 6609–6618.

Shinohara, T., Agehara, S., Yoo, K.S., Leskovar, D.I., 2011. Irrigation and Nitrogen Management of Artichoke: Yield, Head Quality, and Phenolic Content. Hortscience 46, 377–386.

Soliva-Fortuny, R., Balasa, A., Knorr, D., Martin-Belloso, O., 2009. Effects of pulsed electric fields on bioactive compounds in foods: a review. Trends Food Sci. Tech. 20, 544–556.

Souci, S.W., Fachmann, W., Kraut, H., 2008. Food composition and nutrition tables Taylor & Francis. Germany, Stuttgart.

Stewart, A.J., Chapman, W., Jenkins, G.I., Graham, I., Martin, T., Crozier, A., 2001. The effect of nitrogen and phosphorus deficiency on flavonol accumulation in plant tissues. Plant Cell Environ. 24, 1189–1197.

Toor, R.K., Savage, G.P., Lister, C.E., 2006. Seasonal variations in the antioxidant composition of greenhouse grown tomatoes. J. Food Compos. Anal. 19, 1–10.

Treutter, D., 2010. Managing Phenol Contents in Crop Plants by Phytochemical Farming and Breeding—Visions and Constraints. Int. J. Mol. Sci. 11, 807–857.

Vallejo, F., Tomas-Barberan, F.A., Garcia-Viguera, C., 2003. Effect of climatic and sulphur fertilization conditions, on phenolic compounds and vitamin C, in the inflorescences of eight broccoli cultivars. Eur. Food Res. Technol. 216, 395–401.

Wang, C., Lu, J., Zhang, S., Wang, P., Hou, J., Qian, J., 2011. Effects of Pb stress on nutrient uptake and secondary metabolism in submerged macrophyte *Vallisneria natans*. Ecotoxicol. Environ. Saf. 74, 1297–1303.

Wang, S.Y., Lin, H.S., 2000. Antioxidant activity in fruits and leaves of blackberry, raspberry, and strawberry varies with cultivar and developmental stage. J. Agric. Food Chem. 48, 140–146.

Wang, S.Y., Zheng, W., 2001. Effect of plant growth temperature on antioxidant capacity in strawberry. J. Agric. Food Chem. 49, 4977–4982.

Watzl, B., Leitzmann, C., 2005. Bioaktive Substanzen in Lebensmitteln Hippokrates Verlag. Germany, Stuttgart.

Wang, S.Y., Bunce, J.A., Maas, J.L., 2003. Elevated carbon dioxide increases contents of antioxidant compounds in field-grown strawberries. J. Agric. Food Chem. 51, 4315–4320.

Wilkens, R.T., Spoerke, J.M., Stamp, N.E., 1996. Differential responses of growth and two soluble phenolics of tomato to resource availability. Ecology 77, 247–258.

Yamane, T., Jeong, S.T., Goto-Yamamoto, N., Koshita, Y., Kobayashi, S., 2006. Effects of temperature on anthocyanin biosynthesis in grape berry skins. Am. J. Enol. Vitic. 57, 54–59.

Yuri, A., Schmitt, E., Feucht, W., Treutter, D., 1990. Metabolism of Prunus tissues affected by Ca^{2+}-deficiency and addition of prunin. J. Plant Physiol. 135, 692–697.

Zagoskina, N.V., Goncharuk, E.A., Alyavina, A.K., 2007. Effect of cadmium on the phenolic compounds formation in the callus cultures derived from various organs of the tea plant. Russ. J. Plant Physiol. 54, 237–243.

Zornoza, P., Esteban, R.M., 1984. Flavonoids content of tomato plants for the study of the nutrtional status. Plant Soil 82, 269–271.

Plant Polyphenol Profiles as a Tool for Traceability and Valuable Support to Biodiversity

2

Laura Siracusa, Giuseppe Ruberto
Istituto del CNR di Chimica Biomolecolare, Catania, Italy

CHAPTER OUTLINE HEAD

2.1 Introduction

The impressive and large chemical diversity expressed in the plant kingdom is due to the high capacity developed by plant genomes, namely the high diversity of genes able to codify many metabolic enzymes. This is confirmed by the studies to date on genome sequencing carried out on different plant species which have shown that plants possess more genes than other living organisms, such as mammals and bacteria. A result of this large biodiversity is represented by the more than 200,000 secondary metabolites isolated to date from plant species. Furthermore, given that only a small

Polyphenols in Plants. http://dx.doi.org/10.1016/B978-0-12-397934-6.00002-4

portion of over 400,000 worldwide plant species has been analyzed from a phyto-chemical point of view, the number of secondary metabolites is undoubtedly much higher (Yonekura-Sakakibara and Saito, 2009; Marcel *et al.*, 2010). Metabolomics, defined as the exhaustive analysis of all metabolites of an organism, represents, in the post-genomics era, the last step of the other "omics" such as transcriptomics (mRNA) and proteomics (proteins). The last two are concerned with the genotype information of a living system, but metabolomics can also be considered to be intrin-sically connected with the phenotype concept. In fact, most of the thousands of phy-tochemical studies performed in these last decades have shown, more or less directly, plant chemistry (the secondary metabolite profiles) is strictly connected with the relationship of the organism and its environment (Fukusaki and Kobayashi, 2005; Ivanišević *et al.*, 2011).

From an experimental point of view today, it is practically impossible to carry out a complete metabolomics analysis of an organism, because the singular metabo-lites are too chemically different, in terms of polarity and molecular weight, and because of their large range of concentrations. No analytical instrument, despite being highly sophisticated, is able to perform a similar analytical approach. Para-doxically, transcriptomics and proteomics, notwithstanding, deal with much more complex chemical entities (mRNA, proteins), and show relatively "easier" experi-mental protocols owing to "less" chemical diversity. For these reasons one of the most adopted metabolomic studies concerns the analysis of the so-called metabolite profiling, namely the quantitative and qualitative characterization of related com-pounds or of a particular metabolic pattern. As previously mentioned, there are over 200,000 known plant metabolites: 25,000 are terpenoids, 12,000 are alkaloids, and 8000 are phenolics (Croteau *et al.*, 2000). Phenolics, which together with the others can be considered ubiquitous in plants, comprise different chemical classes, namely flavonoids, hydroxybenzoic and hydroxycinnamic acids, gallotannins, proanthocy-anidins, stilbenoids, and lignans. Flavonoids, with over 4000 substances, are largely the most represented phenolic compounds (Harborne *et al.*, 1999; Ignat *et al.*, 2011), which in turn are classified as anthocyanins, flavones, isoflavones, flavanones, flavo-nols, and flavanols (Tsao and Yang, 2003).

In recent years we have focused our phytochemical studies on Sicilian and Medi-terranean flora. Wild plants and horticultural products represent our target, with the aim of exploiting the former as new crops, and valorizing the latter as healthy food for the human diet. The guideline for both aims is to certify a "geographical typical-ness" for the aforementioned plant species, which can be achieved through the study and standardization of their metabolic profiles. Thus, the disclosure of the polyphe-nolic pattern may be recognized as one of the most powerful tools to reach this goal. Finally, one of the challenges of the post-genomic era is represented by the study of biodiversity in order to understand the biological diversity of organisms in different applied fields such as biomedicine, food, and environment. Furthermore, the protec-tion of biodiversity is a major issue in the sustainable development of modern soci-ety. These studies are possible due to the enormous progress achieved in recent years by the next-generation technologies that offer opportunities unimaginable until now.

2.2 Traceability: definition, importance and state of the art in the area concerned

2.2.1 Definition

Global food safety policies have been stipulated by governmental authorities and a new series of regulations has been created and adopted all over the world, with particular incidence in the EU (European Union) and the USA, more as a consequence of several food crises such as dioxin contamination and BSE (bovine spongiform encephalophaty) rather than a natural development in consumers' demands. The Codex Alimentarius Commission (1999) generally defined traceability as "the ability to trace the history, application, or location of an entity by means of recorded identifications"; in 2002, the European Commission (EC) gave a more detailed definition by stating that "it is the ability to trace and follow a food, a feed, food-producing animal of substance intended to be, or expected to be incorporated into a food or a feed, through all stages of production, processing and distribution." This "one-step-up/one step down" approach has also been followed by the International Standards Organization (ISO) which gave its own definition of traceability in 2007. In addition to the general regulations, sector-specific legislation applies to certain categories of food products (fruit and vegetables, fish, honey, olive oil) so that consumers can identify their origin and authenticity.

2.2.2 The importance of traceability for consumers and food/feed businesses

The development of traceability tools and quality policies stem from producer and consumer concerns (Raspor, 2005). Food markets have become more globalized, with the result that consumers have become more concerned about the origin and safety of the food they eat and, at the same time more enthusiastic about high-quality food with a clear regional identity, despite several differences still existing internationally (Kehagia *et al.*, 2007). The demand for quality products also benefits producers: tracing all components of food offered allows the minimization of food safety risks and an increased confidence in food products. In fact, traceability is not only a way to trace the origin of a certain food step by step, but also a tool for responding to potential risks that can arise in food and feed, to ensure that all food products are safe to eat (Du Plessis and du Rand, 2012; Kehagia *et al.*, 2007). Both consumers and producers also benefit from the application of tools for certification based on international standards, such as the registration of Protected Geographical Indications (PGI) and Protected Designation of Origin (PDO). PDO is a brand used for feeds/foods with a strong regional identity that are produced, processed, and prepared in a specific area by using specific techniques, whilst PGI deals with products closely linked to a geographical area in at least one of the stages of production, processing, and preparation (European Commission, 2004). For both certifications, these eligibility conditions concern the reputation of a product and are intrinsically related to the concept of traceability.

2.3 Traceability markers

2.3.1 Definition and features

As already mentioned, implementation of food traceability systems has become a top priority. Identification of the geographical origin and authenticity of food products represent an important goal to ensure organoleptic and nutritional characteristics to consumers, to combat fraudulent practices, and to control product adulteration. This is particularly important when considering agricultural products for human/animal nutrition. Reports on analytical methods for determining the geographical origin and history of foods/feeds have been increasing since the 1980s; the initial focus was more on processed products such as wine, honey, tea, olive oil, and orange juice, while later studies also examined fresh products such as onions, potatoes, pistachios, and garlic, mainly because the dimension of world-wide trade in fresh agricultural products has increased year by year (Luykx and van Ruth, 2008). All analytical strategies reported as suitable for traceability determination are based on two fundamental factors: the accuracy, selectivity, efficiency, reliability, and reproducibility of the instrumentation used, and the nature of analytes considered. In fact, chemometric analysis of the data provided by analytical instruments can be a valuable support to establish links to the history of a food product only if suitable analytical *markers* are used. An analytical marker is generically defined as a substance (or group of substances) which is of interest for analytical purposes, or more precisely, "a specific (bio)chemical owning particular molecular features that make it useful for measuring" (Zolg and Langen, 2004). In order to be adapted for food analysis, an analytical marker has to fulfill several requirements: it has to be stable during sample processing and storage; its presence must be easily determined, and its concentration in the matrix has to be sufficient for measurement; its possible variability along the batch must be easy to follow; its identification must be reliable; and its corresponding reference material must be commercially available at certified high purity levels. A proper qualitative and quantitative analysis of a food/feed matrix, whatever the method used, defines the so-called analytical "fingerprint" of a given product and ultimately its *identity*.

2.3.2 Analytical methodologies commonly used for food and feed traceability

One of the most common methods used to determine the history of plant and other food products relies on DNA-based analytical markers. DNA is stable, durable, and ubiquitous in cells; application of several DNA markers, thanks mainly to the increasing volume of information available on plant genomes, has allowed rapid progress in the investigations of generic variability, parentage assignment, identification of species and varieties. DNA-based markers include restriction fragment length polymorphisms (RFLPs), random amplification of polymorphic DNAs (RAPDs), amplified fragment length polymorphisms (AFLPs) and microsatellites or simple sequence repeats (SSRs) (for a short overview on the use of DNA based markers on this topic, see Varshney *et al.*, 2005; Lopez-Vizcón and Ortega, 2012; Costa *et al.*, 2012; Vietina *et al.*, 2011). Analysis of the isotopic ratios and elemental composition of foods are other attractive methods for product authentication. Evaluation of natural abundance isotope ratios provides information either on plant type or animal diet and geographical origin (Di Paola-Naranjo *et al.*, 2011; Camin *et al.*, 2011). Recently, isotopic parameters have been added to the Protected Denomination of Origin (PDO) technical specification of Grana Padano Cheese. Elemental composition analysis has also been successfully used to discriminate geographical origins of products such as wheat (Zhao *et al.*, 2012), onions (Furia *et al.*, 2011), wine (Di Paola-Naranjo *et al.*, 2011; Coetzee *et al.*, 2005), oranges (Camin *et al.*, 2011), and rice (Ariyama *et al.*, 2012). Data coming either from isotopic ratios or elemental composition change significantly with plant genotype, but also depend on several environmental and geological factors, *thus providing an unique link of a product with its territory,* which is also the ultimate goal of PDO and PGI brands. Another way to highlight the importance of genotype and the influence of environment, storage, and processing method of food and feed products is to measure the effects of all these factors in plant metabolism, and precisely on the production of the so-called secondary metabolites of plants.

2.4 Polyphenols as traceability markers

2.4.1 Plant secondary metabolism and phenolics

As mentioned earlier in this text, given that plants are sessile, they have evolved many strategies for survival, including the capability to produce over 200,000 highly varied but specialized metabolites (Yonekura-Sakakibara and Saito, 2009). All these molecules belong to the secondary metabolism of plants, that is, a pool of metabolic pathways not included in the standard charts. Whether or not secondary metabolism is essential to plants is a subject of debate; a wide collection of literature data suggest that secondary metabolites have a much more restricted distribution than primary metabolites in the whole plant kingdom; that is, they are often found only in one plant species or a taxonomically related group of species (Mazid *et al.*, 2011). Although they appear to have no obvious functions in cell growth and plant development, their defence functions against biotic, abiotic, and man-related stress are widely recognized (Mazid *et al.*, 2011; Ignat *et al.*, 2011; Rohde *et al.*, 2012). The majority of secondary metabolites are assembled from quite similar precursor units; their biogenesis can be traced back from relatively few initiator primary metabolites such as acetate, mevalonate, and shikimate (Dewick, 2002). Structurally speaking, products of plant secondary metabolism are highly variegate: of the above-mentioned 200,000 compounds, over 25,000 are chemically classified as terpenoids, 12,000 as alkaloids, and 8000 as phenolic compounds (Yonekura-Sakakibara and Saito, 2009). The group of secondary metabolites generally called "phenolics" counts for a wide variety of molecules having in common the presence of at least one hydroxyl group on an aromatic ring of their scaffold (Figure 2.1); the group of polyphenols is further

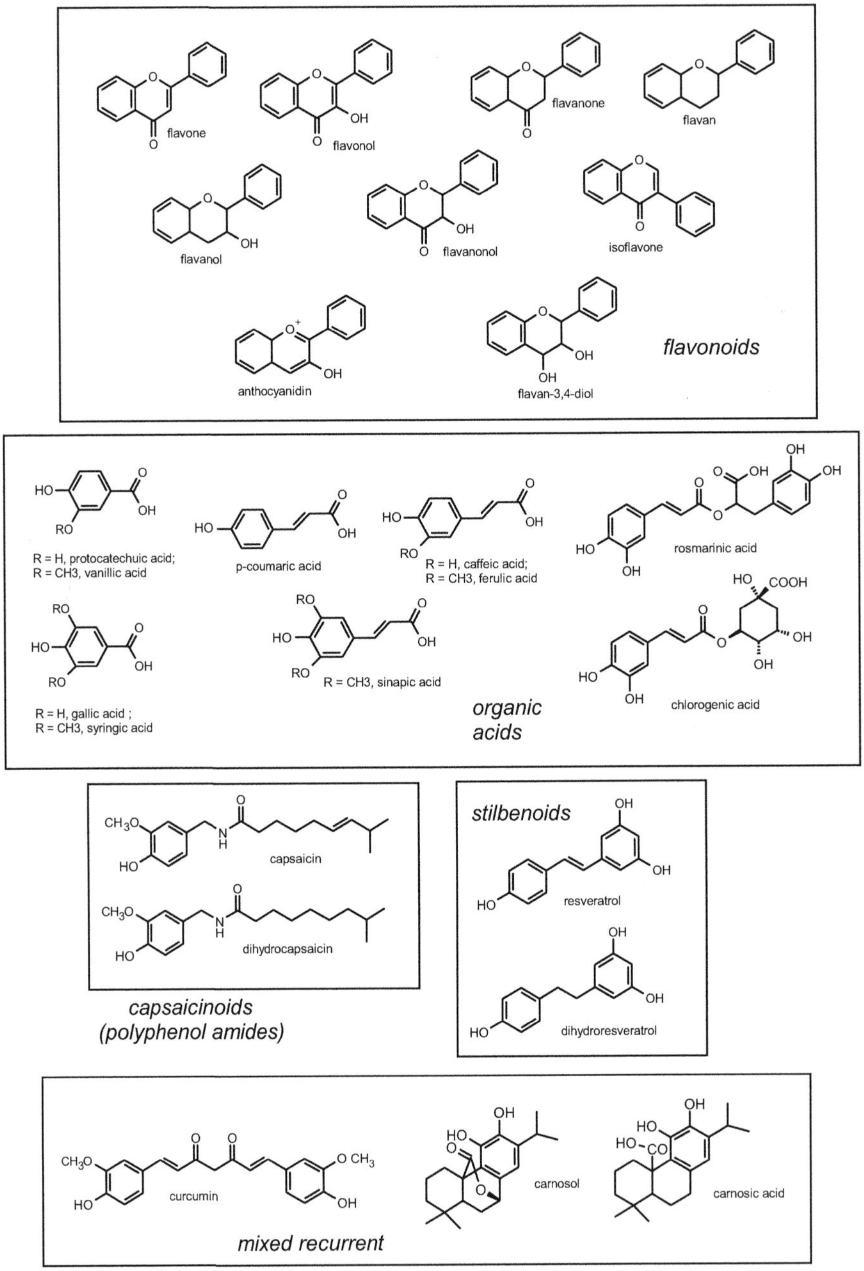

FIGURE 2.1

General chemical structures of some polyphenols described in the text.

divided into several classes by means of structural criteria: from simple phenolic acids and alcohols, to cinnamic acids, flavonoids, stilbenes, chalcones, coumarins, tannins, and lignans.

Phenolics are of considerable physiological and morphological importance for plants, as they may act as phytoalexins, antifeedants, contributors to plant pigmentation and reproduction, UV-light protectors, and antioxidants. As food components, polyphenols are important determinants in the color, sensory, and nutritional quality of fruits, vegetables, and other plants; they have been reported to possess excellent properties as food preservatives as well as having a primary role in protection against a number of pathological and degenerative disturbances (Ignat *et al.*, 2011).

2.4.2 Factors affecting the presence and amount of polyphenols in plants

Generally speaking, the abundance of polyphenols in plants is affected by two main factors: genetics and environment. As secondary metabolites, polyphenol presence and content broadly varies within the plant kingdom. Identification of the genetic basis of this diversity has seen dramatic progress in the last decade; this includes analysis of genetic factors underlying metabolite accumulation and metabolic regulation, studies on the heredity of metabolism, and structure/function studies on genetic polymorphism (Fernie, 2007). A group of structurally related molecules is often found only in one plant species or a taxonomically related group of species, even if the mere recording of a compound or type of compound in apparently unrelated *taxa* is no indication of their systematic affinity, because often the biosynthesis of the structures takes different routes (Reynolds, 2007). The natural inclination to biosynthesis of a specific class of phenolics in plants can be summarized by two explicative examples, both broadly reported in the literature. The first example is the predisposition of the components of the family Fabaceae (also known as Leguminosae) to synthesize a subclass of polyphenols known as isoflavonoids. These molecules belong to an important group of secondary metabolites derived from the phenylpropanoid pathway; in nature, they provide a wide range functions, such as resistance against pathogen attack and signalling for the induction of nod genes (Du *et al.*, 2010). About 1600 isoflavonoids have been identified, of which only 225 are from non-leguminous families (Du *et al.*, 2010; Lapĉîk, 2007). The isoflavonoid pattern dramatically differs in each species, as in the case of alfalfa (*Medicago sativa* L.), whose most abundant compound is medicarpin, and soybean (*Glycine max* (L.) Merr.), where genistein, daidzein, and glycitein are the main isoflavonoids present (Figure 2.2). This compositional diversity has been proposed as support for taxonomic differentiation, as in the case of the genus *Cicer* (Stevenson and Veitch, 1998). A second example is the well-documented characteristic of the members of the *Citrus* genus (Rutaceae) to accumulate in their fruits at specific stages of growth high amounts of another class of flavonoids, namely flavanones (Ortuño *et al.*, 1997), which are often considered as taxonomic markers for the genus itself. Furthermore, within the genus, each *Citrus* species is characterized by a particular flavanone pattern (Mouly *et al.*, 1997;

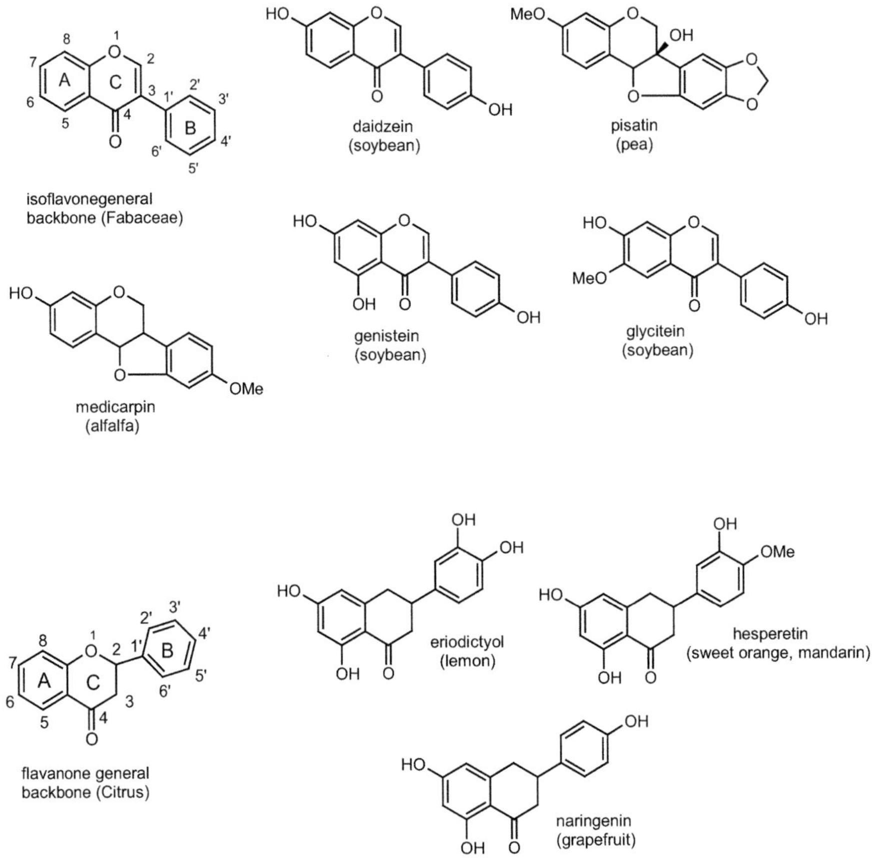

FIGURE 2.2

Chemical structures and corresponding vegetable font of polyphenols described in Section 2.4.2. See text for further details.

Mata Bilbao *et al.*, 2007) which is used for authentication purposes (Mouly *et al.*, 1994; Peterson *et al.*, 2006); for example, *C. reticulata* Blanco (mandarin) and *C. sinensis* Osbeck (sweet orange) possess high amounts of the flavanone glucoside hesperidin, regardless of variety (Gattuso *et al.*, 2007). *C. limon* (L.) Burm. (lemon), apart from hesperidin, is characterized by the considerable presence of another flavanone, eriodictyol; whilst *C. paradisi* Macf. (grapefruit) differs in composition from the other species, narigenin and its derivatives being the characterizing compounds (Figure 2.2) (Gattuso *et al.*, 2007; Peterson *et al.*, 2006; Liu *et al.*, 2012).

Quite recently, the use of flavanones and their derivatives for the authentication of citrus species and their products has been extended to other polyphenols (Mouly *et al*, 1997). A third example is the production of a defined subclass of flavonoids (anthocyans and flavonols, with few examples of dy-hydroflavonols) in onion (*Allium cepa* L.) (Slimestad *et al.*, 2007), and fruits (Robards and Antolovich, 1997).

As previously mentioned, being secondary metabolites, the presence and content of polyphenols in plants depends indeed on genetics, but it is also triggered by external factors. To protect themselves, plants possess an innate immunity that involves different layers of defence responses. Plants accumulate an armoury of antimicrobial secondary metabolites. Some of these defences are preformed (phytoanticipins) and others are activated after recognition of pathogen elicitors (phytoalexins) (Stotz *et al.*, 1999; Templeton and Lamb, 1988; Bednarek and Osbourn, 2009). Phytoalexins are a group of chemically diverse compounds including terpenes, sulfur- and nitrogen-containing metabolites, and also polyphenols (Mazid *et al.*, 2011; Dixon and Paiva, 1995); the analysis of variations in levels and composition of phenolic compounds, supported by robust statistic tools (Robards and Antolovich,1997, and references therein) and preferably coupled with genomics (Fernie, 2007) is a powerful method of monitoring system responses to external factors. Numerous papers can be found in the literature on this particular topic; examples include tea (Tounekti *et al.*, 2013), potato (Reyes and Cisneros-Zevallos, 2003), tomato (Atkinson *et al.*, 2011), and pepper (Wahyuni *et al.*, 2013) among others. Strong variations in polyphenol composition were also observed when particular cultural practices were applied (Siracusa *et al.*, 2012; Incerti *et al.*, 2009; Riggi *et al.*, 2013; Mogren *et al.*, 2006; Tounetki *et al.*, 2013; Venkatesan and Ganapathy, 2004; Downey *et al.*, 2006). While these kinds of studies will shortly lead to a deep understanding of gene-metabolite networks in plants, for authentication and traceability purposes (provided by the great variability of soil, environment and geology of the planet) they contribute to the assessment of the uniqueness of a product grown in a restricted area with specific cultural practices.

2.5 Examples of the use of polyphenols as markers
2.5.1 General

In this paragraph, a collection of examples will be given dealing with the broad applicability of polyphenol analysis and profiling of edible plants, food, and food products according to the previously mentioned concepts of genetic and induced variability. The subsections are organized in order to highlight the principle for which "a food must be traced from its origin to the table," that is, one of the descriptions of traceability. The first examples will therefore deal with the use of polyphenols as chemotaxonomic markers; then a look throughout the food chain will be given, from the assessment of geographical origin and distribution of edible plants, food and feed products, to the analysis of polyphenol content during post-harvest treatments, storage, and processing. All papers and reports regarding species and varietal attribution will be cited in Section 2.5.1, whilst those dealing with ecotype determination will be cited in Section 2.5.2. An easy-to-view scheme of these sections is also given in Table 2.1.

Table 2.1 Examples of the Use of Polyphenols as Markers

Application (section in the text)	Product	References
Taxonomic classification (2.5.1)	Thyme	Boros et al., 2010
	Coffee	Andrade et al., 1998
	Banana	Pothavorn et al., 2010
	Berries	Plessi et al., 2007
	Apple	Chen et al., 2011
	Chicory	Mascherpa et al., 2012
	Plum	Treutter et al., 2012
	Artichoke	Lombardo et al., 2010
	Tea	Li et al., 2010
Designation of origin (2.5.2)	Tea	Li et al., 2010
	Coffee	Andrade et al., 1998
	Durum	Dinelli et al., 2009
Geographical discrimination (2.5.2)	Oregano	Tuttolomondo et al., 2013
	Pepper	Wahyuni et al., 2013
	Plantain	Zubair et al., 2012
	Tomato	Vallverdù-Queralt et al., 2011
	Eggplant	Singh et al., 2009; Luthria et al., 2010
	Onion	Riggi et al., 2013
	Barley/malt	Zimmermann and Galensa, 2007
	Hazelnut	Locatelli et al., 2011
Post-harvest treatment, storage, processing	Grapes/wine	Xu et al., 2011; Downey et al., 2006; Minussi et al., 2003; Kallithraka et al., 2006; Mazza et al., 1999; Radovanović et al., 2010; Sanz et al., 2012
	Apple juice	Karaman et al., 2010; Renard et al., 2011; Oszmiański et al., 2011
	Honey	Kaškoniene and Venskutonis, 2010; Ferreres et al., 1994, 1998; Tomás-Barberán et al., 2001; Andrade et al., 1997; Cherchi et al., 1994; Dimitrova et al., 2007; Hadjmohammadi et al., 2009
	Bakery products	Bedini et al., 2012

See Section 2.5 and given references for further details.

At this point, it is useful to properly define the meaning of "ecotype," especially in comparison with "variety." In evolutionary ecology, an ecotype describes a genetically distinct *geographic* variety or population within species, which is adapted to specific environmental conditions. Typically, ecotypes exhibit phenotypic differences, such as in morphology or physiology, stemming from environmental heterogeneity; they are also capable of interbreeding with other geographically adjacent ecotypes without loss of fertility. With respect to varieties, they do not own a genetic and systematic identity.

2.5.2 Taxonomic classification

Instinctive chemosystematics has always been a feature of living things. Observation of the immediate characteristics of plants (color, scent, and taste), for instance, ultimately lead the inquisitive person to a study of the chemical components. The state of research up until 1975 has been discussed in detail in the robust work of Fairbrothers *et al.* (1975), in which both "micromolecules" (pigments, glucosinolates, terpenes, and phenolics) and "macromolecules," embracing serology, amino acids, and nucleic acid sequences, were compared in relation to a number of taxonomic groups. Given the metabolic diversity within the plant kingdom, it is not surprising that chemosystematics has been extensively used in order to sort out taxonomic problems, and continues to be used, despite sometimes having only partial success (Reynolds, 2007). Several examples of the use of polyphenols as chemotaxonomic markers for classification purposes include the distinction of different thyme (Boros *et al.*, 2010) and coffee species (Andrade *et al.*, 1998), classification of banana clones (Pothavorn *et al.*, 2010), and varietal unambiguous attribution, such as in the case of berries (Plessi *et al.*, 2007), apple (Chen *et al.*, 2011), chicory (Mascherpa *et al.*, 2012), plum (Treutter *et al.*, 2012), artichoke (Lombardo *et al.*, 2010), and tea (*Camellia sinensis* L.) (Li *et al.*, 2010). In the last example, the authors examined 89 different wild, hybrid, and cultivated tea trees from China and Japan for their polyphenol content, to properly distinguish between two varieties (*C. sinensis* var. *sinensis* and var. *assamica*), to investigate their phenetic relationship, and to ultimately find the site of origin of this plant (Li *et al.*, 2010).

2.5.3 Origin and geographical discrimination

The determination of the origin of a plant, on its own or as food/feed component, is the first step in the traceability mechanism. We have already mentioned the work of Li *et al.* (2010) on Chinese and Japanese tea trees; in this case, the analysis of polyphenols gave a valuable support to historical data and allowed us to find the area of origin of the plant (as in the case of coffee (Andrade *et al.*, 1998), in which the authors are confident that the analysis of the hydroxycinnamic acids present in the two varieties considered could be related to the botanical origin of coffee). An interesting investigation was conducted by Dinelli *et al.* (2009), who examined 10 varieties of durum wheat genotypes, including old and modern ones, to evaluate their differences in terms of polyphenol profile and content. The results obtained clearly showed that, even if there were not significant differences in total phenolic content between the old and the modern varieties, they differ remarkably in their profiles, with the old varieties showing the richest number of phenolic compounds present. This data could also contribute to individuating a putative progenitor of the modern varieties. Examples of the use of phenolic fingerprints to discriminate among ecotypes are quite recurrent in literature; this is likely due to the growing trend to characterize a product preferably in relation with its territory. Recent papers on geographical discrimination based on the study of polyphenol profile include spices and edible plants such as oregano (Tuttolomondo *et al.*, 2013), the already cited pepper (Wahyuni *et al.*, 2013), plantain (Zubair *et al.*, 2012), tomato (Vallverdù-Queralt *et al.*, 2011), eggplant (Singh *et al.*, 2009; Luthria *et al.*, 2010), and onion (Riggi *et al.*, 2013). Zimmermann and Galensa (2007) used a subgroup of polyphenols, namely proanthocyanidins, to discriminate among barley vareties and the corresponding malts; whilst Locatelli and co-workers (2011) were able to differentiate the "Tonda Gentile Trilobata" hazelnuts from Piedmont (Italy), a product covered by PGI designation, from other hazelnuts belonging to different cultivars and/or geographic origins.

2.5.4 "From farm to fork:" post-harvest treatments, storage and processing

According to the definition of PDO and PGI brands cited in Section 2.2, the determination of the locality of origin and the attribution of a defined regional identity for a given food/feed is just one of the mandatory steps of the traceability chain. Further steps are the control of post-harvest treatments, storage and processing conditions, and preparation for food products. Polyphenols have been used as a traceability tool for each of these steps, as in the case of one of the most famous products known worldwide: wine. An enormous number of papers have been published in the last 20 years on the polyphenolic composition of grapes and grape-derived products, including wine (Xu *et al.*, 2011; Downey *et al.*, 2006); the reason for such a persistent interest resides on both the commercial value of wine as a product and the widely recognized beneficial role of polyphenols present in grapes. These include flavonoids (flavanols, flavanones, isoflavones, flavonols, and anthocyanins), quinones, but also cinnamic acids, phenolic acids, hydrolysable tannins, and stilbenes (Xu *et al.*, 2011). It has been widely demonstrated that, as for other species, polyphenol content and composition in grapes vary among species as well as in different cultivars within each species, are highly affected by environmental conditions such as soil composition and climate, and undergo specific variations also with post-harvest handling, processing, and storage conditions (Downey *et al.*, 2006; Minussi *et al.*, 2003; Kallithraka *et al.*, 2006; Mazza *et al.*, 1999; Radovanović *et al.*, 2010). Polyphenolic profile has also been recently used to identify the wood used in cooperage (Sanz *et al.*, 2012). Another example of the use of phenolic profile in the traceability chain is the production of apple juice, in which polyphenols have been used as markers for comparison purposes (Karaman *et al.*, 2010), to test the best crushing/pressing conditions during processing (Renard *et al.*, 2011), and to evaluate the potential applicability of enzyme preparation in juice production (Oszmiański *et al.*, 2011). Honey represents another product for which traceability is of pivotal importance. In fact, with the globalization of the honey market, nowadays involving approximately 150 countries, the identification of honey origin together with the proof of its authenticity has become an important issue. As honey is a very complex matrix, the search for reliable chemical markers indicating its floral and/or geographical origin has been a challenge for researchers in the last two decades (Kaškoniene and Venskutonis, 2010). The composition of honey depends not only on the nectar-providing plant species, but also on other factors such as bee species, geographic area, season, mode of storage, and harvest technology. Analysis of polyphenols in honey has been recognized as a valuable tool to distinguish honey coming from different floral sources; for example, the flavanol kaempferol is an indicator for rosemary honey (Ferreres *et al.*, 1994, 1998), quercetin for sunflower honey (Tomás-Barberán *et al.*, 2001), whilst naringenin and luteolin were suggested as markers of lavender honey (Andrade *et al.*, 1997). The hydroxy-cinnamates like caffeic acid, ferulic acid, and p-coumaric acid have been found in chestnut honey (Cherchi *et al.*, 1994). Characteristic flavonoids of propolis like pinocembrin, pinobanksin, and chrysin were also found in most European honey samples (Tomás-Barberán *et al.*, 2001). Other than floral sources, the geographical origin of honey is another important factor affecting its phenolic composition (Dimitrova *et al.*, 2007; Hadjmohammadi *et al.*, 2009). An outstanding application in the food industry for quality control purposes has been recently proposed by Bedini and co-workers (2012); in their work, the analysis of polyphenols (cathechin and epicathechins) has been used to constantly check a standardized bitter taste production quality in biscuits, thus avoiding the adoption of the human panel test approach.

2.6 Concluding remarks

The study of a metabolic profile as described here for polyphenols represents, as previously mentioned, a powerful investigation tool of plant species and their processing products. Several implications of these studies can be highlighted: firstly, being a metabolic profile (metabolomics), the final result of the genes' expression (transcriptomics – proteomics) of a living organism, its analysis allows researchers to establish whether the organism has been subjected to spontaneous or induced genetic modifications; secondly, the metabolic profile is intimately connected with the ecology of the living organism, namely the vital relationships of its existence and evolution in a given environment. We would like to conclude that these kinds of studies would assume a much more effective role and value if there was a multidisciplinary will to take care of their development.

Acknowledgments

We wish to thank Consiglio Nazionale delle Ricerche (CNR, Rome, Italy) for financial support. The authors also express their thanks to the numerous colleagues of other Italian and international institutions, quoted in the reference list, for their continued and efficacious collaboration in this field.

References

Andrade, P., Ferreres, F., Gil, M.I., Tomás-Barberán, F.A., 1997. Determination of phenolic compounds in honeys with different floral origin by capillary zone electrophoresis. Food Chem. 60, 79–84.

Andrade, P.B., Leitão, R., Seabra, R.M., Oliveira, M.B., Ferreira, M.A., 1998. 3,4-Dimethoxycinnamic acid levels as a tool for differentiation of *Coffea canephora* var. *robusta* and *Coffea arabica*. Food Chem. 61, 511–514.

Ariyama, K., Shonozaki, M., Kawasaki, A., 2012. Determination of the geographical origin of rice by chemometrics with strontium and lead isotope ratios and multi-element concentrations. J. Agric. Food Chem. 60, 1628–1634.

Atkinson, N.J., Dew, T.P., Orfila, C., Urwin, P.E., 2011. Influence of combined biotic and abiotic stress on nutritional quality parameters in tomato (*Solanum lycopersicum*). J. Agric. Food. Chem. 59, 9673–9682.

Bedini, A., Zanolli, V., Zanardi, S., Bersellini, U., Dalcanale, E., Suman, M., 2013. Rapid and simultaneous analysis of xanthines and polyphenols as bitter taste markers in bakery products by FT-NIR spectroscopy. Food Analytical Methods 6 (1), 17–27.

Bednarek, P., Osbourn, A., 2009. Plant–microbe interactions: chemical diversity in plant defence. Science 324, 746–747.

Boros, B., Jakabova, S., Dörnyei, A., Horvath, G., Pluhar, Z., Kilar, F., et al., 2010. Determination of polyphenolic compounds by liquid chromatography-mass spectrometry in *Thymus* species. J. Chromatogr. A. 1217, 7931–8072.

Camin, F., Perini, M., Bontempo, L., Fabroni, S., Faedi, W., Magnani, S., et al., 2011. Potential isotopic and chemical markers for characterizing organic fruits. Food Chem. 125, 1072–1082.

Chen, N.-N., Zhao, S.-C., Deng, L.-G., Guo, C.-Y., Mao, J.-S., Zheng, H., et al., 2011. Determination of five polyphenols by HPLC/DAD and discrimination of apple varieties. Chromatographia 73, 595–598.

Cherchi, A., Spanedda, L., Tuberoso, C., Cabras, P., 1994. Solid-phase extraction and HPLC determination of organic acid in honey. J. Chromatogr. 669, 59–64.

Coetzee, P.P., Steffens, F.E., Eiselen, R.J., Augustyn, O.P., Balcaen, L., Vanhaecke, F., 2005. Multi-element analysis of south African wines by ICP-MS and their classification according to geographical origin. J. Agric. Food Chem. 53, 5060–5066.

Costa, J., Mafra, I., Oliveira, M.B.P.P., 2012. Advances in vegetable oil authentication by DNA-based markers. Trends Food Sci. Technol. 26, 43–55.

Croteau, R., Kutchan, T.M., Lewis, N.G., 2000. Natural products (secondary metabolites). In: Buchanan, B.B., Gruissem, W., Joneas, R.L. (Eds.), Biochemistry and Molecular Biology of Plants. American Society of Plant Physiologists, Rockville, MD, USA, pp. 1250–1268.

Dewick, P.M., 2002. Medicinal natural products: a biosynthetic approach, second ed. Published by John Wiley & Sons Ltd, Baffins Lane, Chichester, West Sussex, England.

Dimitrova, B., Gevrenova, R., Anklam, E., 2007. Analysis of phenolic acids in honeys of different floral origin by solid-phase extraction and high-performance liquid chromatography. Phytochem. Anal. 18, 24–32.

Dinelli, G., Segura-Carretero, A., Di Silvestro, R., Marotti, I., Fu, S., Benedettelli, S., et al., 2009. Determination of phenolic compounds in modern and old varieties of durum wheat using liquid chromatography coupled with time-of-flight mass spectrometry. J. Chromatogr. A. 1216, 7229–7240.

Di Paola-Naranjo, R.D., Baroni, M.V., Podio, N.S., Rubinstein, H.R., Fabani, M.P., Badini, R.G., et al., 2011. Fingerprints for main varieties of Argentinean wines: terroir differentiation by inorganic, organic, and stable isotopic analyses coupled to chemometrics. J. Agric. Food Chem. 59, 7854–7865.

Dixon, R.A., Paiva, N.L., 1995. Stress-induced phenylpropanoid metabolism. Plant. Cell 7, 1085–1097.

Downey, M.O., Dokoozlian, N.K., Krstic, M.P., 2006. Cultural practice and environmental impacts on the flavonoid composition of grapes and wine: A review of recent research. Am. J. Enology Viticulture 57, 257–268.

Du, H., Huang, Y., Tang, Y., 2010. Genetic and metabolic engineering of isoflavonoid biosynthesis. Appl. Microbiol. Biotechnol. 86, 1293–1312.

Du Plessis, H.J., du Rand, G.E., 2012. The significance of traceability in consumer decision making towards Karoo lamb. Food Res. Int. 47, 210–217.

European Commission – Working Document of the Commission Services, Aug. 2004. Guide to Community Regulation, second ed.

European Community Regulation 178, 2002. Laying down the general principles and requirements of food law, establishing the European food safety authority and laying down procedures in matters of food safety. Off. J. Eur. Communities L31/I–L31/24.

Fairbrothers, D.E., Mabry, T.J., Scogin, R.L., Tirner, B.L., 1975. The bases of angiosperm phylogeny: chemotaxonomy. Annals of the Missouri Botanical Garden 62, 765–800.

Fernie, A.R., 2007. The future of metabolic phytochemistry: larger numbers of metabolites, higher resolution, greater understanding. Phytochemistry 68, 2861–2880.

Ferreres, F., Blazquez, M.A., Gil, M.I., Tomás-Barberán, F.A., 1994. Separation of honey flavonoids by micellar electrokinetic capillary chromatography. J. Chromatogr. A. 669, 268–274.

Ferreres, F., Juan, T., Perez-Arquillue, C., Herrera-Marteache, A., Garcia-Viguera, C., Tomás-Barberán, F.A., 1998. Evaluation of pollen as a source of kaempferol in rosemary honey. J. Sci. Food Agric. 77, 506–510.

Fukusaki, E., Kobayashi, A., 2005. Plant metabolomics: potential for practical operations. J. Biosci. Bioeng. 100, 347–354.

Furia, E., Naccarato, A., Sindona, G., Stabile, G., Tagarelli, A., 2011. Multielement fingerprinting as a tool in origin authentication of PGI food products: Tropea red onion. J. Agric. Food Chem. 59, 8450–8457.

Gattuso, G., Barreca, D., Gargiulli, C., Leuzzi, U., Caristi, C., 2007. Flavonoid composition of *Citrus* juices. Molecules 12, 1641–1673.

Hadjmohammadi, M.R., Nazari, S., Kamel, K., 2009. Determination of flavonoid markers in honey with SPE and LC using experimental design. Chromatographia 69, 1291–1297.

Harborne, J.B., Baxter, H., Moss, G.P., 1999. Phytochemical dictionary: Handbook of bioactive compounds from plants, second ed. Taylor and Francis, London.

Ignat, I., Volf, I., Popa, V.I., 2011. A critical review of methods for characterization of polyphenolic compounds in fruits and vegetables. Food Chem. 126, 1821–1835.

Incerti, A., Navari-Izzo, F., Parossi, A., Izzo, R., 2009. Seasonal variations in polyphenols and lipoic acid in fruits of tomato irrigated with sea water. J. Sci. Food Agric. 89, 1326–1331.

International Organization for Standardization (ISO), 2007. Standard 2005:2007 Traceability in the feed and food chain – general principles and basic requirements for system design and implementation.

Ivanišević, J., Thomas, O.P., Lejeusne, C., Chevaldonné, P., Pérez, T., 2011. Metabolic fingerprinting as an indicator of biodiversity: towards understanding inter-specific relationship among *Homoscleromorpha* sponges. Metabolomics 7, 289–304.

Kallithraka, S., Tsoutsouras, E., Tzourou, E., Lanaridis, P., 2006. Principal phenolic compounds in Greek red wines. Food Chem. 99, 784–793.

Karaman, Ş., Tütem, E., Sözgen Başkan, K., Apak, R., 2010. Comparison of total antioxidant capacity and phenolic composition of some apple juices with combined HPLC-CUPRAC assay. Food Chem. 120, 1201–1209.

Kaškoniene, V., Venskutonis, P.R., 2010. Floral markers in honey of various botanical and geographic origins: a review. Comprehensive Reviews in Food Science and Food Safety 9, 620–634.

Kehagia, O., Chrysochou, P., Chryssochoidis, G., Krystallis, A., Linardakis, M., 2007. European consumers' perceptions, definitions and expectations of traceability and the importance of labels, and the differences in these perceptions by product type. Sociologia Ruralis 47, 400–416.

Lapčík, O., 2007. Isoflavonoids in non-leguminous taxa: A rarity or a rule? Phytochemistry 68, 2909–2916.

Li, J.H., Nesumi, A., Shimizu, K., Sakata, Y., Liang, M.Z., He, Q.Y., et al., 2010. Chemosystematics of tea trees based on tea leaf polyphenols as phenetic markers. Phytochemistry 71, 1342–1349.

Liu, Y., Heying, E., Tanumihardjo, S.A., 2012. History, global distribution, and nutritional importance of *Citrus* fruits. Comprehensive Reviews in Food Science and Food Safety 11, 530–545.

Locatelli, M., Coïsson, J.D., Travaglia, F., Cereti, E., Garino, C., D'Andrea, M., et al., 2011. Chemotype and genotype chemometrical evaluation applied to authentication and traceability of "Tonda Gentile Trilobata" hazelnuts from Piedmont (Italy). Food Chem. 129, 1865–1873.

Lombardo, S., Pandino, G., Mauromicale, G., Knödler, M., Carle, R., Schieber, A., 2010. Influence of genotype, harvest time and plant part on polyphenolic composition of globe artichoke [*Cynara cardunculus* L. var. *scolymus* (L.) Fiori]. Food Chem. 119, 1175–1181.

Lopez-Vizcón, C., Ortega, F., 2012. Detection of mislabelling in the fresh potato retail market employing microsatellite markers. Food Control 26, 575–579.

Luthria, D., Singh, A.P., Wilson, T., Vorsa, N., Banuelos, G.S., Vinyard, B.T., 2010. Influence of conventional and organic agricultural practices on the phenolic content in eggplant pulp: Plant-to-plant variation. Food Chem. 121, 406–411.

Luykx, D.M.A.M., van Ruth, S.M., 2008. An overview of analytical methods for determining the geographical origin of food products. Food Chem. 107, 897–911.

Macel, M., van Dam, N.M., Keurentjes, J.J.B., 2010. Metabolomics: the chemistry between ecology and genetics. Mol. Ecol. Res. 10, 583–593.

Mascherpa, D., Carazzone, C., Marrubini, G., Gazzani, G., Papetti, A., 2012. Identification of phenolic constituents in *Cichorium endivia* var. *crispum* and var. *latifolium* salads by high-performance liquid chromatography with diode array detection and electrospray ionization tandem mass spectrometry. J. Agric. Food Chem. 60, 12142–12150.

Mata Bilbao, M.L., Andrés-Lacueva, C., Jáuregui, O., Lamuela-Raventós, R.M., 2007. Determination of flavonoids in a Citrus fruit extract by LC-DAD and LC-MS. Food Chem. 101, 1742–1747.

Mazid, M., Khan, T.A., Mohammad, F., 2011. Role of secondary metabolites in defence mechanism of plants. Biol. Med. 3 (Special Issue), 232–249.

Mazza, G., Fukumoto, L., Delaquis, P., Girard, B., Ewert, B., 1999. Anthocyanins, phenolics, and color of Cabernet Franc, Merlot, and Pinot noir wines from British Columbia. J. Agric. Food Chem. 47, 4009–4017.

Minussi, R.C., Rossi, M., Bologna, L., Cordi, L., Rotilio, D., Pastore, G.M., et al., 2003. Phenolic compounds and total antioxidant potential of commercial wines. Food Chem. 82, 409–416.

Mogren, L.M., Olsson, M.E., Gertsson, U.E., 2006. Quercetin content in field cured onions (*Allium cepa* L.): effects of cultivar, lifting time, and nitrogen fertilizer level. J. Agric. Food. Chem. 54, 6185–6191.

Mouly, P.P., Arzouyan, C.R., Gaydou, E.M., Estienne, J.M., 1994. Differentiation of Citrus juices by factorial discriminant analysis using liquid chromatography of flavanone glycosides. J. Agric. Food Chem. 42, 70–79.

Mouly, P.P., Gaydou, E.M., Faure, R., Estienne, J.M., 1997. Blood orange juice authentication using cinnamic acid derivatives. Variety differentiations associated with flavanone glycoside content. J. Agric. Food Chem. 45, 373–377.

Ortuño, A., Reynaldo, I., Fuster, M.D., Botía, J., García Puig, D., Sabater, F., et al., 1997. Citrus cultivars with high flavonoid contents in the fruits. Sci. Hortic. 68, 231–236.

Oszmiański, J., Wojdyło, A., Kolniak, J., 2011. Effect of pectinase treatment on extraction of antioxidant phenols from pomace, for the production of puree-enriched cloudy apple juices. Food Chem. 127, 623–631.

Peterson, J.J., Dwyer, J.T., Beecher, G.R., Bhagwat, S.A., Gebhardt, S.E., Haytowitz, D.B., et al., 2006. Flavanones in oranges, tangerines (mandarins), tangors, and tangelos: a compilation and review of the data from the analytical literature. J. Food Composition and Analysis 19, S66–S73.

Plessi, M., Bertelli, D., Albasini, A., 2007. Distribution of metals and phenolic compounds as a criterion to evaluate variety of berries and related jams. Food Chem. 100, 419–427.

Pothavorn, P., Kitdamrongsont, K., Swangpol, S., Wongniam, S., Atawongsa, K., Svasti, J., et al., 2010. Sap phytochemical compositions of some bananas in Thailand. J. Agric. Food Chem. 58, 8782–8787.

Radovanović, B.C., Radovanović, A.N., Souquet, J.-M., 2010. Phenolic profile and free radical-scavenging activity of Cabernet Sauvignon wines of different geographical origins from the Balkan region. J. Sci. Food. Agric. 90, 2455–2461.

Raspor, P., 2005. Bio-markers: traceability in food safety issues. Acta. Biochim. Pol. 3, 659–664.

Renard, C.M.G.C., Le Quéré, J.-M., Bauduin, R., Symoneaux, R., Le Bourvellec, C., Baron, A., 2011. Modulating polyphenolic composition and orgaleptic properties of apple juices by manipulating the pressing conditions. Food Chem. 124, 117–125.

Reyes, L.F., Cisneros-Zevallos, L., 2003. Wounding stress increases the phenolic content and antioxidant capacity of purple-flesh potatoes. J. Agric. Food Chem. 51, 5296–5300.

Reynolds, T., 2007. The evolution of chemosystematics. Phytochemistry 68, 2887–2895.

Riggi, E., Avola, G., Siracusa, L., Ruberto, G., 2013. Flavonol content and biometrical traits as a tool for the characterization of "Cipolla di Giarratana": a traditional Sicilian onion landrace. Food Chem. 140, 810–816.

Robards, K., Antolovich, M., 1997. Analytical chemistry of fruit bioflavonoids. The Analyst 122, 11R–34R.

Rohde, S., Gochfeld, D.J., Ankisetty, S., Avula, B., Schupp, P.J., Slattery, M., 2012. Spatial variability in secondary metabolites of the Indo-Pacific sponge Stylissa massa. J. Chem. Ecol. 38, 463–475.

Sanz, M., Fernández de Simón, B., Cadahía, E., Esteruelas, E., Muñoz, A.M., Hernández, M.T., et al., 2012. Polyphenolic profile as a useful tool to identify the wood used in wine aging. Anal. Chim. Acta. 732, 33–45.

Singh, A.P., Luthria, D., Wilson, T., Vorsa, N., Singh, V., Banuelos, G.S., et al., 2009. Polyphenols content and antioxidant capacity of eggplant pulp. Food Chem. 114, 955–961.

Siracusa, L., Patanà, C., Avola, G., Ruberto, G., 2012. Polyphenols as chemotaxonomic markers in Italian "long storage" tomato genotypes. J. Agric. Food Chem. 60, 309–314.

Slimestad, R., Fossen, T., Vågen, I.M., 2007. Onions: a source of unique dietary flavonoids. J. Agric. Food Chem. 55, 10067–10080.

Stevenson, P.C., Veitch, N.C., 1998. The distribution of isoflavonoids in Cicer. Phytochemistry 48, 995–1001.

Stotz, H.U., Kroymann, J., Mitchell-Olds, T., 1999. Plant-insect interactions. Curr. Opin. Plant. Biol. 2, 268–272.

Templeton, M.D., Lamb, C.J., 1988. Elicitors and defence gene activation. Plant. Cell Environ. 11, 395–401.

Tomás-Barberán, F.A., Martos, I., Ferreres, F., Radovic, B.S., Anklam, E., 2001. HPLC flavonoid profiles as markers for the botanical origin of European unifloral honeys. J. Sci. Food Agric. 81, 485–496.

Tounekti, T., Joubert, E., Hernández, I., Munné-Bosch, S., 2013. Improving the polyphenol content of tea. Crit. Rev. Plant Sci. 32, 192–215.

Treutter, D., Wang, D., Farag, M.A., Argueta Baires, G.D., Rühmann, S., Neumüller, M., 2012. Diversity of phenolic profiles in the fruits skin of *Prunus domestica* plums and related species. J. Agric. Food Chem. 60, 12011–12019.

Tsao, R., Yang, R., 2003. Optimization of a new mobile phase to know the complex and real polyphenolic composition: Towards a total phenolic index using high-performance liquid chromatography. J. Chromatogr. A. 1018, 29–40.

Tuttolomondo, T., La Bella, S., Licata, M., Virga, G., Leto, C., Saija, A., et al., 2013. Biomolecular characterization of wild Sicilian oregano – Phytochemical screening of essential oils and extracts – Evaluation of their antioxidant activities. Chem. Biodivers 10, 411–433.

Vallverdú-Queralt, A., Medina-Remón, A., Martinez-Huélamo, M., Jauregui, O., Andres-Lacueva, C., Lamuela-Raventos, R.M., 2011. Phenolic profile and hydrophilic antioxidant capacity as chemotaxonomic markers of tomato varieties. J. Agric. Food Chem. 59, 3994–4001.

Varshney, R.K., Graner, A., Sorrels, M.E., 2005. Genic microsatellite markers in plants: features and applications. Trends Biotechnol. 23, 48–55.

Venkatesan, S., Ganapathy, M.N.K., 2004. Impact of nitrogen and potassium fertilizer application on quality of CTC teas. Food Chem. 84, 325–328.

Vietina, M., Agrimonti, C., Marmiroli, M., Bonas, U., Marmiroli, N., 2011. Applicability of SSR markers to the traceability of monovarietal olive oils. J. Sci. Food Agric. 91, 1381–1391.

Wahyuni, Y., Ballester, A.R., Tikunov, Y., de Vos, R.C.H., Pelgrom, K.T.B., Maharijaya, A., et al., 2013. Metabolomics and molecular marker analysis to explore pepper (*Capsicum* sp.) biodiversity. Metabolomics 9, 130–144.

Xu, Y., Simon, J.E., Welch, C., Wightman, J.D., Ferruzzi, M.G., Ho, L., et al., 2011. Survey of polyphenol constituents in grapes and grape-derived products. J. Agric. Food Chem. 59, 10586–10593.

Yonekura-Sakakibara, K., Saito, K., 2009. Functional genomics for plant natural product biosynthesis. Nat. Prod. Rep. 26, 1466–1487.

Zhao, H., Guo, B., Wei, Y., Zhang, B., 2012. Effects of wheat origin, genotype, and their interaction on multielement fingerprints for geographical traceability. J. Agric. Food Chem. 60, 10957–10962.

Zimmermann, B.F., Galensa, R., 2007. One for all—all for one: proof of authenticity and tracing of foods with flavonoids. Analysis of proanthocyanidins in barley and malt. Eur. Food Res. Technol. 224, 385–393.

Zolg, J.W., Langen, H., 2004. How industry is approaching the search for new diagnostic markers and biomarkers. Mol. Cell. Proteomics 3, 345–354.

Zubair, M., Nybom, H., Ahnlund, M., Rumpunen, K., 2012. Detection of genetic and phytochemical differences between and within populations of *Plantago major* L. (plantain). Sci. Hortic. 136, 9–16.

Stress and Polyphenols in Plants

Phenolic Compounds and Saponins in Plants Grown Under Different Irrigation Regimes

3

Ana Maria Gómez-Caravaca[*†], **Vito Verardo**[‡], **Antonio Segura-Carretero**[*†], **Alberto Fernández-Gutiérrez**[*†], **Maria Fiorenza Caboni**[‡§]

Department of Analytical Chemistry, University of Granada, Granada, Spain, †Functional Food Research and Development Centre (CIDAF), Armilla (Granada), Spain, ‡Inter-Departmental Centre for Agri-Food Industrial Research (CIRI Agroalimentare), University of Bologna, Italy, §Department of Agro-Food Sciences and Technologies, Alma Mater Studiorum Università di Bologna, Bologna, Italy

CHAPTER OUTLINE HEAD

3.1 Introduction

Plants produce a wide range of metabolites during their growth and life. These metabolites can be divided into primary and secondary metabolites, even though it is not easy to assign many of them to an unique class of metabolites. Primary metabolites are defined as those compounds that have essential roles associated with photosynthesis, respiration, growth, and development. Compounds such as carbohydrates, lipids,

Polyphenols in Plants. http://dx.doi.org/10.1016/B978-0-12-397934-6.00003-6

nucleotides, amino acids, and organic acids are included in this group. Conversely, secondary metabolites are natural products that are not essential for basic growth and development, but are frequently involved in environmental adaptation to both biotic and abiotic stresses. These compounds are structurally diverse and many of them are distributed among a very limited number of species within the plant kingdom and so can be diagnostic in chemotaxonomic studies (Crozier *et al.*, 2006; Seigler, 1998).

Secondary metabolites are of great importance for plants because they have been shown to have a key role in protecting plants from herbivores and microbial infection, as attractants for pollinators and seed-dispersing animals, as allelopathic agents, UV protectants, etc. They also contribute to the specific odors, tastes, and colors in plants (Bennett and Wallsgrove, 1994) and many of them have interesting applications in pharmacology, chemical industry, novel products, agriculture, and forestry (Cannes do Nascimento and Fett-Neto, 2010).

Since secondary metabolism is dispensable for growth and development, its components can be continuously modified and adapted to the demands of an incessantly changing environment. This high plasticity is mirrored by its attributes: they are unique (some of them are characteristics of certain plants), diverse, and have an adaptative function (Hartmann, 2007).

As has been mentioned before, the secondary metabolites are synthesized mainly to fight against biotic and abiotic stresses. The biotic stress can be caused by bacteria, fungi, plants, insects, nematodes, mammals, and/or birds (Kliebenstein, 2012); while the abiotic stress is produced by high and low temperature, drought, alkalinity, salinity, UV stress, etc. (Seigler, 1998). Some secondary metabolites are continuously produced in plants, whereas others are newly formed in response to stress signals. Agriculturally, the most damaging abiotic stress is water deficit leading to drought (Chaves *et al.*, 2003). Exposure to water or osmotic stress triggers the production of active oxygen species that can be extremely harmful to plant cells, causing oxidative damage and inactivation of enzymes.

Although secondary metabolites are chemically very complex, most of them can be grouped into major classes based on their chemical structure and biosynthetic origin: (a) shikimate-phenylpropanoid pathway (from where the phenolic compounds originate); (b) terpenoid pathway; and (c) lypoxygenase/hydroperoxide lyase pathway (Chen *et al.*, 2009).

Phenolic compounds and terpenoids are two of the major groups of secondary metabolites in plants. Thus, in this chapter, we will present an outline of the main abiotic factors that affect the secondary metabolism of phenolic compounds and a family of terpenoids, named saponins, and special attention will be paid to the effect produced due to water-deficit or drought.

3.2 Secondary metabolites: phenolic compounds and saponins

The phenolic compounds family includes a large number of secondary plant products which differ in chemical structure and reactivity, ranging from simple compounds to highly polymerized compounds and derives from phenylalanine and, to a lesser extent in some plants, also from tyrosine (Shahidi and Naczk, 1995). Phenolics are characterized by having at least one aromatic ring with one or more hydroxyl groups attached. In excess of 8000 phenolic structures have been reported and they are widely dispersed throughout the plant kingdom (Strack, 1997). Plants contain a large variety of phenolic derivatives and, the groups of phenolics most relevant for human health are phenolic acids, flavonoids (flavones, flavanones, flavonols, 3-flavanols, isoflavones, anthocyanins) lignans, and stilbenes (Hooper and Cassidy, 2006) as reported in Figure 3.1.

Phenolic compounds have attracted great interest in recent years because of evidence of their beneficial health effects on humans, their potent antioxidant properties, their abundance in the diet, and their credible effects in the prevention of various oxidative-stress-associated diseases (Manach *et al.*, 2004). The ingestion of foods rich in these substances has been associated in humans and experimental animals with reduction in dyslipidemia and atherosclerosis, endothelial dysfunction and hypertension, platelet activation and thrombosis, and inflammatory processes associated with induction and perpetuation of cardiovascular diseases (Fraga *et al.*, 2010).

Saponins are a diverse family of secondary metabolites produced by many plants species. Saponins are glycosides and are synthesized from mevalonic acid via the isoprenoid pathway. They can be classified into two groups based on the nature of their aglycone skeleton. The first group consists of the steroidal saponins, which are almost exclusively present in the monocotyledonous angiosperms. The second group, and the most common one, consists of the triterpenoid saponins, which occur mainly in the dicotyledonous angiosperms (Bruneton, 1995).

Saponins are characterized by their surfactant properties (saponin derived from the Latin *sapo*, which means soap) and give stable, soap-like foams in aqueous solution (see Figure 3.2).

Most saponins have haemolytic properties and are toxic to most cold-blooded animals. Despite that, if taken orally by warm-blooded species, saponins have only a weak toxicity, which is probably attributed to low absorption rates (Dini *et al.*, 2001a, 2001b).

The natural role of these molecules in plants is likely to be in conferring protection against attack by potential pathogens (Morrissey and Osbourn, 1999). Therefore, due to their toxicity to various organisms, saponins can be used as antimolluscicidal, antifungal, antiyeast, antibacterial, antimicrobial, antiparasitic and antitumorals. Also, they present anti-inflammatory activity (Sparg *et al.*, 2004).

They also have a variety of commercial applications for a diversity of purposes including drugs and medicines, precursors for hormone synthesis, adjuvants, foaming agents, sweeteners, taste modifiers, and cosmetics (Osbourn, 2003).

3.3 Factors that influence plants' secondary metabolism

Plants are exposed to numerous and diverse environmental stresses that alter their basic metabolism during their development and life. To successfully complete their life cycle, plants have to face many environmental constraints, and if these parameters exceed a critical limit, plants are said to be under stress.

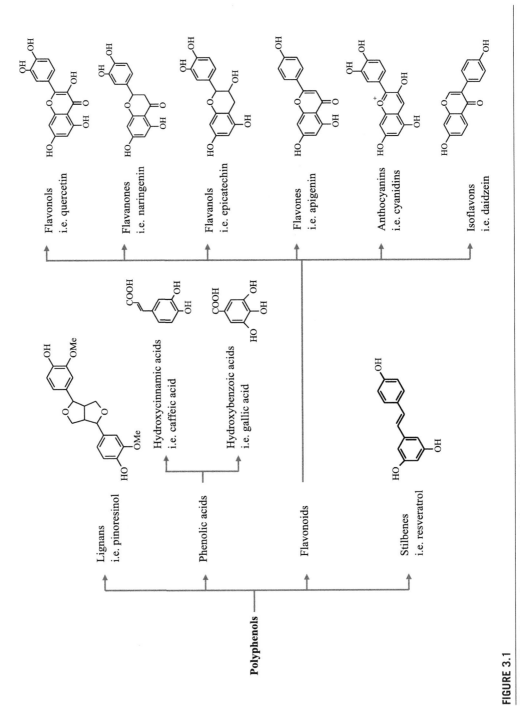

FIGURE 3.1

Main families of phenolic compounds found in plants.

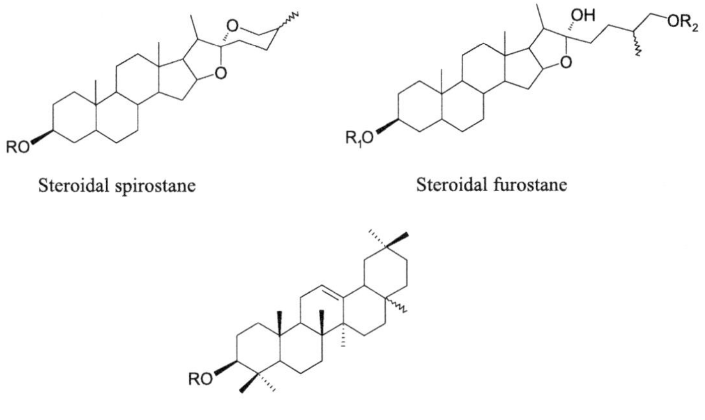

FIGURE 3.2

Skeletons of the different aglycone saponins.

Phytochemical studies have indicated that stresses result in the overproduction of reactive oxygen species (ROS). Thus, an antioxidant serves to establish a delicate balance between the antioxidant scavenging system and ROS at their site of production (Gechev *et al.*, 2006) and introduce tolerance to stress. This balance requires a complex antioxidant system that includes glutathione, ascorbate, and powerful antioxidants such as polyphenols.

Most plants constitutively synthesize low concentrations of polyphenols. However, accumulation of phenolic compounds in plants can be induced by abiotic and biotic stresses, such as UV radiation, high light intensity, temperature, salinity, supply of water, wounding, low nutrients, and pathogen attack (Ramakrishna and Ravishankar, 2011). In fact, phenolics have been reported to accumulate under extreme stress conditions, accompanied by increased concentrations of phenylalanine ammonia-lyase but decreased activities of peroxidase polyphenyl lyase (Bartwal *et al.*, 2013).

In the case of saponins, the content of these molecules is very high in some plants, reaching a high percentage of the dry weight. Nevertheless, the concentration of saponins can be significantly influenced by many external factors. The external factors comprise environmental stimuli, both abiotic and biotic, including temperature, soil fertility, the availability of water and light, feeding of phytophagous insects or other herbivorous animals, competition with neighboring plants, and interactions with pathogens and parasites such as bacteria, fungi, viruses, and nematodes (Szakiel *et al.*, 2011).

As has been reported in the literature, biotic and abiotic factors influence the production of phenolics and saponins and they have a bearing on enhancing the potential to overproduce them for further applications. Thus, significant efforts are being made in the qualitative and quantitative analysis of phenolics and saponins in different plant varieties and cultivars of particular importance, with the aim of increasing the content of compounds with pharmacological or nutritional value, or decreasing the level of toxins or antinutrients (Ramakrishna and Ravishankar, 2011; Shim *et al.*, 2003; Pecetti *et al.*, 2006; Schwarzbach *et al.*, 2006).

3.4 Influence of the irrigation regimes on secondary metabolites

One of the most important environmental conditions that affects plants is the quantity of water they receive during their growth and development. Plants respond to water stress by dramatically complex mechanisms, from genetic molecular express, biochemical metabolism through individual plant physiological processes to ecosystem levels. One of these mechanisms consists of acquiring drought resistance via altering metabolic pathways to survive under severe stress (i.e., increased antioxidant metabolism) (Xu *et al.*, 2010).

Most studies compare the concentration of the different secondary metabolites between stressed and unstressed plants, but it has to be taken into account that just one stress factor can be covered comprehensively. However, it is important to know that it is nearly impossible to avoid the interference of other factors, i.e., the lower water availability is very often related to higher salt concentrations in the soil. Despite that, results could be scientifically inconclusive for the effects of single factors on the accumulation of secondary plant products. Because of that, the influence of the irrigation regimes on secondary metabolites of plants (phenolic compounds and saponins) has been the aim of this study.

3.4.1 Phenolic compounds

Phenolic compounds act as potential antioxidant compounds by donating electrons to guaiacol peroxidases for detoxification of high amounts of H_2O_2 produced under stress conditions (Decker, 1997).

Regarding the different irrigation regimes that can be used during plant growth, there are many studies that focus on the phenolic content variation under the diverse situations of water stress. Situations of drought stress and different irrigation regimes and their influence on phenolic content have been reported in several constituents of the plant kingdom, such as, olive oil, tea, wine, tomato, potato, cereals, medicinal plants, etc.

3.4.1.1 Olive fruit and olive oil

It is known that the olive tree is drought-resistant; in fact, it is usually grown in the Mediterranean region, an area with limited water resources. However, it has been seen that olive trees respond positively to irrigation and that controlled water deficit is useful for olive oil production. Furthermore, olive fruit and olive oil quality seems to improve under situations of water stress. This is in part due to antioxidant compounds such as polyphenols that confer higher oxidative stability on the fruit and oil.

Olive fruit and olive oil phenolic compounds are highly affected by the activity of an enzyme responsible for the phenilpropanoid pathway. This enzyme is the L-phenylalanine ammonia lyase (PAL), and Tovar *et al.* have reported that there is a significant correlation between PAL and phenolic compounds, and that the activity of this enzyme varied as a result of changes in water status (Tovar *et al.*, 2002a). As the PAL activity is greater under higher water stress conditions, this will lead to superior phenolic contents in the olive flesh and, therefore, in the virgin olive oils obtained.

Many studies corroborate the affirmation carried out by Tovar *et al.* (2002a) and confirm that water deficit and drought provoke an increase of phenolic compounds in olive fruits (Patumi *et al.*, 2002; Marsilio *et al.*, 2006) and olive oil (Motilva *et al.*, 2000; Gómez-Rico *et al.*, 2007; Servili *et al.*, 2007; Dabbou *et al.*, 2011).

Major phenolic compounds of olive oil such as oleuropein aglycone and its derivatives have been reported to increase under water stress situations (Tovar *et al.*, 2001; Marsilio *et al.*, 2006; Dabbou *et al.*, 2011) and, because of that, these compounds could be a stress index for olive tree (Marsilio *et al.*, 2006). Indeed, the concentration of lignans as (+)-1-acetoxypinoresinol and (+)-1-pinoresinol were slightly affected by the different irrigation regimes (Servili *et al.*, 2007; Dabbou *et al.*, 2011).

The control of the irrigation regime to increment the phenolic compounds content in olive oil has become of great interest because these compounds affect a positive sensory attribute of olive oil, namely bitterness and pungency, and improve the oxidative stability of olive oil (Tovar *et al.*, 2002b).

Despite these findings, there are some authors that have found a decrease of phenolic compounds in olive oils from trees under severe drought (Greven *et al.*, 2009; Dabbou *et al.*, 2010). Probably, other factors can simultaneously affect olive trees, altering the final phenolic content.

3.4.1.2 Solanaceae

The solanaceae family constitutes some of the most important crops for human nutrition, such as tomato and potato.

Some authors have demonstrated that polyphenolic profiles of potato are to some extent correlated to variations in gene expression profiles. Drought-induced changes of the expression of the gene profile were also directly related to variations of the phenolic content. Therefore, the control of the gene expression plays an essential role in the polyphenolic biosynthetic pathway. Although, the range of induction or repression of gene expression in potatoes under water stress is highly cultivar-dependent (André *et al.*, 2009).

The previous theory also seems to be fulfilled in the case of the tomato. Some studies have found that chlorogenic acid and its derivatives and flavonoids diminish in a generalized way under water stress in different cultivars; however, there are some other cultivars more water-stress tolerant that produce a higher content of phenolics under water deficit (Sánchez-Rodríguez *et al.*, 2010, 2011, 2012; Atkinson *et al.*, 2011). Because of that, the cultivar and the activity of its enzymes (PAL) influence the production of phenolics. It has been demonstrated that PAL-deficient mutants are sensitive to environmental stress, demonstrating the role of these antioxidants in plant adaptation (Gitz *et al.*, 2004).

3.4.1.3 Grapevine and wine

Grapevine (*Vitis vinifera* L.) is known to adapt well to dry and semi-dry regions; but despite the drought tolerance, grapevine growth under low water conditions will present limitations in development, abnormal ripening and reduced berry quality. Because of that, studies under different irrigation regimes have been carried out to improve the quality of wines obtained.

Most studies have demonstrated that irrigation of grapevines incite an increase in phenolic content, but it has also been found that in order to increase the concentration of phenolic compounds in wines, water stress should be applied during the pre-veraison period (Koundouras *et al.*, 2009; Intrigliolo and Castel, 2010; Zarrouk *et al.*, 2012).

3.4.1.4 Tea

The tea plant is a perennial and, because of that, encounters a large number of environmental stresses throughout its life. Among the main climatic variables that influence the growth of tea plant is the soil water deficit.

According to Chakraborty *et al.*, water stress results in an accumulation of phenolic compounds after a period of 4–8 days followed by a decline of them under prolonged water stress (Chakraborty *et al.*, 2002).

Other studies have reported that low soil water content diminishes phenolic content; however, the decrease is less extensive in drought-tolerant clones of tea. This might provide a basis for clonal selection, improvement and/or management of tea for better yield and quality (Cheruiyot *et al.*, 2007). Moreover, epicatechin and catechin could be used as indicators in predicting drought tolerance in tea (Cheruiyot *et al.*, 2008), and the rehydration of tea plants after elevated water stress shows a high recovery of antioxidant properties (Upadhyaya *et al.*, 2008). Furthermore, Hernández *et al.* (2006) have found that flavonoid accumulation occurs under water deficit.

3.4.1.5 Cereals and pseudocereals

Cereal grains contribute more than 60% of the total world food demand; because of that, the study of the influence of drought stress on the crop productivity is of great interest.

Wheat leaves from different genotypes have shown that water stress produced an increase of total phenolic content (TPC) (Hameed *et al.*, 2012). Similar results were found for rice cultivars and, in this case, it was evident that the increment of phenolic compounds was much higher in the varieties with a higher tolerance to drought stress (Basu *et al.*, 2010).

Regarding maize, it has been reported that its resistant genotypes observed an increase in the TPC, while other less drought-tolerant genotypes have a decrease of these compounds. Thus, the increased level and synthesis during drought can be an indicator of the resistance of maize to water stress (Hura *et al.*, 2008). Besides, despite the decrease of TPC, water stress significantly increases the flavonoids content in maize (Ali *et al.*, 2010). Analogous results have been found by Dicko *et al.* (2005) for sorghum.

Quinoa, a pseudocereal that has been reported to be highly drought-tolerant, also shows wide increments of phenolic compounds under severe water deficit (Gómez-Caravaca *et al.*, 2012).

3.4.1.6 Medicinal plants

Generally, medicinal plants grow freely in the countryside and, thus, most of them are well-adapted to environmental stresses. Different studies about the effect of drought on phenolic compounds in these kinds of plants have been carried out and they have shown that under water deficit medicinal plants accumulate higher concentrations of phenolics (Table 3.1).

Table 3.1 Irrigation Regimes and their Influence on the Phenolic Content in Medicinal Plants

Plant	Treatment	Influence on Phenolics	References
Echinacea purpurea	Two drydown periods were imposed during the reproductive stage for two consecutive seasons. Some plants received drought stress during the initial flowering and other plants during the seed production period.	Total phenolics increased an average of 67.1% for all treatments.	Gray *et al.*, 2003
Ligustrum vulgare	Plants supplied with either 100% or 40% of the daily evapotranspiration demand during 8 weeks.	Drought-induced decrease phenolic compounds content, but increase the CO_2-based polyphenol accumulation.	Tattini *et al.*, 2004
Hypericum brasiliense	Drought stress was applied gradually by watering with half of the weight loss per day during 15 days.	Phenolic compounds concentration increase under water deficit.	De Abreu and Mazzaferra, 2005
Salvia officinalis	Plants were subjected to different water levels: 100%, 50%, and 25% during two months.	Significant increase of phenolic compounds: four-fold under medium water deficit and three-fold under severe water deficit.	Bettaieb *et al.*, 2011
Cuminum cyminum	Plants were subjected to different water levels: 100%, 50%, and 25% and two harvests (2010 and 2011) were carried out.	Significant increase of phenolic compounds: 43.77% under medium water deficit and 15.28% under severe water deficit.	Rebey *et al.*, 2012
Mentha pulegium	Six weeks after sowing, plants were grown under soil moisture corresponding to 100, 75, 50, and 25% field capacity.	The phenolic contents significantly increased in the roots and leaves under water deficit stress.	Hassanpour *et al.*, 2012
Erica multiflora	The drought treatment was applied for 2- to 3-month periods during the spring and autumn growing seasons by covering the vegetation with waterproof, transparent covers.	No significant differences in total phenolics concentrations were found among treatments in spring, whereas in summer the concentrations of these compounds were higher under drought.	Nogués *et al.*, 2012

3.4.1.7 Other plants

Other plants have also been studied to understand the influence of the irrigation regimes on their phenolic content.

Bouchereau *et al.* (1996) studied three spring genotypes of rapeseed by inducing drought at three different stages of development of the plant: at the early vegetative growth period (T1), at a larger stage (T2) and, during flowering (T3). It was found that the phenolic content was highly sensitive to the stage of the application of the water shortage treatment. At T1 and T3 the accumulation of phenolic compounds was strongly inhibited; however, at T2 no significant differences were observed.

Populus nigra L., a rapid growing plant, was subjected to drought stress and it was observed that total phenolic glycoside concentrations were 89% higher in the drought-stressed relative compared to the well-watered plants (Hale *et al.*, 2005).

A study to elucidate whether or not the degree of deficit irrigation affects strawberry fruit quality and biochemistry has also been carried out. The results showed that total phenolics were approximately between 1.3–1.4-fold higher in fruit form plants under water deficit than those well watered. Thus, the increment of phenolics and other healthfulness phytochemicals makes manipulation of water a viable prospect for increasing fruit quality (Terry *et al.*, 2007).

Tolerant and sensitive cotton genotypes, grown in the Aegean region of Turkey, have also been compared at normal (field capacity) and limited ($\frac{1}{3}$ field capacity) water supply. Chlorogenic acid isomers and flavonoids were identified in HPLC pattern of polyphenols. At normal water supply, the tolerant genotype was distinguished by a higher content of all polyphenol types compared with the sensitive genotype. In plants subjected to water deficit, a decline of all polyphenol compounds was observed. However, this response was less pronounced in the tolerant than in the sensitive genotype. Despite the stress conditions imposed, the tolerant plants maintained a more effective defence system (Yildiz-Aktas *et al.*, 2009).

Lettuce (*Lactuca sativa* L.), another major crop within Europe, has also been studied to see the influence of water on phenolics. Baslam *et al.* (2012) studied plants grown under four different water regimes: optimal irrigation (field capacity (FC)), a water regime equivalent to $\frac{2}{3}$ of FC, a water regime equivalent to ½ of FC and a cyclic drought (CD). The results showed that phenolics were not affected by water stress or were slightly increased, but nothing could be concluded because the variety influenced the behaviour of the phenolic content (Baslam and Goicoechea, 2012). Fresh-cut lettuce is also of interest to the food industry, because of the effects of five drip irrigation treatments (excess 50%, excess 25%, control, deficit 25%, and deficit 50%) on the quality and shelf-life of fresh-cut iceberg lettuce were examined in six different harvest dates during three consecutive years. After cutting, lettuce phenolic compounds (mainly caffeic acid derivatives) increased at second day of storage, but the increase was more noticeable in the less irrigated samples because of the low polyphenol oxidase (PPO) activity and therefore less consumption of PPO substrates. Thus, the quality and shelf-life of the fresh-cut lettuce was better preserved by reduction in irrigation quantity during the growing period. This result may provide an opportunity to lettuce growers for improving sustainability in growing areas highly dependent on water (Luna *et al.*, 2012).

The isoflavones of soybean have several uses for food and nutraceutical applications. Therefore, cultivation involved drip irrigation and rainfed conditions, for 2 years. Yield components and hypocotyl and cotyledon isoflavone concentrations were measured in stems and branches separately. Unexpectedly, the response to water supply was almost negligible. However, the most important finding was a 24.4% cotyledon (1.54 *versus* 1.96 mg/g) and 4.2% hypocotyl (14.9 *versus* 14.3 mg/g) greater isoflavone concentration in seeds of branches compared with those of the main stem (Vameralia *et al.*, 2012).

3.4.2 Saponins

The current understanding considers saponins to be derived from intermediates of the phytosterol pathway, and predominantly enzymes belonging to the multigene families of oxidosqualene cyclases (OSCs), cytochromes P450 (P450s) and family 1 UDP-glycosyltransferases (UGTs) are thought to be involved in their biosynthesis. However, the total number of identified genes in saponin biosynthesis remains low as the complexity and diversity of these multigene families impede gene discovery based on sequence analysis and phylogeny (Augustin *et al.*, 2011).

Concerning the influence of the different irrigation regimes on the concentration of these compounds there is not much information in the literature.

Ayet *et al.* (1997) observed the saponins contained in lentils after germination and treatment watering daily or on alternate days. The authors detected that 6-day germinated seeds contained the maximum level of soysapogenol B when alternating watering was applied. Because of that, the saponin levels during germination could be explained by their implication in the defence system of the plants.

Four different levels of water stress, mild water stress (W1, 65% FC), moderate water stress (W2, 50% FC), severe water stress (W3, 35% FC) and the control (CK, 80% FC), were applied to study the effects of water stress on the growth and total saponin content of fibrous root of *Liriope muscari* (Decne.) Bailey by pot culture. The total saponin yield of fibrous root of the control treatment was significantly higher than those of the mild and severe water treatments, while the total saponin contents of fibrous root had no significant difference among the different water stress treatments (Hailan *et al.*, 2011).

Quinoa (*Chenopodium quinoa*) is the plant that has been studied more in depth regarding its content of saponins under different irrigation regimes. Soliz-Guerrero *et al.* (2002) found that soil water deficit treatment affected the saponin content; high deficit promoted low saponin content for the two cultivars. Saponin content depended on the growing stage for all treatments and cultivars. The highest saponin content was found at blooming and decreased at the grain filling stage for treatments and cultivars. Later, Martínez *et al.* (2009) carried out research with two landraces from Chile and it was observed that saponin content was different between landraces and stable within the same order of magnitude between the harvests. However, conclusions about the effects of the irrigation conditions on saponin content could not be provided. Other authors were in agreement with Soliz-Guerrero *et al.* (2002), and their studies reported the irrigation deficit was associated with a minor accumulation of saponins in quinoa seeds and that samples not subject to drought stress showed the highest content of saponins (Gómez-Caravaca *et al.* 2012; Pulvento *et al.*, 2012).

3.5 **Concluding remarks**

Phenolic compounds and saponins, as secondary metabolites, are highly influenced by environmental stresses such as water scarcity.

As can be seen, results reported in the literature show that phenolic content depends on the activity of some enzymes, and these enzymes depend on the genotype of the plants. Therefore, in general, highly drought-tolerant species present an increment of phenolic compounds while less drought-tolerant species can remain without changes or present a decrease in the concentration of phenolic compounds.

Saponin content under different irrigation regimes has not been widely studied and also the enzymes involved in their biosynthesis are not known in depth. However, in highly drought-tolerant species, such as quinoa, saponins seem to diminish under water stress conditions.

Hence, further studies in this direction are needed to help growers to understand the balance between enhancing phytochemical content by regulating irrigation and maintaining good quality of the product.

References

Ali, Q., Ashraf, M., Anwar, F., 2010. Seed composition and seed oil antioxidant activity of maize under water stress. J. Am. Oil. Chem. Soc. 87, 1179–1187.

André, C.M., Schafleitner, Roland., Legay, S., Lefèvre, I., Alvarado Aliaga, C.A., Nomberto, G., et al., 2009. Gene expression changes related to the production of phenolic compounds in potato tubers grown under drought stress. Phytochemistry 70, 1107–1116.

Atkinson, N.J., Dew, T.P., Orfila, C., Urwin, P.E., 2011. Influence of combined biotic and abiotic stress on nutritional quality parameters in tomato (*Solanum lycopersicum*). J. Agric. Food Chem. 59, 9673–9682.

Augustin, J.M., Kuzina, V., Andersen, S.B., Bak, S., 2011. Molecular activities, biosynthesis and evolution of triterpenoid saponins. Phytochemistry 72, 435–457.

Ayet, G., Burbano, C., Cuadrado, C., Pedrosa, M.M., Robredo, L.M., Muzquiz, M., et al., 1997. Effect of germination, under different environmental conditions, on saponins, phytic acid and tannins in lentils (*Lens culinaris*). J. Sci. Food. Agric. 74, 273–279.

Bartwal, A., Mall, R., Lohani, P., Guru, S.K., Arora, S., 2013. Role of secondary metabolites and brassinosteroids in plant defense against environmental stresses. J. Plant Growth Regul. 32, 216–232.

Baslam, M., Goicoechea, N., 2012. Water deficit improved the capacity of arbuscular mycorrhizal fungi (AMF) for inducing the accumulation of antioxidant compounds in lettuce leaves. Mycorrhiza 22, 347–359.

Basu, S., Roychoudhury, A., Saha, P.P., Sengupta, D.N., 2010. Differential antioxidative responses of indica rice cultivars to drought stress. Plant Growth Regul. 60, 51–59.

Bennett, R.N., Wallsgrove, R.M., 1994. Secondary metabolites in plant defence mechanisms. New Phytol. 127, 617–633.

Bettaieb, I., Hamrouni-Sellami, I., Bourgou, S., Limam, F., Marzouk, B., 2011. Drought effects on polyphenol composition and antioxidant activities in aerial parts of *Salvia officinalis* L. Acta. Physiol. Plant. 33, 1103–1111.

Bouchereau, A., Clossais-Besnard, N., Bensaoud, A., Leport, L., Renard, M., 1996. Water stress effects on rapeseed quality. Eur. J. Agronomy 5, 19–30.

Bruneton, J., 1995. Pharmacognosy, Phytochemistry. Medicinal Plants. Lavoisier Publishing, Paris.

Cannes do Nascimento, N., Fett-Neto, A.G., 2010. Plant secondary metabolism and challenges in modifying its operation: an overview. In: Fett-Neto, A.G. (Ed.), Plant Secondary Metabolism Engineering, Methods in Molecular Biology. Humana Press Inc., New York, pp. 1–13.

Chakraborty, V., Dutta, S., Chakraborty, B.N., 2002. Responses of tea plant to water stress. Biol. Plant. 45 (4), 557–562.

Chaves, M.M., Maroco, J.P., Pereira, J.S., 2003. Understanding plant responses to drought—From genes to the whole plant. Functional Plant Biol. 30, 239–264.

Chen, F., Liu, C.-J., Tschaplinski, T.J., Zhao, N., 2009. Crit. Rev. Plant Sci. 28, 375–392.

Cheruiyot, E.K., Mumera, L.M., Ng'etich, W.K., Hassanali, A., Wachira, F., 2007. Polyphenols as potential indicators for drought tolerance in tea (*Camellia sinensis* L.). Biosci. Biotechnol. Biochem. 71, 2190–2197.

Cheruiyot, E.K., Mumera, L.M., Ng'etich, W.K., Hassanali, A., Wachira, F., Wanyoko, J.K., 2008. Shoot epicatechin and epigallocatechin contents respond to water stress in tea [Camellia sinensis (L.) O. Kuntze]. Biosci. Biotechnol. Biochem. 72 (5), 1219–1226.

Crozier, A., Clifford, M.N., Ashihara, H., 2006. Plant Secondary Metabolites: Occurrence, Structure and Role in the Human Diet. Blackwell Publishing Ltd, Oxford, United Kingdom.

Dabbou, S., Chehab, H., Faten, B., Dabbou, S., Esposto, S., Selvaggini, R., et al., 2010. Effect of three irrigation regimes on Arbequina olive oil produced under Tunisian growing conditions. Agric. Water Manag. 97, 763–768.

Dabbou, S., Dabbou, S., Chehab, H., Brahmi, F., Taticchi, A., Servili, M., et al., 2011. Chemical composition of virgin olive oils from Koroneiki cultivar grown in Tunisia with regard to fruit ripening and irrigation regimes. Int. J. Food Sci. Technol. 46, 577–585.

De Abreu, I.N., Mazzafera, P., 2005. Effect of water and temperature stress on the content of active constituents of *Hypericum brasiliense* Choisy. Plant Physiol. Biochem. 43, 241–248.

Decker, E.A., 1997. Phenolics: prooxidants or antioxidants? Nutr. Rev. 55, 396–407.

Dicko, M.H., Gruppen, H., Barro, C., Traore, A.S., Van Berkel, W.J.H., Voragen, A.G.J., 2005. Impact of phenolic compounds and related enzymes in sorghum varieties for resistance and susceptibility to biotic and abiotic stresses. J. Chem. Ecol. 31 (11), 2671–2688.

Dini, I., Schettino, O., Simioli, T., Dini, A., 2001a. Studies on the constituents of *Chenopodium quinoa* seeds: isolation and characterization of new triterpene saponins. J. Agric. Food Chem. 49, 741–746.

Dini, I., Tenore, G.C., Schettino, O., Dini, A., 2001b. New oleanane saponins in *Chenopodium quinoa*. J. Agric. Food Chem. 49, 3976–3981.

Fraga, C.G., Galleano, M., Verstraeten, S.V., Oteiza, P.I., 2010. Basic biochemical mechanisms behind the health benefits of polyphenols. Mol. Aspects. Med. 31, 435–445.

Gechev, T.S., Van Breusegem, F., Stone, J.M., Denev, I., Laloi, C., 2006. Reactive oxygen species as signals that modulate plant stress responses and programmed cell death. Bioessays 28, 1091–1101.

Gitz, D.C., Lui-Gitz, L., McClure, J.W., Huerta, A.J., 2004. Effects of PAL inhibitor on phenolic accumulation and UV-B tolerance in *Spirodela intermedia* (Koch.). J. Exp. Bot. 55, 919–927.

Gómez-Caravaca, A.M., Iafelice, G., Lavini, A., Pulvento, C., Caboni, M.F., Marconi, E., 2012. Phenolic compounds and saponins in quinoa samples (*Chenopodium quinoa* Willd.) grown under different saline and nonsaline irrigation regimens. J. Agric. Food Chem. 60, 4620–4627.

Gómez-Rico, A., Salvador, M.D., Moriana, A., Pérez, D., Olmedilla, N., Ribas, F., et al., 2007. Influence of different irrigation strategies in a traditional *Cornicabra* cv. Olive orchard on virgin olive oil composition and quality. Food Chem. 100, 568–578.

Gray, D., Pallardy, S.G., Garrett, H.E., Rottinghaus, G.E., 2003. Acute drought stress and plant age effects on alkamide and phenolic acid content in purple coneflower roots. Planta Med. 69, 50–55.

Greven, M., Neal, S., Green, S., Dichio, B., Clothier, B., 2009. The effects of drought on the water use, fruit development and oil yield from young olive trees. Agric. Water Manag. 96, 1525–1531.

Hailan, F., Wei, H., Chengzhen, W., Jian, L., Can, C., Jinhu, H., et al., 2011. Effects of water stress on growth and total saponin content of *Liriope muscari* (Decne.) Bailey. Chinese J. Appl. Environ. Biol. 17 (3), 345–349.

Hale, B.K., Herms, D.A., Hansen, R.C., Clausen, T.P., Arnold, D., 2005. Effects of drought stress and nutrient availability on dry matter allocation, phenolic glycosides, and rapid induced resistance of poplar to two lymantriid defoliators. J. Chem. Ecol. 31 (11), 2601–2620.

Hameed, A.M., Goher, M., Iqbal, N., 2012. Drought induced programmed cell death and associated changes in antioxidants, proteases and lipid peroxidation in wheat leaves. 10.1007/s10535-012-0286-9.

Hartmann, T., 2007. From waste products to ecochemicals: Fifty years research of plant secondary metabolism. Phytochemistry 68, 2831–2846.

Hassanpour, H., Khavari-Nejad, R.A., Niknam, V., Najafi, F., Razavi, K., 2012. Effects of penconazole and water deficit stress on physiological and antioxidative responses in pennyroyal (*Mentha pulegium* L.). Acta. Physiol. Plant. 34, 1537–1549.

Hernández, I., Alegre, L., Munné-Bosch, S., 2006. Enhanced oxidation of flavan-3-ols and proanthocyanidin accumulation in water-stressed tea plants. Phytochemistry 67, 1120–1126.

Hooper, L., Cassidy, A., 2006. A review of the health care potential of bioactive compounds. J. Sci. Food. Agric. 86, 1805–1813.

Hura, T., Hura, K., Grzesiak, S., 2008. Contents of total phenolics and ferulic acid, and PAL activity during water potential changes in leaves of maize single-cross hybrids of different drought tolerance. J. Agronomy Crop Sci. 194, 104–112.

Intrigliolo, D.S., Castel, J.R., 2010. Response of grapevine cv. "Tempranillo" to timing and amount of irrigation: water relations, vine growth, yield and berry and wine composition. Irrigation Sci. 28, 113–125.

Kliebenstein, D.J., 2012. Plant Defense Compounds: Systems Approaches to Metabolic Analysis. Annu. Rev. Phytopathol. 50, 155–173.

Koundouras, S., Hatzidimitriou, E., Karamolegkou, M., Dimopoulou, E., Kallithraka, S., Tsialtas, J.T., et al., 2009. Irrigation and rootstock effects on the phenolic concentration and aroma potential of *Vitis vinifera* L. cv. Cabernet Sauvignon grapes. J. Agric. Food Chem. 57, 7805–7813.

Luna, M.C., Tudela, J.A., Martínez-Sánchez, A., Allende, A., Marín, A., Gil, M.I., 2012. Long-term deficit and excess of irrigation influences quality and browning related enzymes and phenolic metabolism of fresh-cut iceberg lettuce (*Lactuca sativa* L.). Postharvest. Biol. Technol. 73, 37–45.

Manach, C., Scalbert, A., Morand, C., Remesy, C., Jimenez, L., 2004. Polyphenols: food sources and bioavailability. Am. J. Clin. Nutr. 79, 727–747.

Martínez, E.A., Veas, E., Jorquera, C., San Martín, R., Jara, P., 2009. Re-introduction of quinoa into arid Chile: cultivation of two lowland races under extremely low irrigation. J. Agronomy Crop Sci. 195, 1–10.

Marsilio, V., d'Andria, R., Lanza, B., Russi, F., Iannucci, E., Lavini, A., et al., 2006. Effect of irrigation and lactic acid bacteria inoculants on the phenolic fraction, fermentation and sensory characteristics of olive (*Olea europaea* L. cv. Ascolana tenera) fruits. J. Sci. Food. Agric. 86, 1005–1013.

Morrissey, J.P., Osbourn, A.E., 1999. Fungal resistance to plant antibiotics as a mechanism of pathogenesis. Microbiol. Mol. Biol. Rev. 63, 708–724.

Motilva, M.J., Tovar, M.J., Romero, M.P., Alegre, S., Girona, J., 2000. Influence of regulated deficit irrigation regimes applied to olive trees (Arbequina cultivar) on oil yield and oil composition during the fruit ripening period. J. Sci. Food. Agric. 80 (14), 2037–2043.

Nogués, I., Peñuelas, J., Llusià, J., Estiarte, M., Munnè-Bosch, S., Sardans, J., et al., 2012. Physiological and antioxidant responses of *Erica multiflora* to drought and warming through different seasons. Plant Ecol. 213, 649–661.

Osbourn, A.E., 2003. Saponins in cereals. Phytochemistry 62, 1–4.

Patumi, M., d'Andria, R., Marsilio, V., Fontanazza, G., Morelli, G., Lanza, B., 2002. Olive and olive oil quality after intensive monocone olive growing (*Olea europaea* L., cv. Kalamata) in different irrigation regimes. Food Chem. 77, 27–34.

Pecetti, L., Tava, A., Romani, M., De Benedetto, M.G., Corsi, P., 2006. Variety and environmental effects on the dynamics of saponins in lucerne (*Medicago sativa* L.). Eur. J. Agronomy 25, 187–192.

Pulvento, C., Riccardi, M., Lavini, A., Iafelice, G., Marconi, E., d'Andria, R., 2012. Yield and quality characteristics of quinoa grown in open field under different saline and non-saline irrigation regimes. J. Agronomy Crop Sci. 198, 254–263.

Ramakrishna, A., Ravishankar, G.A., 2011. Influence of abiotic stress signals on secondary metabolites in plants. Plant Signal. Behav. 6 (11), 1720–1731.

Rebey, I.B., Jabri-Karoui, I., Hamrouni–Sellami, I., Bourgou, S., Limam, F., Marzouk, B., 2012. Effect of drought on the biochemical composition and antioxidant activities of cumin (*Cuminum cyminum* L.) seeds. Industrial Crops and Products 36, 238–245.

Sánchez-Rodríguez, E., Rubio-Wilhelmi, M.M., Cervilla, L.M., Blasco, B., Rios, J.J., Rosales, M.A., et al., 2010. Genotypic differences in some physiological parameters symptomatic for oxidative stress under moderate drought in tomato plants. Plant Sci. 178, 30–40.

Sánchez-Rodríguez, E., Moreno, D.A., Ferreres, F., Rubio-Wilhelmi, M.M., Ruiz, J.M., 2011. Differential responses of five cherry tomato varieties to water stress: Changes on phenolic metabolites and related enzymes. Phytochemistry 72, 723–729.

Sánchez-Rodríguez, E., Ruiz, J.M., Ferreres, F., Moreno, D.A., 2012. Phenolic profiles of cherry tomatoes as influenced by hydric stress and rootstock technique. Food Chem. 134, 775–782.

Schwarzbach, A., Schreiner, M., Knorr, D., 2006. Effect of cultivars and deep freeze storage on saponin content of white asparagus spears (*Asparagus officinalis* L.). Eur. Food Res. Technol. 222, 32–35.

Seigler, D.S., 1998. Plant Secondary Metabolism. Kluwer Academic Publishers, Boston MA.

Servili, M., Esposto, S., Lodolini, E., Selvaggini, R., Taticchi, A., Urbani, S., et al., 2007. Irrigation effects on quality, phenolic composition and selected volatiles of virgin olive oils cv Leccino. J. Agric. Food Chem. 55, 6609–6618.

Shahidi, F., Naczk, M., 1995. Food Phenolics: Sources, Chemistry, Effects, Applications. Technolomic Pub. Co. Inc., Lancaster, PA.

Shim, S.J., Jun, W.J., Kang, B.H., 2003. Evaluation of nutritional and antinutritional components in Korean wild legumes. Plant Foods Hum. Nutr. 58, 1–11.

Solíz-Guerrero, J.B., de Rodriguez, D.J., Rodríguez-García, R., Angulo-Sánchez, J.L., Méndez-Padilla, G., 2002. Quinoa saponins: concentration and composition analysis. In: Janick, J., Whipkey, A. (Eds.), Trends in new crops and new uses. ASHS Press, Alexandria, pp. 110–114.

Sparg, S.G., Light, M.E., van Staden, J., 2004. Biological activities and distribution of plant saponins. J. Ethnopharmacol. 94, 219–243.

Strack, D., 1997. Phenolic metabolism. In: Dey, P.M., Harborne, J.B. (Eds.), Plant Biochemistry. Academic Press, London, pp. 387–416.

Szakiel, A., Pączkowski, C., Henry, M., 2011. Influence of environmental abiotic factors on the content of saponins in plants. Phytochem. Rev. 10, 471–491.

Tattini, M., Galardi, C., Pinelli, P., Massai, R., Remorini, D., Agati, G., 2004. Differential accumulation of flavonoids and hydroxycinnamates in leaves of *Ligustrum vulgare* under excess light and drought stress. New Phytol. 163, 547–561.

Terry, L.A., Chope, G.A., Giné Bordonaba, J., 2007. Effect of water deficit irrigation and inoculation with botrytis cinerea on strawberry (*Fragaria x ananassa*) fruit quality. J. Agric. Food. Chem. 55, 10812–10819.

Tovar, M.J., Romero, M.P., Motilva, M.J., 2001. Changes in the phenolic composition of olive oil from young trees (*Olea europaea* L. cv. Arbequina) grown under linear irrigation strategies. J. Agric. Food. Chem. 49, 5502–5508.

Tovar, M.J., Romero, M.P., Girona, J., Motilva, M.J., 2002a. L-Phenylalanine ammonia-lyase activity and concentration of phenolics in developing fruit of olive tree (*Olea europaea* L. Cv. Arbequina) grown under different irrigation regimes. J. Sci. Food. Agric. 82, 892–898.

Tovar, M.J., Romero, M.P., Alegre, S., Girona, J., Motilva, M.J., 2002b. Composition and organoleptic characteristics of oil from Arbequina olive (*Olea europaea* L.) trees under deficit irrigation. J. Sci. Food. Agric. 82, 1755–1763.

Upadhyaya, H., Panda, S.K., Dutta, B.K., 2008. Variation of physiological and antioxidative responses in tea cultivars subjected to elevated water stress followed by rehydration recovery. Acta. Physiol. Plant 30, 457–468.

Vameralia, T., Barionb, G., Hewidyc, M., Mosca, G., 2012. Soybean isoflavone patterns in main stem and branches as affected by water and nitrogen supply. Eur. J. Agronomy 41, 1–10.

Xu, Z., Zhou, G., Shimizu, H., 2010. Plant responses to drought and rewatering. Plant Signal. Behav. 5, 649–654.

Yildiz-Aktas, L., Dagnon, S., Gurel, A., Gesheva, E., Edreva, A., 2009. Drought tolerance in cotton: involvement of non-enzymatic ROS-scavenging compounds. J. Agronomy Crop Sci. 195, 247–253.

Zarrouk, O., Francisco, R., Pinto-Marijuan, M., Brossa, R., Santos, R.R., Pinheiro, C., et al., 2012. Impact of irrigation regime on berry development and flavonoids composition in Aragonez (Syn. Tempranillo) grapevine. Agric. Water Manag. 114, 18–29.

Lichen Phenolics: Environmental Effects

4

Gajendra Shrestha, Larry L. St Clair

Department of Biology and the M.L. Bean Life Science Museum, Brigham Young University, Provo, UT, USA

CHAPTER OUTLINE HEAD

4.1 Introduction

Lichens are symbiotic systems consisting of a filamentous fungus and a photosynthetic partner (an eukaryotic alga and/or cyanobacterium). However, more recent studies indicate a more diverse and complex assemblage of symbiotic partners (Hodkinson and Lutzoni, 2009; Selbmann *et al.*, 2010). Lichens are found in a diversity of habitats; with distributions ranging from the tropics to the polar regions (Brodo *et al.*, 2001). Current estimates suggest that the number of lichen-forming fungal species may be as high as 18,500 taxa worldwide (Boustie and Grube, 2005; Feuerer and Hawksworth, 2007). Lichens produce an abundance of secondary compounds, and more than 1000 secondary metabolites have been identified from intact lichens and aposymbiotically cultured mycobionts (Molnar and Farkas, 2010). Similar to higher plants and other non-lichenized fungi, lichens produce both primary and secondary metabolites. Primary lichen substances are intracellular in origin and are synthesized independently by both symbionts and play specific roles in cellular metabolism. In contrast, secondary metabolites are almost exclusively of fungal origin (Culberson and Armaleo 1992; Hamada *et al.*, 1996; Stocker-Wörgötter and Elix 2002).

Polyphenols in Plants. http://dx.doi.org/10.1016/B978-0-12-397934-6.00004-8

However, the complicated metabolic interaction between the mycobiont and photobiont is fundamentally important to the production of secondary compounds. There are some cases where the photobiont, especially cyanobacteria, are known to independently produce secondary metabolites (Cox *et al.*, 2005; Yang *et al.*, 1993). Fungal secondary metabolites are typically transported outside the fungal hyphae and deposited as crystals in the upper cortex and medullary layer of the lichen thallus (Fahselt, 1994). Most secondary metabolites are generated from a few key intermediate products from primary metabolic pathways (Deduke *et al.*, 2012). There are three specific pathways from which lichen secondary compounds are derived: acetyl-polymalonyl, mevalonic acid, and shikimic acid pathways. The acetyl-polymalonyl pathway produces aliphatic acids, esters, and polyketide derived aromatic compounds; while the mevalonic acid pathway generates terpenes and steroids, with pulvinic acid derivatives coming from the shikimic acid pathway. In lichens, most secondary metabolites are produced by the acetyl-polymalonyl pathway (Elix and Stocker-Worgotter, 2008).

Lichen phenolic compounds are distinctly different from vascular plant phenolics. Lichen phenolics are mainly depsides, depsidones, dibenzenofurans, whereas vascular plant phenolics include tannins, lignins, and flavonoids. Orsellinic acid is the basic unit in the biosynthesis of lichen phenolics (Gaucher and Shepherd, 1968). Lichen phenols are generally secreted by the fungal partner and deposited as crystals on the surface of the cell wall of the fungal hyphae (Hyvärinen *et al.*, 2000). Lichen phenols are primarily acetate-polymalonate-derived with the exception of pulvinic acid derivatives which are synthesized via the shikimic acid pathway (Mosbach, 1969). Lichen phenolics are composed of two monocyclic phenols joined either by an ester bond as in depsides or by both ester and ether bonds in depsidones or a furane heterocycle bond as found in dibenzofurans, such as usnic acid (Blanch *et al.*, 2001). Phenolic compounds, including depsides, depsidones, dibenzofurans, and pulvinic acid derivatives, have attracted much attention recently due to their potential antibiotic, anticancer, antiviral, anti-proliferative, and plant growth inhibitory activity. (Huneck, 1999; Molnar and Farkas, 2010). Besides the potential human benefits as pharmaceuticals, other important roles for lichen phenolics include protecting the algal layer from intense solar radiation, especially in the ultraviolet spectrum (Waring, 2008). They also play an important role as antiherbivory agents (Lawrey, 1986).

4.2 Effects of environmental factors on the production of lichen phenolics

Production and regulation of phenolic compounds in lichenized fungi is complex and variously influenced by environmental factors, including light (Armaleo *et al.*, 2008; Waring, 2008), UV exposure (Bjerke *et al.*, 2005b), elevation (Swanson *et al.*, 1996), temperature fluctuations (Bjerke *et al.*, 2004; Culberson *et al.*, 1983), and seasonality (Ravinskaya and Vainstein, 1975; Swanson *et al.*, 1996). In this chapter we summarize the research investigating the influence of the various environmental parameters on the production of phenolic compounds by lichens.

4.2.1 **Light**

Phenolic levels in lichens vary widely between individual thalli and under different habitat conditions. Although numerous studies have attempted to correlate the variation of phenolic concentrations in lichens under different light levels, the results have often been contradictory. Armaleo *et al.* (2008) reported a positive correlation between the amount of light available to *Parmotrema hypotropum* annually and the concentration of atranorin in the thallus; however, there was a decrease in norstictic acid concentration under the same conditions. In contrast, Stephenson and Rundel (1979) did not find any correlation between atranorin levels and light intensity for *Letharia vulpina*. Bjerke and Dahl (2002) and Rundel (1969) reported a positive correlation between thallus concentration of usnic acid and light intensity in *Ramalina siliquosa* under culture conditions but there was no correlation between light intensity and salazinic acid concentration in the same species. Hamada (1982) and Culberson *et al.* (1983) also found reduced production of phenolics with variation in light intensity for *Cladonia cristatella*. Reasons for conflicting patterns in production of phenolic compounds in response to varying light conditions in lichens remain unclear. Bjerke *et al.* (2005b) and Bjerke *et al.* (2004) have suggested that these conflicting results may be due, at least in part, to the interactive effects of multiple variables across differing habitat conditions.

4.2.2 **UV**

Lichen communities can be a prominent component in many high elevation/latitude ecosystems (Billings, 1974) where UV levels are significantly higher (Caldwell *et al.*, 1989). Understanding how lichen phenolic concentrations vary in response to UV exposure may provide important insights into how lichens thrive in these harsh environments. One of the primary functions of lichen secondary compounds is to protect the algal layer from intense light levels, especially in the ultraviolet range (Waring, 2008). For example, lichen-derived phenolic compounds including, depsides, depsidones, usnic acid, and pulvinic acid derivatives are effective UV absorbing compounds (McEvoy *et al.*, 2007). Extensive field experiments have assessed the effects of UV on the concentration of lichen phenolics with a pattern of increasing concentrations under high UV-A conditions (BeGora and Fahselt, 2001; Swanson and Fahselt, 1997). On the other hand, UV-B radiation appears to have a negative effect on phenolic concentrations (BeGora *et al.*, 2001; Swanson *et al.*, 1997; Swanson *et al.*, 1996). Increased levels of phenolics under high UV-A conditions may be due to greater production of photosynthates which in turn support increased production of phenolic compounds (BeGora *et al.*, 2001); while lower production of phenolics in the presence of UV-B may be due to the production of extracellular degradative enzymes or electronic transitions within molecules caused by exposure to UV-B (Fessenden and Fessenden, 1986). Waring (2008) compared the abundance of UV-protective compounds in lichens growing on trees in open areas, at the forest edge, and in the forest interior and consistently found higher abundance of UV-protective compounds in lichens on trees occurring in open areas. He suggested that this may be due to the fact that the trunks of trees in open areas are more directly exposed to UV light; therefore, the attached bark lichens produce more UV-protective compounds. In addition to the positive and negative effects of UV light on the phenolic content of some lichens, other studies found no significant effects. For example, *Flavocetraria nivalis* and *Nephroma arcticum* did not show any differences in phenolic content between thalli exposed to enhanced UV-B radiation and ambient UV-B (Bjerke *et al.*, 2005b)

4.2.3 **Geographical and seasonal variation**

Various studies have shown that differences in geographic location and changing seasons can influence the concentration of phenolics in some lichen species. Studies examining variation in microclimatic conditions have also shown differences in the production of phenolic compounds. Swanson *et al.* (1996) reported that the content of the phenolics in *Umbilicaria americana* was significantly higher in samples from sites near Ontario, Canada when compared with specimens from New York, USA. Despite similarities in climatic conditions, Hamada (1982) found significant variation in the salazinic acid content of *Ramalina siliquosa* thalli from two sites in Japan. Similarly, Bjerke *et al.* (2004) showed variation in the usnic acid content of *Flavocetraria. nivalis* thalli collected from 25 sites in northwestern Spitsbergen, Norway. A significant decrease in the quantity of salazinic acid from southern to northern latitudes has been documented for *Xanthoparmelia viriduloumbrina* (Deduke *et al.*, 2012). In addition to the influence of geographic location, the phenolic content of lichens has also been reported to vary by season. For example, Swanson *et al.* (1996) documented the highest levels of phenolics for *Umbilicaria americana* during the winter months. BeGora and Fahselt (2001) also reported the lowest levels of usnic acid for *Cladonia mitis* during the spring and summer with the highest concentrations occurring in the late winter. The lower levels of usnic acid reported for summer months may be due to drought and heat stress which have a depressing effect on metabolic activity (Bjerke *et al.*, 2005a). Another reason could be a function of higher levels of UV or the ratio of UV-B to UV-A during the summer (Madronich, 1993). However, Bjerke *et al.* (2005a) found increased levels of usnic acid in *Flavocetraria nivalis* in the late spring and early summer and generally lower levels during the autumn and winter which they believe may be due to higher levels of precipitation during the spring and summer months. However at one of their sites in central west Greenland, they detected remarkably high levels of usnic acid during the late autumn and early winter. They suggest this pattern may be due to decreasing temperatures resulting in lower evaporation rates and increasing availability of water during the autumn which stimulated production of usnic acid.

4.2.4 **Elevation**

Studies have also demonstrated that the production and/or content of phenolics in lichens varies with changes in elevation. For example, it has been reported for *Umbilicaria americana* that phenolic production decreases with increasing elevation (Swanson *et al.*, 1996). Similarly, Vatne *et al.* (2011) found decreased production of carbon-based secondary compounds (stictic acid, cryptostictic acid, and norstictic acid) in *Lobaria pulmonaria* with increasing elevation. Swanson *et al.* (1996) suggested that the degradation of phenolic compounds by UV-B may be a possible explanation for the decrease in phenolics with increasing elevation. However, Vatne *et al.* (2011) questioned this explanation and suggested that other environmental factors, especially temperature, may have a greater influence on the speed of biosynthetic pathways and thus a greater effect on phenolic concentrations in lichens. In contrast, Rubio *et al.* (2002) found increasing levels of rhizocarpic acid in *Acarospora schleicheri* with increasing elevation in spite of higher levels of UV-B. Rubio's explanation for this pattern, suggests that UV-B light may be stimulating the production of rhizocarpic acid. However, the direct effects of elevation on the production of phenolic compounds by lichens may in fact be complicated by the dynamic interaction of various environmental factors (levels of UV light, temperature, and precipitation) as they vary across an elevation gradient.

4.2.5 **Temperature**

Variation in temperature has also been shown to influence the levels of phenolic compounds in lichens. For example, the concentration of salazinic acid in *Ramalina siliquosa* was positively correlated with mean annual temperature (Hamada, 1982). Under semi-natural conditions, the amount of gyrophoric acid and methyl gyrophorate was reported to increase by an average of 6.7 and 22.3 times, respectively in *Peltigera* didactyla with an increase in temperature of 3°C (Bjerke *et al.*, 2003). Hamada (1991) observed a steady increase in the concentration of usnic acid and 4-*O*-dimethylbarbatic acid in mycobiont cultures from *Ramalina siliquosa* when the temperature of the culture was increased to 12°C for usnic acid and 15°C for 4-*O*-dimethylbarbatic acid, with higher temperatures resulting in a decrease in the content of both compounds. In contrast, the content of 4-*O*-dimethylbarbatic acid in the intact lichen increased with a temperature up to 19°C (Hamada, 1984), suggesting that the optimal temperature for the production of lichen substances is generally lower in the isolated mycobiont when compared with the intact lichen. Increased concentrations in lichen phenolics may be due to increased photosynthetic activity related to increased temperature. However, in a controlled phytotron experiment, Culberson *et al.* (1983) found the concentrations of barbatic acid, obtrusatic acid, and didymic acid in *Cladonia cristatella* were higher at lower temperatures. Similarly, Bjerke *et al.* (2004) found that in *Flavocetraria nivalis*, the concentration of usnic acid was highest at the sites with the lowest temperatures. Tundra lichens are typically metabolically active at near-zero or subzero temperatures (Kappen *et al.*, 1996). Thus, *Flavocetraria nivalis*, a common tundra lichen, is most likely metabolically active under conditions of lower temperatures and relatively strong solar radiation. With high solar radiation comes the possibility of irreversible damage to algal cells. Hence, Bjerke *et al.* (2004) speculated that increased production of usnic acid by the mycobiont at lower temperatures would likely prevent UV-related damage to the photobiont and thus better support the level of photosynthetic activity required to sustain increase production of phenolic compounds.

4.2.6 Culture media

Various experiments have been conducted using cultured lichen mycobionts. It has been shown that the type of growth media, available nutrients, presence or absence of the algal partner, and pH of the media may potentially influence the production of secondary metabolites. Lichenized fungi have been found to produce different secondary metabolites when cultured aposymbiotically. Deduke *et al.* (2012) proposed two explanations for this phenomenon: 1) artificial media may trigger induction of an alternate pathway, and 2) the availability of certain trace elements, carbohydrates, or a slightly different pH may alter enzyme activity resulting in the production of a different group of secondary compounds. However, nutrient media containing the disaccharide sucrose have been shown to produce lichen substances typical of intact specimens of *Lobaria spathulata* (Stocker-Wörgötter and Elix, 2002). In contrast, Cordeiro *et al.* (2004) found that in *Ramalina peruviana* sekikaic acid is produced when grown on solid medium but atranorin (not detected in any voucher specimens) was found in addition to sekikaic acid when grown in liquid culture. Similarly, Verma *et al.*, (2012) found that Bold's Basal Media (BBM) was most suitable for increasing the production of salazinic acid, sekikaic acid, and usnic acid in the lichens *Ramalina nervulosa* and *Ramalina pacifica*. pH also plays an important role in the production of phenolics by lichens. In aposymbiotic cultures of the lichen-forming fungal genera *Umbilicaria* and *Lasallia*, Stocker-Wörgötter (2001) demonstrated that diagnostic phenolic compounds are produced only when they are grown on an acidic medium (potato-dextrose-agar). Similarly, Hamada (1989) found that aposymbiotic *Ramalina siliquosa* cultures produced the highest amount of depsides when the culture medium was at pH 6.5; which is also the optimal pH suggested by Ahmadjian (1961) for the growth of isolated lichen mycobionts. These studies demonstrate that differences in culture conditions, i.e., types of carbohydrates, pH of the medium etc., can trigger different pathways leading to the production of different metabolites by lichen mycobionts.

4.3 Role of photobionts in phenolics production

In lichens, secondary metabolites are produced exclusively by the fungal partner (Fahselt, 1994; Culberson and Armaleo, 1992; Hamada *et al.*, 1996; Kon *et al.*, 1997; and Stocker-Wörgötter and Elix, 2002). However, the metabolic interactions between the mycobiont and photobiont are essential to the production of lichen secondary chemistry. The important contributions of the photobiont have been documented using aposymbiotic axenic cultures of the mycobiont where the metabolites typical of the intact lichen are either produced at a reduced concentration or not at all. Molina *et al.* (2003) reported that intact thalli of *Physconia distorta* produced several phenolic acids including, atranorin, chloroatranorin, and usnic acid. On the other hand, the main phenolic metabolite extracted from axenic cultures of the mycobiont was malonprotocetraric acid, produced by a completely different biosynthetic pathway than the phenolics extracted from the intact lichen. Variation in phenolic compounds between intact thalli and aposymbiotic mycobiont cultures may be attributed to different carbon sources. Lichenized fungi with a green algal photobiont use ribitol as a carbon source while in axenic cultures glucose is the main carbon source. Differences in carbon sources can trigger different pathways ultimately leading to the production of distinct groups of phenolic compounds between cultured mycobionts and intact lichen thalli of the same species (Molina *et al.*, 2003).

4.4 **Conclusions**

Lichens produce various kinds of phenolic compounds using different metabolic pathways. Lichen phenolic compounds serve a variety of biological functions, including protection of lichen thalli from various environmental stresses such as light intensity, water potential changes, and other factors associated with natural fluctuations in environmental conditions (Deduke *et al.*, 2012). Due to issues associated with global climate change, it has been predicted that there will be an increase in average surface temperature, intensity and length of drought periods, along with an increase in global water vapor, evaporation, and precipitation (Meehl *et al.*, 2007). The influence of various environmental factors on the production and concentration of different lichen phenolics may be further complicated by these potential changes related to global climate change. Research has also shown contradictory results with the effects of environmental conditions on phenolic production by lichens. This condition is largely due to the fact that lichens tend to be very sensitive to microclimatic variation. A slight change in environmental conditions can significantly affect the distribution of lichens which ultimately influences the production of phenolics in terms of both variety and quantity. Furthermore, production of phenolics in lichens is inevitably affected by the dynamic interaction between a suite of environmental parameters. Sorting out this kind of synergistic interaction by examining the influence of a single environmental factor is likely simplistic at best. In addition, phenolics are produced by different pathways; therefore, different combinations of regulatory genes could mediate divergent responses under the same environmental conditions (Armaleo *et al.*, 2008). Lichens are also important primary producers in many ecosystems especially in the Arctic. Hence, it is important to understand the relationships between the production of lichen phenolics and the predicted changes in environmental conditions associated with global climate change.

References

Ahmadjian, V., 1961. Studies on Lichenized Fungi. Bryologist 64, 168–179.

Armaleo, D., Zhang, Y., Cheung, S., 2008. Light might regulate divergently depside and depsidone accumulation in the lichen *Parmotrema hypotropum* by affecting thallus temperature and water potential. Mycologia 100, 565–576.

BeGora, M.D., Fahselt, D., 2001. Usnic Acid and Atranorin Concentrations in Lichens in Relation to Bands of UV Irradiance. Bryologist 104, 134–140.

Billings, W.D., 1974. Arctic and alpine vegetation: plant adaptations to cold summer climates. In: Ives, J.D.a.R.G.B. (Ed.), Arctic and Alpine Environments. William Clowes and Sons Ltd., Great Britian, pp. 403–443.

Bjerke, J.W., Dahl, T., 2002. Distribution patterns of usnic acid-producing lichens along local radiation gradients in West Greenland. Nova Hedwigia 75, 487–506.

Bjerke, J.W., Zielke, M., Solheim, B., 2003. Long-term impacts of simulated climatic change on secondary metabolism, thallus structure and nitrogen fixation activity in two cyanolichens from the Arctic. New Phytol. 159, 361–367.

Bjerke, J.W., Joly, D., Nilsen, L., Brossard, T., 2004. Spatial trends in usnic acid concentrations of the lichen *Flavocetraria nivalis* along local climatic gradients in the Arctic (Kongsfjorden, Svalbard). Polar Biol. 27, 409–417.

Bjerke, J.W., Elvebakk, A., Dominguez, E., Dahlback, A., 2005a. Seasonal trends in usnic acid concentrations of Arctic, alpine and Patagonian populations of the lichen *Flavocetraria nivalis*. Phytochemistry 66, 337–344.

Bjerke, J.W., Gwynn-Jones, D., Callaghan, T.V., 2005b. Effects of enhanced UV-B radiation in the field on the concentration of phenolics and chlorophyll fluorescence in two boreal and arctic–alpine lichens. Environ. Exp. Bot. 53, 139–149.

Blanch, M., Blanco, Y., Fontaniella, B., Legaz, M.E., Vicente, C., 2001. Production of phenolics by immobilized cells of the lichen *Pseudevernia furfuracea*: the role of epiphytic bacteria. Int. Microbiol. 4, 89–92.

Boustie, J., Grube, M., 2005. Lichens — a promising source of bioactive secondary metabolites. Plant Genetic Resources 3, 273–287.

Brodo, I.M., Sharnoff, S.D., Sharnoff, S., 2001. Lichens of North America Yale University Press/New Haven and London.

Caldwell, M.M., Teramura, A.H., Tevini, M., 1989. The changing solar ultraviolet climate and the ecological consequences for higher plants. Trends Ecol. Evol. 4, 363–367.

Cordeiro, L.M., Iacomini, M., Stocker-Worgotter, E., 2004. Culture studies and secondary compounds of six *Ramalina* species. Mycol. Res. 108, 489–497.

Cox, P.A., Banack, S.A., Murch, S.J., Rasmussen, U., Tien, G., Bidigare, R.R., 2005. Diverse taxa of cyanobacteria produce β-N-methylamino-l-alanine, a neurotoxic amino acid. Proc. Natl. Acad. Sci. U. S. A. 102, 5074–5078.

Culberson, C.F., Culberson, W.L., Johnson, A., 1983. Genetic and environmental effects of growth and production of secondary compounds in *Cladonia cristatella*. Biochem. Syst. Ecol. 11, 77–84.

Culberson, C.F., Armaleo, D., 1992. Induction of a complete secondary-product pathway in a cultured lichen fungus. Exp. Mycol. 16, 52–63.

Deduke, C., Timsina, B., Piercey-Normore, M.D., 2012. Effect of Environmental Change on Secondary Metabolite Production in Lichen-Forming Fungi. In: Silvern, S.S.Y.a.S.E. (Ed.), International Perspectives on Global Environmental Change. InTech, pp. 197–230.

Elix, J.A., Stocker-Wörgötter, E., 2008. Biochemistry and secondary metabolites. In: III, T.H.N. (Ed.), Lichen Biology. second ed. Cambridge University Press.

Fahselt, D., 1994. Secondary Biochemistry of Lichens. Symbiosis 16, 117–165.

Fessenden, R.J., Fessenden, J.S., 1986. Organic chemistry, third ed. Brooks/Cole Publishing, Monterey, California.

Feuerer, T., Hawksworth, D., 2007. Biodiversity of lichens, including a world-wide analysis of checklist data based on Takhtajan's floristic regions. Biodiv. Conserv. 16, 85–98.

Gaucher, G.M., Shepherd, M.G., 1968. Isolation of orsellinic acid synthase. Biochem. Biophys. Res. Commun. 32, 664–671.

Hamada, N., 1982. The effect of temperature on the content of the medullary depsidone salazinic acid in *Ramalina siliquosa* (lichens). Can. J. Bot. 60, 383–385.

Hamada, N., 1984. The Content of Lichen Substances in *Ramalina Siliquosa* Cultured at Various Temperatures in Growth Cabinets. Lichenologist 16, 96–98.

Hamada, N., 1989. The Effect of Various Culture Conditions on Depside Production by an Isolated Lichen Mycobiont. Bryologist 92, 310–313.

Hamada, N., 1991. Environmental Factors Affecting the Content of Usnic Acid in the Lichen Mycobiont of *Ramalina siliquosa*. Bryologist 94, 57–59.

Hamada, N., Miyagawa, H., Miyawaki, H., Inoue, M., 1996. Lichen Substances in Mycobionts of Crustose Lichens Cultured on Media with Extra Sucrose. Bryologist 99, 71–74.

Hodkinson, B., Lutzoni, F., 2009. A microbiotic survey of lichen-associated bacteria reveals a new lineage from the Rhizobiales. Symbiosis 49, 163–180.

Huneck, S., 1999. The Significance of Lichens and Their Metabolites. Naturwissenschaften 86, 559–570.

Hyvärinen, M., Koopmann, R., Hormi, O., Tuomi, J., 2000. Phenols in reproductive and somatic structures of lichens: a case of optimal defence? Oikos 91, 371–375.

Kappen, L., Schroeter, B., Scheidegger, C., Sommerkorn, M., Hestmark, G., 1996. Cold resistance and metabolic activity of lichens below 0_C. Adv. Space. Res. 18, 119–128.

Kon, Y., Kashiwadani, H., Wardlaw, J.D., Elix, J.A., 1997. Effects of culture conditions on dibenzofuran production by cultured mycobionts of lichens. Symbiosis 23, 97–106.

Lawrey, J.D., 1986. Biological role of lichen substances. Bryologist 89, 111–122.

Madronich, S., 1993. The atmosphere and UV-B radiation at ground level. In: Young, A.R. (Ed.), Environmental UV photobiology. Plenum Press, New York, pp. 1–40.

McEvoy, M., Gauslaa, Y., Solhaug, K.A., 2007. Changes in pools of depsidones and melanins, and their function, during growth and acclimation under contrasting natural light in the lichen *Lobaria pulmonaria*. New Phytologist 175, 271–282.

Meehl, G.A., Stocker, T.F., Collins, W.D., Friedlingstein, P., Gaye, A.T., Gregory, J.M., Kitoh, A., Knutti, R., Murphy, J.M., Noda, A., Raper, S.C.B., Watterson, I.G., Weaver, A.J., Zhao, Z.C., 2007. Global Climate Projections. In: Solomon, S., Qin, D., Manning, M., Chen, Z., Marquis, M., Averyt, K.B., Tignor, M., Miller, H.L. (Ed.), Climate Change 2007: The Physical Science Basis. Contribution of Working Group I to the Fourth Assessment Report of the Intergovernmental Panel on Climate Change. Cambridge University Press, Cambridge, United Kingdom and New York, NY, USA.

Molina, M.C., Crespo, A., Vicente, C., Elix, J.A., 2003. Differences in the composition of phenolics and fatty acids of cultured mycobiont and thallus of *Physconia distorta*. Plant Physiol. Biochem. 41, 175–180.

Molnar, K., Farkas, E., 2010. Current results on biological activities of lichen secondary metabolites: a review. Z. Naturforsch. C. 65, 157–173.

Mosbach, K., 1969. Zur Biosynthese von Flechtenstoffen, Produkten einer symbiotischen Lebensgemeinschaft. Angewandte Chemie 81, 233–244.

Ravinskaya, A.P., Vainshtein, A., 1975. Influence of certain ecological factors on the content of various compounds in lichens. Ékologiya 3, 82–85.

Rubio, C., Fernández, E., Hidalgo, M.E., Quilhot, W., 2002. Effects of solar UV-B radiation in the accumulation of rhizocarpic acid in a lichen species from alpine zones of Chile. Boletín de la Sociedad Chilena de Química 47, 67–72.

Rundel, P.W., 1969. Clinal Variation in the Production of Usnic Acid in *Cladonia subtenuis* along Light Gradients. Bryologist 72, 40–44.

Selbmann, L., Zucconi, L., Ruisi, S., Grube, M., Cardinale, M., Onofri, S., 2010. Culturable bacteria associated with Antarctic lichens: affiliation and psychrotolerance. Polar Biol. 33, 71–83.

Stephenson, N.L., Rundel, P.W., 1979. Quantitative variation and the ecological role of vulpinic acid and atranorin in thallus of *Letharia vulpina*. Biochem. Syst. Ecol. 7, 263–267.

Stocker-Wörgötter, E., 2001. Experimental studies of the lichen symbiosis: DNA-analysis, differentiation and secondary chemistry of selected mycobionts, artificial resynthesis of two- and tripartite symbioses. Symbiosis 30, 207–227.

Stocker-Wörgötter, E., Elix, J.A., 2002. Secondary chemistry of cultured mycobionts: formation of a complete chemosyndrome by the lichen fungus of Lobaria spathulata. Lichenologist 34, 351–359.

Swanson, A., Fahselt, D., 1997. Effects of ultraviolet on polyphenolics of *Umbilicaria americana*. Can. J. Bot. 75, 284–289.

Swanson, A., Fahselt, D., Smith, D., 1996. Phenolic levels in *Umbilicaria americana* in relation to enzyme polymorphism, altitude and sampling date. Lichenologist 28, 331–339.

Vatne, S., Asplund, J., Gauslaa, Y., 2011. Contents of carbon based defence compounds in the old forest lichen *Lobaria pulmonaria* vary along environmental gradients. Fungal Ecol. 4, 350–355.

Verma, N., Behera, B.C., Joshi, A., 2012. Studies on nutritional requirement for the culture of lichen *Ramalina nervulosa* and *Ramalina pacifica* to enhance the production of antioxidant metabolites. Folia. Microbiol. 57, 107–114.

Waring, B., 2008. Light Exposure Affects Secondary Compound Diversity in Lichen Communities in Monteverde, Costa Rica. PennScience 6, 11–13.

Yang, X., Shimizu, Y., Steiner, J.R., Clardy, J., 1993. Nostoclide I and II, extracellular metabolites from a symbiotic cyanobacterium, *Nostoc* sp., from the lichen *Peltigera canina*. Tetrahedron Lett. 34, 761–764.

Plant systems of polyphenol modification

CHAPTER

Modulation of Plant Endogenous Antioxidant Systems by Polyphenols

5

Ramón Rodrigo, Matías Libuy

Molecular and Clinical Pharmacology Program, Institute of Biomedical Sciences,
Faculty of Medicine, University of Chile, Santiago, Chile

CHAPTER OUTLINE HEAD

5.1 Introduction

Reactive oxygen species (ROS) may be generated in plants, as in other cell types, through several metabolic pathways leading to univalent reductions of ground state oxygen (Apel and Hirt, 2004). Under physiological conditions, there is a balance between their production and the scavenging mechanisms capable of reducing the active oxygen derivates with a concomitant oxidation of substrates and/or antioxidant molecules. However, this balance may be challenged by several factors, thereby putting the plant into conditions of oxidative stress. Accordingly, low temperatures are closely associated with enhancement of ROS production (Arora *et al.*, 2002; Fowler and Thomashow, 2002), which is consistent with the reported increased ROS in plants exposed to liquid nitrogen (Johnston *et al.*, 2007; Uchendu *et al.*, 2010). Thus, plant species of pharmacological importance such as *Hypericum perforatum L.* showed increased biomarkers of oxidative stress following cryopreservation, an

Polyphenols in Plants. http://dx.doi.org/10.1016/B978-0-12-397934-6.00005-X

effect that persisted for at least 2 months (Skyba *et al.*, 2010). In addition, the most toxic substance to restrict plant growth is salt (Zhu, 2002), as it is capable of altering a broad range of metabolic processes such as hormone production and causing photosynthetic inhibition, which results in excessive generation of ROS in stressed plants (Allakhverdiev and Murata, 2008). It has been shown that UV-B treatment results in a decrease in the light-saturated rate of CO_2 assimilation, accompanied by decreases in carboxylation velocity and ascorbic acid content (Allen *et al.*, 1997). Limited CO_2 assimilation due to UV-B leads to excessive production of ROS which, in turn, causes oxidative damage in plants (Han *et al.*, 2009; Strid *et al.*, 1994).

In recent years there has been a remarkable increment in scientific articles dealing with oxidative stress. Several reasons justify this trend: knowledge about reactive oxygen and nitrogen species metabolism; definition of markers for oxidative damage; evidence linking chronic diseases and oxidative stress; identification of flavonoids and other dietary polyphenol antioxidants present in plant foods as bioactive molecules; and data supporting the idea that health benefits associated with fruits, vegetables, and red wine in the diet are probably linked to the polyphenol antioxidants they contain (Urquiaga and Leighton, 2000).

The aim of this chapter is to present an update of the mechanisms whereby polyphenols can modulate the activity of the plant antioxidant defense system.

5.2 Plant oxidative stress

The following environmental pathophysiological stimuli, are involved in plant oxidative damage (Figures 5.1 and 5.2):

5.2.1 Heat

High temperature is now a major concern to plant biologists due to the threat to productivity in global agriculture. Heat affects photosynthesis respiration, water relations, and membrane stability and also modulates levels of hormones and primary and secondary metabolites. ROS production constitutes a major plant response to heat stress (Whaid *et al.*, 2007). ROS generation is a symptom of cellular injury due to high temperature (Liu and Huang, 2000), causing the autocatalytic peroxidation of membrane lipids and pigments thus leading to the loss of membrane semi-permeability and modifying its functions (Xu *et al.*, 2006). Superoxide radicals are regularly synthesized in the chloroplast and mitochondrion and some quantities are also produced in microbodies. ROS such as •OH and O_2•– can damage chlorophyll, protein, DNA, lipids and other important macromolecules, thus fatally affecting plant metabolism and limiting growth and yield (Sairam and Tyagi, 2004). Heat stress also affects the organization of microtubules by splitting and/or elongation of spindles, formation of microtubule asters in mitotic cells, and elongation of phragmoplast microtubules (Smertenko *et al.*, 1997). These injuries eventually lead to starvation, inhibition of growth, reduced ion flux, production of toxic compounds, and ROS (Schöffl *et al.*, 1998; Wahid *et al.*, 2007).

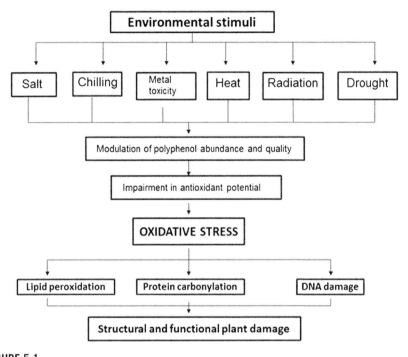

FIGURE 5.1

Role of changes in polyphenol content and quality in plant oxidative damage.

It has been reported that protection against oxidative stress is an important component in determining the survival of a plant under heat stress. Studies on heat-acclimated *versus* non-acclimated cool season turfgrass species suggested that the former had lower production of ROS as a result of enhanced synthesis of ascorbate and glutathione (Xu *et al.*, 2006). Available data suggest that some signaling molecules may cause an increase in the antioxidant capacity of cells (Gong *et al.*, 1997).

5.2.2 **Salt**

The most toxic substance to restrict plant growth is salt (Zhu, 2002). Primarily, salinity stress results in excessive ROS generation (Tanou *et al.*, 2009; Hernández *et al.*, 2000). Furthermore, a high concentration of NaCl causes an inhibition of plant growth, reduces the content of photosynthetic pigments, increases lipid peroxidation, and enhances the entire antioxidant defense. Biomolecules could be damaged by ROS (Mittler, 2002), thus inducing the alternation of cellular redox state or homeostasis (Imlay, 2003; Potters *et al.*, 2010). Salt stress can lead to stomatal closure that reduces carbon dioxide (CO_2) availability in the leaves and inhibits carbon fixation that, in turn, causes the exposure of chloroplasts to excessive excitation energy and an over-reduction of the photosynthetic electron transport system leading to enhanced

ROS generation. A low chloroplastic CO_2/O_2 ratio also favors photorespiration leading to increased ROS production (Hernández *et al.*, 2000). Elevated CO_2 mitigates the oxidative stress caused by salinity, involving lower ROS generation and a better maintenance of redox homeostasis as a consequence of higher assimilation rates and lower photorespiration (Pérez-López *et al.*, 2009). Salinity-induced ROS disrupt normal metabolism through denaturing proteins and nucleic acids in several plant species (Hernández *et al.*, 2000). Plants with high antioxidant levels have been reported to have higher resistance to oxidative damage (Pérez-López *et al.*, 2009; D'Amico *et al.*, 2003; Chaparzadeh *et al.*, 2004; Pérez-López *et al.*, 2010).

5.2.3 Radiation

The effect of radiation on plant homeostasis acts particularly on photosynthesis through enhanced UV-B exposure that inhibits photosynthetic rate. It has been shown that UV-B treatment results in the decrease in the light-saturated rate of CO_2 assimilation, accompanied by decreases in carboxylation velocity and ascorbic acid content (Allen *et al.*, 1997). Studies have observed a marked decrease in the ratios of variable to maximum chlorophyll fluorescence yield and in the quantum yield of photosynthetic oxygen evolution in pea and rice leaves (He *et al.*, 1993). Also, limited CO_2 assimilation due to UV-B leads to excessive ROS production, which in turn causes oxidative damage in plants (Strid *et al.*, 1994). Other studies have suggested that UV-B exposure increases NADPH–oxidase activity. Therefore, plants must adapt to the deleterious effects of UV-B radiation because they are dependent on sunlight for photosynthesis and, therefore, cannot avoid exposure to UV-B radiation (Rao *et al.*, 1996). Photosynthetic activity is inhibited in plant tissues due to an imbalance between light capture and its utilization under drought stress (Foyer and Noctor, 2000). Dissipation of excess light energy in the PSII core and antenna leads to generation of ROS that are potentially dangerous under drought stress conditions (Fenta *et al.*, 2012).

5.2.4 Drought

Oxidative stress imposed by ROS under drought conditions profoundly affects plant growth and development (Lee *et al.*, 2012). ROS production might be higher in several ways due to drought stress. Inhibition of CO_2 assimilation, coupled with the changes in photosystem activities and photosynthetic transport capacity under drought stress results in accelerated production of ROS via the chloroplast Mehler reaction (Asada, 1999). During drought stress, CO_2 fixation is limited due to stomatal closure, which, in turn, leads to reduced NADP+ regeneration through the Calvin cycle. Consequently, there is a higher leakage of electrons to O_2 by the Mehler reaction. Some studies reported 50% more leakage of photosynthetic electrons to the Mehler reaction in drought stressed wheat plants, compared to unstressed plants. Under drought stress, the photorespiratory pathway is also enhanced (Biehler *et al.*, 1996). Other studies have estimated that photorespiration is likely to account for over

70% of total H_2O_2 production under drought stress conditions (Noctor *et al.*, 2002). The $O_2\bullet-$ initiates a chain reaction leading to the production of more toxic radical species, which may cause damage far in excess of the initial reaction products. Under drought stress one of the real threats towards the chloroplast is the production of the $\bullet OH$ in the thylakoids.

5.2.5 Chilling

It has been reported that chilling leads to the overproduction of ROS by exacerbating imbalance between light absorption and light use by inhibiting Calvin-Benson cycle activity (Logan *et al.*, 2006), enhancing photosynthetic electron flux to O_2 and causing an over reduction of the respiratory electron transport chain (Hu *et al.*, 2008). Chilling stress also causes significant reductions in *rbc*L and *rbc*S transcripts, leading to higher electron flux to O_2. H_2O_2 accumulation in chloroplast was negatively correlated with the initial ascorbic acid content and photosynthetic rate. Chilling-induced oxidative stress, lipid peroxidation, and protein carbonylation is a significant factor in relation to chilling injury in plants (Fryer, 1992; Prasad, 1997; Yong *et al.*, 2008). Protein carbonyl content, an indication of oxidative damage, was increased two-fold in maize seedlings when exposed to chilling temperatures (Prasad, 1997). Lipoxygenase activity as well as lipid peroxidation was increased in maize leaves during low temperatures, suggesting that lipoxygenase-mediated peroxidation of membrane lipids contributes to the oxidative damage occurring in chill-stressed maize leaves (Fryer, 1992).

5.2.6 Toxicity by redox-active metals

All materials of biological origin contain small amounts of transition metals. Transition metals, e.g., Fe, Cu, Co, which possess two or more valence states with a suitable oxidation-reduction potential affect both the speed of autoxidation and the direction of hydroperoxide (Grosch, 1982). Plant growth and metabolism may be drastically affected by increasing levels of metals into the environment, leading to severe losses in crop yields (Mishra and Dubey, 2005). Consequently, metals can lead to oxidative damage to different cell constituents (Gallego *et al.*, 2002). Under metal stress condition, net photosynthesis (Phn) decreases due to damage to the photosynthetic metabolism, including photosynthetic electron transport (Phet) (Vinit-Dunand *et al.*, 2002). For example, copper has been shown to negatively affect components of both the light reactions (Vinit-Dunand *et al.*, 2002) and CO_2-fixation reactions (Moustakas *et al.*, 1994). These alterations lead to ROS overproduction. The induction of ROS production due to metals, such as cadmium and zinc, in *Nicotiana tabacum* L. cv. Bright Yellow 2 (TBY- 2) cells in suspension cultures showed properties comparable to the elicitor-induced oxidative burst in other plant cells (Zróbek-Sokolnik *et al.*, 2009). Redox-active metals, such as iron, copper, and chromium, undergo redox cycling producing ROS, whereas redox-inactive metals, such as lead, cadmium, mercury, and others, deplete cells' major antioxidants (Shah *et al.*, 2001). If metal-induced production of ROS is not adequately counterbalanced by cellular anti-oxidants, it produces oxidative damage of lipids, proteins, and nucleic acids (Mishra and Dubey, 2005). Significant enhancement in lipid peroxidation and decline in protein thiol contents were observed when rice seedlings were subjected to Al, Ni, and Mn toxicity (Srivastava and Dubey, 2011).

5.3 **Antioxidant defense system**

Plants are exposed to various environmental conditions likely to affect their growth, development, reproduction, and quality of their products for medicinal, industrial, and other uses. The ROS may behave as mediators of the mechanisms of these effects. To avoid the damage caused by ROS, plants have evolved enzymatic and nonenzymatic systems to scavenge these reactive species (Alsocher et al., 2002). The endogenous antioxidant system includes enzymatic and nonenzymatic components. Among the enzymatic components should be mentioned the activity of peroxidases (POD), superoxide dismutase (SOD), and catalase (CAT), which together with the other enzymes of the ascorbate-glutathione cycle such as ascorbate peroxidase (APX), monodehydroascorbate reductase (MDHAR), dehydroascorbate reductase (DHAR), and glutathione reductase (GR) promote the scavenging of ROS (Cavalvanti et al., 2004). Nonenzymatic components comprise endogenous systems such as reduced glutathione (GSH), exogenous ascorbate (AsA), carotenoids, or tocopherols among others.

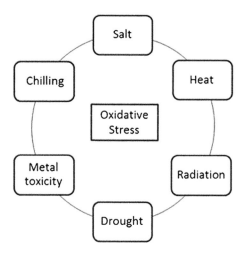

FIGURE 5.2

Prooxidant mechanisms asociated to damage in polyphenols.

5.3.1 **Nonenzymatic antioxidants**

Nonenzymatic components of the antioxidative defense system include the major cellular redox buffers ascorbate (AsA) and glutathione (γ-glutamyl-cysteinyl-glycine, GSH) as well as tocopherol, carotenoids, and phenolic compounds. They interact with numerous cellular components and in addition to crucial roles in defense and as enzyme cofactors, these antioxidants influence plant growth and development by modulating processes from mitosis and cell elongation to senescence and cell death (De Pinto and De Gara, 2004). Mutants with decreased nonenzymatic antioxidant contents have been shown to be hypersensitive to stress (Gao and Zhang, 2008; Semchuk *et al.*, 2009).

Ascorbate (AsA) is a major metabolite in plants and is the most abundant antioxidant. In association with other components of the antioxidant system, it protects plants against oxidative damage resulting from aerobic metabolism, photosynthesis and a range of pollutants (Smirnoff, 1996) and has a key role in defense against oxidative stress caused by enhanced levels of ROS (Shao *et al.*, 2008). It is also considered a powerful antioxidant because of its ability to donate electrons in a number of enzymatic and nonenzymatic reactions. AsA has been shown to play an important role in several physiological processes in plants, including growth, differentiation, and metabolism. It is detected in the majority of cell types, organelles, and apoplast in plants (Shao *et al.*, 2008) and is found to be particularly abundant in photosynthetic tissues (Smirnoff *et al.*, 2004). The biochemical functions of ascorbate can be divided into four categories: (i) Antioxidant, because ascorbate reacts rapidly with superoxide, singlet oxygen, ozone, and hydrogen peroxide. It thus participates in removal of this reactive form of oxygen which is generated during ascorbic metabolism and during exposure to some pollutants and herbicides (Horemans *et al.*, 1994). (ii) Enzyme cofactor. It is a cofactor for a range of hydroxylase enzymes for example, prolyl and lysyl hydroxylases involved in hydroxyproline and hydroxylisine synthesis (Arrigoni *et al.*, 1989). (iii) Electron transport. AsA is well known as an *in vitro* electron donor for photosynthetic and mitochondrial electron transport. (iv) Oxalate and tartrate synthesis. AsA can be cleaved to form oxalate and tartate (Saito, 1996). Most of AsA, more than 90%, is localized in cytoplasm, but unlike other soluble antioxidants a substantial portion is exported to the apoplast, where it is present in millimolar concentration. Apoplastic AsA is believed to represent the first line of defense against potentially damaging external oxidants (Foyer *et al.*, 1991).

Glutathione (GSH) is one of the important nonprotein thiols that play a major role in intracellular defense against ROS-induced oxidative damage (Foyer and Noctor, 2003) In plants, the physiological significance of GSH may be divided into two categories: sulfur metabolism and defense. GSH is the predominant nonprotein thiol (Rennenberg, 1982), and it regulates sulfur uptake at root level (Lappartient and Touraine, 1996). It is used by the GSH S-transferases in the detoxification of xenobiotics (Lamoureux and Rusness, 1993) and is a precursor of the phytochelatins, which are crucial in controlling cellular heavy metal concentrations (Scheller *et al.*, 1987). In addition to its effects on expression of defense genes (Shaul *et al.*, 1996), GSH may also be involved in redox control of cell division (Wingate *et al.*, 1988).

This enzyme has been detected in virtually all cell compartments such as cytosol, chloroplasts, endoplasmic reticulum, vacuoles, and mitochondria (Foyer and Noctor 2003). GSH is synthesized in the cytosol and chloroplasts of plant cells by compartment specific isoforms of γ-glutamyl-cysteinyl synthetase (γ-ECS) and glutathione synthetase (GS). The balance between the GSH and glutathione disulfide (GSSG) is a central component in maintaining cellular redox state. Due to its reducing power, GSH plays an important role in diverse biological processes, including cell growth/division, regulation of sulfate transport, signal transduction, conjugation of metabolites, enzymatic regulation, synthesis of proteins and nucleic acids, synthesis of phytochelatins for metal chelation, detoxification of xenobiotics, and the expression of the stress-responsive genes (Foyer *et al.*, 1991). GSH functions as an antioxidant in many ways. It can react chemically with $O_2 \bullet-$, $\bullet OH$, H_2O_2 and, therefore, can function directly as a free radical scavenger. GSH can protect macromolecules (i.e., proteins, lipids, DNA) either by the formation of adducts directly with reactive electrophiles (glutathiolation) or by acting as a proton donor in the presence of ROS or organic free radicals, yielding GSSG (Asada, 1994). It can participate in regeneration of another potential antioxidant AsA, via the AsA-GSH cycle. GSH recycles AsA from its oxidized to reduced form by the enzyme dehydroascorbate reductase (DHAR) (Loewus, 1988).

Tocopherols (α, β, γ, and δ) are a group of lipophilic antioxidants involved in scavenging of oxygen free radicals and lipid peroxy radicals (Diplock *et al.*, 1989). Tocopherols have two principal oxidation mechanisms (Neely *et al.*, 1988). Firstly, they may be oxidized in a one-electron-transfer reaction to a tocopheryl-radical. Secondly, they may react with singlet oxygen to a hydroperoxide, equivalent to a two-electron-transfer reaction. Both reactions can be reversed, as both the radical as well as the hydroperoxide can be re-reduced to tocopherol by AsA. However, the hydroperoxide is also easily split by mild acidic conditions (as in the chloroplast lumen) to tocopherylquinone and this reaction competes with re-reduction. The cleavage to tocopherylquinone makes the second oxidation mechanism irreversible (Krieger and Trebst, 2006). Relative antioxidant activity of the tocopherol isomers *in vivo* is $\alpha > \beta > \gamma > \delta$ which is due to the methylation pattern and the amount of methyl groups attached to the phenolic ring of the polar head structure (Fukuzawa *et al.*, 1982). Hence, α-tocopherol with its three methyl substituents has the highest antioxidant activity of tocopherols (Kamal-Eldin and Appelqvist, 1996). Tocopherols are synthesized only by photosynthetic organisms and are present in only green parts of plants. The tocopherol biosynthetic pathway utilizes two compounds (homogentisic acid (HGA) and phytyl diphosphate (PDP)) as precursors. At least five enzymes, 4-hydroxyphenylpyruvate dioxygenase (HPPD), homogentisate phytyl transferases (VTE2), 2-methyl-6-phytylbenzoquinol methyltransferase (VTE3), tocopherol cyclase (VTE1), γ-tocopherol methyltransferase (VTE4), are involved in the biosynthesis of tocopherols, excluding the bypass pathway of phytyl-tail synthesis and utilization (Li *et al.*, 2010). Tocopherols are known to protect lipids and other membrane components by physically quenching and chemically reacting with O_2 in chloroplasts, thus protecting the structure and function of PSII (Ivanov and Khorobrykh, 2003).

Carotenoids are colored terpenes synthesized in plants, algae and some yeasts and bacteria. In plants and algae, these lipophilic molecules exert functional roles in hormone synthesis, photosynthesis, photomorphogenesis and photoprotection (Pizarro and Stange, 2009). Additionally, they belong to the group of lipophilic antioxidants and are able to detoxify various forms of ROS (Young, 1991). Carotenoids are found in plants as well as microorganisms. As an antioxidant, they scavenge O_2 to inhibit oxidative damage and quench triplet sensitizer and excited chlorophyll molecules to prevent the formation of O_2 to protect the photosynthetic apparatus. Carotenoids also serve as precursors to signaling molecules that influence plant development and biotic/abiotic stress responses (Li *et al.*, 2008). The ability of carotenoids to scavenge, prevent, or minimize the production of triplet chlorophyll may be accounted for by their chemical specificity. Carotenoids contain a chain of isoprene residues bearing numerous conjugated double bonds which allows easy energy uptake from excited molecules and dissipation of excess energy as heat (Grene *et al.*, 2002).

5.3.2 **Enzymatic antioxidants**

The enzymatic components of the antioxidative defense system comprise of several antioxidant enzymes such as superoxide dismutase (SOD), catalase (CAT), guaiacol peroxidase (GPX), enzymes of ascorbate-glutathione (AsA–GSH) cycle ascorbate peroxidase (APX), monodehydroascorbate reductase (MDHAR), dehydroascorbate reductase, and glutathione reductase (GR) (Li *et al.*, 2008). These enzymes operate in different subcellular compartments and respond in concert when cells are exposed to oxidative stress.

Superoxide Dismutase (SOD) constitutes the first line of defence against ROS in plant cells (Grene *et al.*, 2002). The enzyme SOD belongs to the group of metalloenzymes and catalyzes the dismutation of $O_2^{\bullet-}$ to O_2 and H_2O_2. It is present in most of the subcellular compartments that generate activated oxygen. Three isozymes of SOD copper/zinc SOD (Cu/Zn-SOD), manganese SOD (Mn-SOD), and iron SOD (Fe-SOD) are reported in plants (Fridovich, 1989; Racchi *et al.*, 2001). All forms of SOD are nuclear encoded and targeted to their respective subcellular compartments by an amino terminal targeting sequence (Bowler *et al.*, 1992). MnSOD is localized in mitochondria, whereas Fe- SOD is localized in chloroplasts (Jackson *et al.*, 1978). Cu/Zn-SOD is present in three isoforms, which are found in the cytosol, chloroplast, and peroxisome and mitochondria (Bowler *et al.*, 1992, Kanematsu and Asada, 1989; Bueno *et al.*, 1995; Del Río *et al.*, 1998). Eukaryotic Cu/Zn-SOD is cyanide sensitive and presents as a dimer, whereas the other two (Mn-SOD and Fe-SOD) are cyanide insensitive and may be a dimer or tetramer (Del Río *et al.*, 1998). Specialization of function among the SODs may be due to a combination of the influence of subcellular location of the enzyme and upstream sequences in the genomic sequence. The commonality of elements in the upstream sequences of Fe, Mn and Cu/Zn SODs suggests a relatively recent origin for those regulatory regions (Grene *et al.*, 2002). The evolutionary reason for the separation of SODs with different metal requirements is probably related to the different availability of soluble transition metal compounds in the biosphere in relation to the O_2 content of the atmosphere in different geological eras (Bannister and Parker, 1985). SOD activity has been reported to increase in plants exposed to various environmental stresses, including drought and metal toxicity. Increased activity of SOD is often correlated with increased tolerance of the plant against environmental stresses (Sharma and Dubey, 2005).

Catalase (CAT) is an ubiquitous tetrameric heme-containing enzyme that catalyzes the dismutation of two molecules of H_2O_2 into water and oxygen. It has high specificity for H_2O_2, but weak activity against organic peroxides. Plants contain several types of H_2O_2-degrading enzymes, however, CATs are unique as they do not require cellular reducing equivalent. These enzymes have extremely high maximum catalytic rates but low substrate affinities, since the reaction requires the simultaneous access of two H_2O_2 molecules to the active site (Willekens *et al.*, 1995). CATs have a very fast turnover rate, but a much lower affinity for H_2O_2 than APX. The peroxisomes are major sites of H_2O_2 production. CAT scavenges H_2O_2 generated in this organelle during photorespiratory oxidation, β-oxidation of fatty acids, and other enzyme systems such as XOD coupled to SOD (Corpas *et al.*, 2008). Though there are frequent reports of CAT being present in cytosol, chloroplast, and mitochondria, the presence of significant CAT activity in these is less well established (Mhamdi *et al.*, 2010).

Guaiacol Peroxidase (GPX) is an important group of peroxidases, which oxidize guaiacol (omethoxyphenol) as a commonly used reducing substrate. They are found in cellular cytoplasm and appoplasm fractions and are involved in a range of processes related to plant growth and development (Tayefi-Nasrabadi *et al.*, 2011). This heme containing protein preferably oxidizes aromatic electron donors such as guaiacol and pyragallol at the expense of H_2O_2. These enzymes have four conserved disulfide bridges and contain two structural Ca^{2+} ions (Schuller *et al.*, 1996). Many isoenzymes of GPX exist in plant tissues localized in vacuoles, the cell wall, and the cytosol (Asada, 1992). GPX is associated with many important biosynthetic processes, including lignification of cell wall, biosynthesis of ethylene, wound healing, and defense against abiotic and biotic stresses (Kobayashi *et al.*, 1996). GPXs are widely accepted as a stress "enzyme." GPX can function as effective quencher of reactive intermediary forms of O_2 and peroxyl radicals under stressed conditions. Various stressful conditions of the environment have been shown to induce the activity of GPX. The peroxidase system of higher plants exists in multiple isoforms that are developmentally regulated and highly reactive in response to exogenous stimuli (Vangronsveld and Clijsters, 1994).

5.4 Polyphenol properties

More than 8000 polyphenolic compounds have been identified in various plant species. All plant phenolic compounds arise from a common intermediate, phenylalanine, or a close precursor, shikimic acid. Primarily they occur in conjugated forms, with one or more sugar residues linked to hydroxyl groups, although direct linkages

Table 5.1 Polyphenol Chemical Classification

Category	Compound
Anthocyanins	Cyanidin
	Delphindin
	Malvidin
	Pelargonidin
	Peonidin
	Petunidin
Flavan-3-ols	(+)-catechin
	(–)-epicathechin
Flavonols	Isorhamnetin
	Kaempferol
	Myricetin
	Quercetina
Hydroxybenzoic acids	Ellagic
	Gallic
	3,4-dihydroxybenzoic
	p-hydroxybenzoic
	Protocatechic
	Syringic
	Vanillic
Hydroxycinnamic acids	Caffeic
	Chlorogenic
	p-coumaric
	Ferulic
Procyanidins	procyanidins A2, B1, B2, B3, B4
Stilbenes	Piceid
	Resveratrol

Adapted from Rodrigo, R., et al. (2011) Clin. Chim. Acta. 412, 410–424.

of the sugar to an aromatic carbon also exist. Association with other compounds, like carboxylic and organic acids, amines, lipids and linkage with other phenols is also common (Kondratyuk and Pezzuto, 2004). Polyphenols may be classified into different groups as a function of the number of phenol rings that they contain and on the basis of structural elements that bind these rings to one another. The main classes include phenolic acids, flavonoids, stilbenes and lignans (Spencer, 2009) (Table 5.1).

5.4.1 Mechanisms of protection

Different mechanisms of protection exist in polyphenols. Polyphenols can chelate transition metal ions, can directly scavenge molecular species of active oxygen, and can inhibit lipid peroxidation by trapping the lipid alkoxyl radical. They also modify lipid packing order and decrease fluidity of the membranes (Arora *et al.*, 2002). These changes could strictly hinder diffusion of free radicals and restrict peroxidative reactions. Moreover, it has been shown that, especially, flavonoids and phenylpropanoids are oxidized by peroxidase, and act in H_2O_2-scavenging, phenolic/AsA/POD systems (Sharma *et al.*, 2012). Polyphenols also contain an aromatic ring with $-OH$ or OCH_3 substituents which together contribute to their biological activity, including antioxidant action. They have been shown to outperform well-known antioxidants, AsA, and α-tocopherol, in *in vitro* antioxidant assays because of their strong capacity to donate electrons or hydrogen atoms. (Sharma *et al.*, 2012). The polyphenolic compounds may be responsible for the major portion of the antioxidant activity of leaves. These results are in good agreement with a report by Lee *et al.* (2009), who reported that the antioxidant activity of olive leaves was well correlated with the amount of phenolic compounds and flavonoids. Many previous studies have reported significant correlation between polyphenolic and antioxidant activities in fruits, barley and mushrooms (Lee *et al.*, 2009; Goupy *et al.*, 1999; Leong and Shui, 2002; Choi *et al.*, 2005). AsA, for example, is not only an efficient chemical scavenger of toxic free radicals, but is also a required molecule for the formation of zeaxanthin, the most important quencher of excess light energy (Demming-Adams and Adams, 1992), and for the regeneration of α-tocopherol, an important protectant of thylakoid membranes (Fryer, 1992).

Polyphenols can modulate by acting as antioxidants against the damaging effects of increased ROS levels, for example due to a common condition such as salt stress (Tattini *et al.*, 2006). In this case, the responses of five antioxidant enzymes showed that ascorbate peroxidase, guaiacol peroxidase, and superoxide dismutase activities were the most enhanced after NaCl exposure, catalase moderately, and glutathione reductase the least (Chang *et al.*, 2012). Genomic and proteomic screenings carried out in *Physcomitrella patens* plants showed that they responded to salinity stress by upregulating a large number of genes involved in the antioxidant defense mechanism (Wang *et al.*, 2008), suggesting that the antioxidative system may play a crucial role in protecting cells from oxidative damage following exposure to salinity stress in *P. patens*. Salinity induced oxidative stress and a possible relationship between the status of the components of antioxidative defense system and the salt tolerance in Indica rice (*Oryza sativa* L.) genotypes were studied by Mishra *et al.* (2011). Seedlings of salt-sensitive cultivar showed a substantial increase in the rate of $O_2 \bullet-$ production, elevated levels of H_2O_2, MDA, declined levels of thiol, AsA, and GSH, and lower activity of antioxidant enzymes compared to salt-tolerant seedlings. It was suggested that a higher status of antioxidants AsA and GSH and a coordinated higher activity of the enzymes SOD, CAT, GPX, APX, and GR can serve as the major determinants in the model for depicting salt tolerance in Indica rice seedlings (Mishra *et al.*, 2011). Similarly, study of immediate responses (enzymatic and nonenzymatic) to salinity-induced oxidative stress in two major rice (*Oryza sativa* L.) cultivars, salt sensitive Pusa Basmati 1 (PB) and salt-tolerant Pokkali (PK), revealed a lesser extent of membrane damage (lipid peroxidation), lower levels of H_2O_2, higher activity of the ROS scavenging enzyme, CAT and enhanced levels of antioxidants such as ASA and GSH

Table 5.2 Type of Stimuli and Antioxidant Response

Stimuli	Antioxidant Defense	Reference
Drought	Superoxide dismutase, catalase	[105],[94]
Salt	Catalase, glutathione reductase	[116],[7]
Chilling	Ascorbate peroxidase, superoxide dismutase, glutathione reductase	[33],[43]
Metals	Ascorbate peroxidase, superoxide dismutase, glutathione reductase	[69],[101]
Radiation	Superoxide dismutase,catalase	[31]
Temperature	Superoxide dismutase,catalase	[43]

in PK compared to PB (Vaidyanathan, 2003). NADP-dehydrogenases and peroxidase have been suggested as key antioxidative enzymes in olive plants under salt stress conditions (Valderrama *et al.*, 2006). Indeed, Mittal and Dubey (1991) observed a correlation between peroxidase activity and salt tolerance in rice seedling.

Plants also possess antioxidative enzymatic scavengers SOD, POD, CAT, and APX, and nonenzymatic antioxidants like AsA, GSH, and carotenoids to keep the balance between the production and removal of ROS.

Other examples of polyphenol modulation are:

Flavonoids, which represent a class of phenolic metabolites that have significant antioxidant and chelating properties. Their propensity to inhibit free radical mediated events is governed by their chemical structure. This structure–activity relationship has been well established *in vitro* and has been demonstrated by many studies (Heim *et al.*, 2002; Amić *et al.*, 2007). The isolated flavone glycosides (luteolin 7-*O*-rutinoside, luteolin 7-Oneohesperidoside, luteolin 7-*O*-glucoside, luteolin 4′-*O*-glucoside) have shown to be active antioxidants. The mechanism behind the radical-scavenging activity of flavonoids is thought to be hydrogen atom donation and the structural requirements include an ortho-dihydroxy substitution in the B ring, a C2–C3 double bond and a C4 carbonyl group in the C ring. The free hydroxyl groups on the B ring donate hydrogen to a radical, thus stabilising it and giving rise to a relatively stable flavonoid radical. Flavones with the C2–C3 double bond in conjugation with a C4 carbonyl group are planar; this structural feature allows a charge delocalization throughout the three rings system. In flavonoids with the ortho-dihydroxy (catechol) group, the formation of flavonoid phenoxy radicals may be stabilized by the mesomeric equilibrium to ortho-semiquinone structures (Amić *et al.*, 2007). Aglycones are more potent antioxidants than their corresponding glycosides. *O*-glycosylation of the A or B ring decreases the antiradical power (De Marino, 2012). It has been shown that hesperidin is able to inhibit lipid peroxidation since it has been reported to act as a powerful superoxide, singlet oxygen and hydroxyl radicals scavenger, and to react with peroxyl radicals involving termination of radical chain reactions (Torel *et al.*, 1986).

Neohesperidin and naringin have been reported to exhibit antioxidant and radical scavenger properties and to offer protection against lipid peroxidation which is linked to the presence of effective antioxidant structures including hydroxylation and C2–C3 double bond in conjugation with a 4-oxo function (Di Majo *et al.*, 2005; Hamdan *et al.*, 2011) Moreover, kampferol could participate in the antioxidant activities of bitter orange since it is a powerful antioxidant compound (Yang *et al.*, 2009).

Hydroxybenzoic acids. Gallic acid has been reported as an efficient 2,2-diphenyl-1-picrylhydrazyl (DPPH) antiradical compound which may be attributed to its proton-donating ability (Zhou *et al.*, 2006). This compound is characterized by the presence of galloyl moiety (three –OH groups) in the B ring which contribute to the high scavenging reactive oxygen species (ROS) activity (Villano *et al.*, 2007).

Linoleic acid. Regarding the structure–activity relationship, the presence of a free hydroxyl group on C-5 of 8-*O*-acetylharpagid could be responsible for its higher antioxidant ability, with respect to what was observed for compound Tucardosid (Pacífico *et al.*, 2009). However, it should be noted that the process of linoleic acid peroxidation inhibition could be due to the effect of more and/or combined antioxidant mechanisms (De Graft-Johnson *et al.*, 2007). The effect of these expogenous antioxidants is summarized in Table 5.2.

5.5 **Concluding remarks**

Reactive oxygen species (ROS) may be generated in plants, as in other cell types, through several metabolic pathways leading to univalent reductions of ground state oxygen. Oxidative stress arises from the imbalance prooxidant–antioxidant in favor of the first. In the plant, this balance may be challenged by several factors. Thus, low temperatures are closely associated with enhancement of ROS production. Heat stress affects the organization of microtubules by splitting and/or elongation of spindles, formation of microtubule asters in mitotic cells, and elongation of phragmoplast microtubules. These injuries eventually lead to starvation, inhibition of growth, reduced ion flux, production of toxic compounds and reactive oxygen species (ROS). Under salt stress, antioxidant systems are similar among plants and the defense mechanisms are shared by different plant species; however, salt responses in reprogramming of the whole plant to attain a new metabolic and cellular homeostasis to cope with the upcoming stress are species-specific. Salinity may stimulate the biosynthesis of some polyphenols, but others like hydroxytyrosol are not affected by the salt stress. Photosynthetic activity is inhibited in plant tissues due to an imbalance between light capture and its utilization under drought stress. Oxidative stress imposed by ROS under drought conditions profoundly affects plant growth and development. It has been reported that chilling leads to the overproduction of ROS by exacerbating the imbalance between light absorption and light. Plant growth and metabolism may be drastically affected by increasing levels of metals in the environment, leading to severe losses in crop yields. A better knowledge of the effect of the mechanisms modulating this activity, such as the plant polyphenol composition, could provide new paradigms facilitating the design of studies aimed to modify their abundance toward an improvement of plant development and stability, as well as for their production for industry purposes or potential human health benefits.

References

Allakhverdiev, S.I., Murata, N., 2008. Salt stress inhibits photosystem II and I in cynobacteria. Photosynth. Res. 98, 529–539.

Allen, D.J., Mckee, F., Farage, P.K., Baker, N.R., 1997. Analysis of limitations to CO_2 assimilation on exposure of leaves of two *Brassica napus* cultivars to UV-B. Plant Cell Environ. 20, 633–640.

Alsocher, R.G., Erturk, N., Heath, L., 2002. Role of superoxide dismutases (SODs) in controlling oxidative stress in plants. J. Exp. Bot. 53, 1331–1341.

Amić, D., Davidović-Amić, D., Beslo, D., Rastija, V., Lucić, B., Trinajstic, N., 2007. SAR and QSAR of the antioxidant activity of flavonoids. Curr. Med. Chem. 14, 827–845.

Apel, K., Hirt, H., 2004. Reactive oxygen species: metabolism, oxidative stress, and signal transduction. Annu. Rev. Plant. Biol. 55, 373–399.

Arora, A., Sairam, R.K., Srivastava, G.C., 2002. Oxidative stress and antioxidative system in plants. Curr. Sci. 82, 1227–1238.

Arrigoni, O., Bitonti, M.B., Cozza, R., Innocenti, A.M., Liso, R., Veltri, R., 1989. Ascorbic acid effect on pericycle cell line in *Allium cepa* root. Caryologia 42, 213–216.

Asada, K., 1992. Ascorbate peroxidase: a hydrogen peroxide scavenging enzyme in plants. Physiol. Plantarum 85, 235–241.

Asada, K., 1994. Production and action of active oxygen species in photosynthetic tissues. In: Foyer, C.H., Mullineaux, P.M. (Eds.), "Causes of Photooxidative Stress and Amelioration of Defense Systems in Plants". CRC, Boca Raton, pp. 77–104.

Asada, K., 1999. The water–water cycle in chloroplasts: scavenging of active oxygens and dissipation of excess photons. Plant Mol. Biol. 50, 601–639.

Bannister, J.V., Parker, M.W., 1985. The presence of a copper/zinc superoxide dismutase in the bacterium *Photobacterium leiognathi*: a likely case of gene transfer from eukaryotes to prokaryotes. Proc. Natl. Acad. Sci. 82, 149–252.

Biehler, K., Fock, H., 1996. Evidence for the contribution of the Mehler-peroxidase reaction in dissipating excess electrons in drought-stressed wheat. Plant Physiol. 112, 265–272.

Bowler, C., Van Montagu, M., Inzé, D., 1992. Superoxide dismutase and stress tolerance. Annu. Rev. Plant Phys. 43, 83–116.

Bueno, P., Varela, J., Giménez-Gallego, G., Del Rio, L.A., 1995. Peroxisomal copper, zinc superoxide dismutase. Characterization of the isoenzyme from watermelon cotyledons. Plant Physiol. 108, 1151–1160.

Cavalcanti, F.R., Oliveira, J.T.A., Martins, M.A.S., Viegas, R.A., Silveira, J.A.G., 2004. Superoxide dismutase, catalase and peroxidase activities do not confer protection against oxidative damage in saltstressed cowpea leaves. New Phytol. 163, 563–571.

Chang, I., Cheng, K., Huang, P., Lin, Y., Cheng, L., Cheng, T., 2012. Oxidative stress in greater duckweed (*Spirodela polyrhiza*) caused by long term NaCl exposure. Physiol. Plant 34, 1165–1176.

Chaparzadeh, N., D'Amico, M.L., Khavari-Nejad, R.A., Izzo, R., Navari-Izzo, F., 2004. Antioxidative responses of *Calendula officinalis* under salinity conditions. Plant Physiol. Biochem. 42, 695–701.

Choi, Y., Ku, J., Chang, H., Lee, J., 2005. Antioxidant activities and total phenolics of ethanol extracts from several edible mushrooms produced in Korea. Food Sci. Biotechnol. 14, 703–770.

Corpas, F.J., Palma, J.M., Sandalio, L.M., Valderrama, R., Barroso, J.B., del Río, L.A., 2008. Peroxisomal xanthine oxidoreductase: characterization of the enzyme from pea (*Pisum sativum* L.) leaves. J. Plant Physiol. 165, 1319–1330.

D'Amico, M.L., Izzo, R., Tognoni, F., Pardossi, A., Navari-Izzo, F., 2003. Application of diluted sea water to soilless culture of tomato (*Lycopersicon esculentum*): effects on plant growth, yield, fruit quality and antioxidant capacity. Food Agric. Environ. 1, 112–116.

De Graft-Johnson, J., Koloddziejczyk, K., Krol, M., Krol, B., Nowak, P., Nowak, D., 2007. Ferric-Reducing ability power of selected plant poliphenols and their metabolites: implications for clinical studies on the antioxidant effects of fruits and vegetable consumption. Basic Clin. Pharmacol. Toxicol. 100, 345–352.

De Marino, S., 2012. Antioxidant activity of phenolic and phenylethanoid glycosides from *Teucrium polium* L. Food Chem. 303, 21–28.

De Pinto, M.C., De Gara, L., 2004. Changes in the ascorbate metabolism of apoplastic and symplastic spaces are associated with cell differentiation. J. Exp. Bot. 55, 2559–2569.

Del Río, L.A., Pastori, G.M., Palma, J.M., 1998. The activated oxygen role of peroxisomes in senescence. Plant Physiol. 116, 1195–1200.

Demmig-Adams, B., Adams III, W.W., 1992. Photoprotection and other responses of plants to high light stress. Plant Physiol. 43, 599–626.

Di Majo, D., Giammanco, M., La Guardia, M., Tripoli, E., Giammanco, S., et al., 2005. Flavanones in Citrus fruit: structure–antioxidant activity relationships. Food Res. Int. 38, 1161–1166.

Diplock, T., Machlin, L.J., Packer, L., Pryor, W.A., 1989. Vitamin E: biochemistry and health implications. Ann. Ny Acad. Sci. 570, 372–378.

Fenta, B.A., Kunert, K.J., Driscoll, S.P., Foyer, C.H., 2012. Characterization of Drought-Tolerance Traits in Nodulated Soya Beans: The Importance of Maintaining Photosynthesis and Shoot Biomass Under Drought-Induced Limitations on Nitrogen Metabolism. J. Agron. Crop. Sci. 198, 92–103.

Fowler, S., Thomashow, M.F., 2002. Arabidopsis transcriptome profiling indicates that multiple regulatory pathways are activated during cold acclimation in addition to the CBF cold response pathway. Plant Cell. 14, 1675–1690.

Foyer, C.H., Noctor, G., 2003. Redox sensing and signaling associated with reactive oxygen in chloroplasts, peroxisomes and mitochondria. Physiol. Plantarum. 119, 355–364.

Foyer, C.H., Noctor, G., 2000. Oxygen processing in photosynthesis: regulation and signaling. New Phytol. 146, 359–388.

Foyer, C.H., Lelandais, M., Galap, C., Kunert, K.J., 1991. Effects of elevated cytosolic glutathione reductase activity on the cellular glutathione pool and photosynthesis in leaves under normal and stress conditions. Plant Physiol. 97, 863–872.

Fridovich, I., 1989. Superoxide dismutases. An adaptation to a paramagnetic gas. J. Biol. Chem. 264, 7761–7764.

Fryer, M.J., 1992. The antioxidant effects of thylakoid vitamin E (atocopherol). Plant Cell. Environ. 15, 381–392.

Fukuzawa, K., Tokumura, A., Ouchi, S., Tsukatani, H., 1982. Antioxidant activities of tocopherols on Fe^{2+}-ascorbate induced lipid peroxidation in lecithin liposomes. Lipids 17, 511–514.

Gallego, S., Benavides, M., Tomaro, M., 2002. Involvement of an antioxidant defence system in the adaptive response to heavy metal ions in *Helianthus annuus* L. cells. Plant Growth Regul. 36, 267–273.

Gao, Q., Zhang, L., 2008. Ultraviolet-B-induced oxidative stress and antioxidant defense system responses in ascorbate deficient vtc1 mutants of *Arabidopsis thaliana*. J. Plant Physiol. 165, 138–148.

Gong, M., Chen, S., Song, Y., Li, Z., 1997. Effect of calcium and calmodulin on intrinsic heat tolerance in relation to antioxidant systems in maize seedlings. Aust. J. Plant Physiol. 24, 371–379.

Goupy, P., Hugues, M., Boivin, P., Amiot, M.J., 1999. Antioxidant composition and activity of barley (*Hordeum vulgare*) and malt extracts and of isolated phenolic compounds. J. Sci. Food Agric. 79, 1625–1634.

Grene, R., Erturk, N., Lenwood, S., 2002. Heath Role of superoxide dismutases (SODs) in controlling oxidative stress in plants. J. Exp. Bot. 53, 1331–1341.

Grosch, W., 1982. Lipid degradation products and flavor. In: Morton, I.D., Macleod, A.J. (Eds.), Food flavours Part A, vol. 5. Elsevier, Oxford, pp. 325–385.

Gupta, K.J., Stoimenova, M., Kaiser, W.M., 2005. In higher plants, only root mitochondria, but not leaf mitochondria reduce nitrite to NO, *in vitro* and *in situ*. J. Exp. Bot. 56, 2601–2609.

Hamdan, D., Zaki El-Readi, M., Tahrani, A., Herrmann, F., Kaufmann, D., Farrag, N., et al., 2011. Chemical composition and biological activity of *Citrus jambhiri* Lush. Food Chem. 127, 394–403.

Han, C., Liu, Q., Yang, Y., 2009. Short-term effects of experimental warming and enhanced ultraviolet-B radiation on photosynthesis and antioxidant defense of *Picea asperata* seedlings. Plant Growth Regul. 58, 153–162.

He, J., Huang, L.K., Chow, W.S., Whitecross, M.L., Anderson, J.M., 1993. Effects of supplementary ultraviolet-B radiation on rice and pea plants. Aust. J. Plant Physiol. 20, 129–142.

Heim, K.E., Tagliaferro, A., Bobiya, D., 2002. Flavonoid antioxidants: chemistry, metabolism and structure–activity relationships. J. Nutr. Biochem. 13, 572–584.

Hernández, J.A., Jiménez, A., Mullineaux, P., Sevilia, F., 2000. Tolerance of pea (*Pisum sativum* L.) to long-term salt stress is associated with induction of antioxidant defences. Plant Cell Environ. 43, 853–862.

Horemans, N., Asard, H., Caubergs, R.J., 1994. The role of ascorbate free radical as an electron acceptor to cytochrome b-mediated transplasma membrane electron transport in higher plants. Plant Physiol. 104, 1455–1458.

Hu, W.H., Song, X.S., Shi, K., Xia, X.J., Zhou, Y.H., Yu, J.Q., 2008. Changes in electron transport, superoxide dismutase and ascorbate peroxidase isoenzymes in chloroplasts and mitochondria of cucumber leaves as influenced by chilling. Photosyn. 46, 581–588.

Imlay, J.A., 2003. Pathways of oxidative damage. Ann. Rev. Microbiol. 57, 395–418.

Ivanov, B.N., Khorobrykh, S., 2003. Participation of photosynthetic electron transport in production and scavenging of reactive oxygen species. Antioxid Redox Sign. 5, 43–53.

Jackson, C., Dench, J., Moore, A.L., Halliwell, B., Foyer, C.H., Hall, D.O., 1978. Subcellular localization and identification of superoxide dismutase in the leaves of higher plants. Eur. J. Biochem. 91, 339–344.

Johnston, J.W., Harding, K., Benson, E.E., 2007. Antioxidant status and genotypic tolerance of Ribes *in vitro* cultures to cryopreservation. Plant Sci. 172, 524–534.

Kamal-Eldin, A., Appelqvist, L.A., 1996. The chemistry and antioxidant properties of tocopherols and tocotrienols. Lipids 31, 671–701.

Kanematsu, S., Asada, K., 1989. CuZn-superoxide dismutases in rice: occurrence of an active, monomeric enzyme and two types of isozyme in leaf and non-photosynthetic tissues. Plant Cell Physiol. 30, 381–391.

Kobayashi, K., Kumazawa, Y., Miwa, K., Yamanaka, S., 1996. ε- (γ-Glutamyl) lysine cross-links of spore coat proteins and transglutaminase activity in *Bacillus subtilis*. FEMS Microbiol. Lett. 144, 157–160.

Kondratyuk, T.P., Pezzuto, J.M., 2004. Natural Product Polyphenols of Relevance to Human Health. Pharm Biol. 42, 46–63.

Krieger-Liszkay, A., Trebst, A., 2006. Tocopherol is the scavenger of singlet oxygen produced by the triplet states of chlorophyll in the PSII reaction centre. J. Exp. Bot. 57, 1677–1684.

Lamoureux, G.L., Rusness, D.G., 1993. Glutathione in the metabolism and detoxification of xenobiotics in plants. In: de Kok, L.J., Stulen, I., Rennenberg, H., Brunold, C., Rauser, W.E. (Eds.), "Sulfur Nutrition and Assimilation in Higher Plants". The Hague: SPB Acad, The Netherlands, pp. 221–237.

Lappartient, A.G., Touraine, B., 1996. Demand-driven control of root ATP sulfurylase activity and sulphate uptake in intact Canola. Plant Physiol. 111, 147–157.

Lee, O.H., Lee, B.Y., Lee, J., Lee, H., Son, J., Park, C., et al., 2009. Assessment of phenolics-enriched extract and fractions of olive leaves and their antioxidant activities. Bioresour. Technol. 100, 6107–6113.

Lee, S., Park, C.M., 2012. Regulation of reactive oxygen species generation under drought conditions in Arabidopsis. Plant Signal Behav. 7, 599–601.

Leong, L., Shui, G., 2002. An investigation of antioxidant capacity of fruits in Singapore markets. Food Chem. 76, 69–75.

Li, F., Vallabhaneni, R., Yu, J., Rocheford, T., Wurtzel, E.T., 2008. The maize phytoene synthase gene family: overlapping roles for carotenogenesis in endosperm, photomorphogenesis, and thermal stress tolerance. Plant Physiol. 147, 1334–1346.

Li, Y., Zhou, Y., Wang, Z., Sun, X., Tang, K., 2010. Engineering tocopherol biosynthetic pathway in *Arabidopsis* leaves and its effect on antioxidant metabolism. Plant Sci. 178, 312–320.

Liu, X., Huang, B., 2000. Heat stress injury in relation to membrane lipid peroxidation in creeping bent grass. Crop. Sci. 40, 503–510.

Loewus, F.A., 1988. Ascorbic acid and its metabolic products. In: Preiss, J. (Ed.), "The Biochemistry of Plants". Academic Press, New York, pp. 85–107.

Logan, B.A., Kornyeyev, D., Hardison, J., Holaday, A.S., 2006. The role of antioxidant enzymes in photoprotection. Photosyn. Res. 88, 119–132.

Mhamdi, A., Queval, G., Chaouch, S., Vanderauwera, S., Van Breusegem, F., Noctor, G., 2010. Catalase function in plants: a focus on *Arabidopsis* mutants as stress-mimic models. J. Exp. Bot. 61, 4197–4220.

Mishra, S., Dubey, R.S., 2005. Heavy metal toxicity induced alterations in photosynthetic metabolism in plants. In: Pessarakli, M. (Ed.), Handbook of Photosynthesis. CRC Press, Taylor and Francis Publishing Company, Florida, USA, pp. 845–863.

Mishra, S., Jha, A.B., Dubey, R.S., 2011. Arsenite treatment induces oxidative stress, upregulates antioxidant system, and causes phytochelatin synthesis in rice seedlings. Protoplasma 248, 565–577.

Mittler, R., 2002. Oxidative stress, antioxidants and stress tolerance. Trends Plant Sci. 7, 405–410.

Mittler, R., Dubey, R.S., 1991. Behaviour of peroxidases in rice: changes in enzymatic activity and isoforms in relation to salt tolerance. Plant Physiol. Biochem. 29, 31–40.

Moustakas, M., Lanaras, T., Symeonidis, L., Karataglis, S., 1994. Growth and some photosynthetic characteristics of field grown *Avena sativa* under copper and lead stress. Photosynthetica 30, 389–396.

Neely, W.C., Martin, M., Barker, S.A., 1988. Products and relative reaction rates of the oxidation of tocopherols with singlet molecular oxygen. Photochem. Photobiol. 48, 423–428.

Noctor, G., Veljovic-Jovanovic, S., Driscoll, S., Novitskaya, L., Foyer, C.H., 2002. Drought and oxidative load in the leaves of C3 plants: a predominant role for photorespiration? Ann. Bot. 89, 841–850.

Pacifico, S., D'Abrosca, B., Pascarella, M., Letizia, M., Uzzo, P., Piscopo, V., 2009. Antioxidant efficacy of iridoid and phenylethanoid glycosides from the medicinal plant *Teucrium chamaedris* in cell-free systems. Bioorgan. Med. Chem. 17, 6173–6179.

Pérez-López, U., Robredo, A., Lacuesta, M., Sgherri, C., Muñoz-Rueda, A., Navari- Izzo, F., et al., 2009. The oxidative stress caused by salinity in two barley cultivars is mitigated by elevated CO_2. Physiol. Plant 135, 29–42.

Pérez-López, U., Robredo, A., Lacuesta, M., Sgherri, C., Mena-Petite, A., Navari-Izzo, F., et al., 2010. Lipoic acid and redox status in barley plants subjected to salinity and elevated CO2. Physiol. Plant 139, 256–268.

Pizarro, L., Stange, C., 2009. Light-dependent regulation of carotenoid biosynthesis in plants. Cien. Inv. Agr. 36, 143–162.

Potters, G., Horemans, N., Jansen, M.A.K., 2010. The cellular state in plant stress biology— A charging concept. Plant Physiol. Biochem. 48, 292–300.

Prasad, T.K., 1997. Role of catalase in inducing chilling tolerance in pre-emergent maize seedlings. Plant Physiol. 114, 1369–1376.

Racchi, M.L., Bagnoli, F., Balla, I., Danti, S., 2001. Differential activity of catalase and superoxide dismutase in seedlings and *in vitro* micropropagated oak (*Quercus robur* L.). Plant Cell Rep. 20, 169–174.

Rao, M.V., Paliyath, G., Ormrod, D.P., 1996. Ultraviolet-B- and ozone-induced biochemical changes in antioxidant enzymes of *Arabidopsis thaliana*. Plant Physiol. 110, 125–136.

Rennenberg, H., 1982. Glutathione metabolism and possible biological roles in higher plants. Phytochemistry 21, 2771–2781.

Rodrigo, R., Miranda, A., Vergara. L., 2011. Modulation of endogenous antioxidant system by wine polyphenols in human disease. Clin. Chim. Acta 412, 410–424.

Sairam, R., Tyagi, A., 2004. Physiology and molecular biology of salinity stress tolerance in plants. Curr. Sci. 86, 407–421.

Saito, K., 1996. Formation of l-ascorbic acid and oxalic acid from dglucosone in *Lemna minor*. Phytochemistry 41, 145–149.

Scheller, H.V., Huang, B., Hatch, E., Goldsbrough, P.B., 1987. Phytochelatin synthesis and glutathione levels in response to heavy metals in tomato cells. Plant Physiol. 85, 1031–1035.

Schöffl, F., Prandl, R., Reindl, A., 1998. Regulation of the Heat-Shock Response. Plant Physiol. 117, 1135–1141.

Schuller, D.J., Ban, N., Van Huystee, R.B., McPherson, A., Poulos, T.L., 1996. The crystal structure of peanut peroxidase. Structure 4, 311–321.

Semchuk, N.M., Lushchak, O.V., Falk, J., Krupinska, K., Lushchak, V.I., 2009. Inactivation of genes, encoding tocopherol biosynthetic pathway enzymes, results in oxidative stress in outdoor grown *Arabidopsis thaliana*. Plant Physiol. Bioch. 47, 384–390.

Shah, K., Kumar, R.G., Verma, S., Dubey, R.S., 2001. Effect of cadmium on lipid peroxidation, superoxide anion generation and activities of antioxidant enzymes in growing rice seedlings. Plant Sci. 161, 1135–1144.

Shao, H.B., Chu, L.Y., Lu, Z.H., Kang, C.M., 2008. Primary antioxidant free radical scavenging and redox signaling pathways in higher plant cells. Int. J. Biol. Sci. 4, 8–14.

Sharma, P., Dubey, R.S., 2005. Drought induces oxidative stress and enhances the activities of antioxidant enzymes in growing rice seedlings. Plant Growth Regul. 46, 209–221.

Sharma, P., Jha, A.B., Dubey, R.S., Pessarakli, M., 2012. Reactive Oxygen Species, Oxidative Damage and Antioxidative Defense Mechanism in Plants under Stressful Conditions. J. Bot. 2012, 1–26.

Shaul, O., Mironov, V., Burssens, S., Van Montagu, M., Inze, D., 1996. Two *Arabidopsis* cyclin promoters mediate distinctive transcriptional oscillation on synchronized tobacco 3 Y-2 cells. Proc. Natl. Acad. Sci. 93, 4868–4872.

Skyba, M., Urbanová, M., Kapchina-Toteva, V., Harding, K., Cellárová, E., 2010. Physiological, biochemical and molecular characteristics of cryopreserved *Hypericum perforatum* L. Cryo-Letters 31, 249–260.

Smertenko, A., Draber, P., Viklicky, V., Opatrny, Z., 1997. Heat stress affects the organization of microtubules and cell division in *Nicotiana tabacum* cells. Plant Cell Environ. 20, 1534–1542.

Smirnoff, N., 1996. The Function and Metabolism of Ascorbic Acid in Plants. Ann. Bot. 78, 661–669.

Smirnoff, N., Running, J.A., Gatzek, S., 2004. Ascorbate biosynthesis: a diversity of pathways. In: Asard, H., May, J.M., Smirnoff, N. (Eds.), Vitamin C: ItsFunctions and Biochemistry in Animals and Plants, vol. 4. BIOS Scientific, New York, pp. 7–29.

Spencer, J.P., Abd El Mohsen, M.M., Minihane, A.M., Mathers, J.C., 2009. Biomarkers of the intake of dietary polyphenols: strengths, limitations and application in nutrition research. Br. J. Nutr. 99, 12–22.

Srivastava, S., Dubey, R.S., 2011. Manganese-excess induces oxidative stress, lowers the pool of antioxidants and elevates activities of key antioxidative enzymes in rice seedlings. Plant Growth Regul. 5, 1–16.

Strid, A., Chow, W.S., Anderson, J.M., 1994. UV - B damage and protection at the molecular level in plants. Photosynth. Res. 39, 475–39, 489.

Tanou, G., Job, C., Rajjou, L., Arc, E., Belghazi, M., Diamantidis, G., et al., 2009. Proteomics reveals the overlapping roles of hydrogen peroxide and nitric oxide in the acclimation of citrus plants to salinity. Plant J. 60, 795–804.

Tattini, M., Remorini, D., Pinelli, P., Agati, G., Saracini, E., Traversi, M., et al., 2006. Morpho-anatomical, physiological and biochemical adjustments in response to root zone salinity stress and high solar radiation in two Mediterranean evergreen shrubs, *Myrtus communis* and *Pistacia lentiscus*. New Phytol. 170, 779–794.

Tayefi-Nasrabadi, H., Dehghan, G., Daeihassani, B., Movafegi, A., Samadi, A., 2011. Some biochemical properties of guaiacol peroxidases as modified by salt stress in leaves of salt-tolerant and salt-sensitive safflower (*Carthamus tinctorius* L.cv.) cultivars. Afr. Biotechnol. 10, 751–763.

Torel, J., Cillard, J., Cillard, P., 1986. Antioxidant activity of flavonoids and reactivity with peroxyl radical. Phytochemistry 25, 383–385.

Uchendu, E.E., Muminova, M., Gupta, S., Reed, B.M., 2010. Antioxidant and anti-stress compounds improve regrowth of cryopreserved *Rubus* shoot tips. In vitro Cell. Dev. Biol. Plant 46, 386–393.

Urquiaga, I., Leighton, F., 2000. Plant polyphenol antioxidant and oxidative stress. Biol. Res. 33, 55–64.

Vaidyanathan, H., Sivakumar, P., Chakrabarty, R., Thomas, G., 2003. Scavenging of reactive oxygen species in NaCl stressed rice (*Oryza sativa* L.)—Differential response in salt tolerant and sensitive varieties. Plant Sci. 165, 1411–1418.

Valderrama, R., Corpas, F.J., Carreras, A., 2006. The dehydrogenase-mediated recycling of NADPH is a key antioxidant system against salt-induced oxidative stress in olive plants. Plant. Cell. Environ. 29, 1449–1459.

Vangronsveld, J., Clijsters, H., 1994. Toxic effects of metals. In: Farago, M.E. (Ed.), Plants and the Chemical Elements. Biochemistry Uptake, Tolerance and Toxicity, vol. 3. VCH Publishers, Weinheim, Germany, pp. 150–177.

Villano, D., Fernandez-Pachon, M., Moya, M., Troncoso, A., García-Parrilla, M., 2007. Radical scavenging ability of polyphenolic compounds towards DPPH free radical. Talanta 71, 230–235.

Vinit-Dunand, F., Epron, D., Alaoui-Sosse, B., Badot, P.M., 2002. Effects of copper on growth and on photosynthesis of mature and expanding leaves in cucumber plants. Plant Sci. 163, 53–58.

Wang, X., Yang, P., Gao, Q., 2008. Proteomic analysis of the response to high-salinity stress in *Physcomitrella patens*. Planta 228, 167–177.

Whaid, A., Gelan, S., Ashraf, M., Foolad, M., 2007. Heat tolerance in plants: an overview. Environ. Exp. Bot. 61, 199–223.

Willekens, H., Inzé, D., Van Montagu, M., Van Camp, W., 1995. Catalase in plants. Mol. Breed 1, 207–228.

Wingate, V.P.M., Lawton, M.A., Lamb, C.J., 1988. Glutathione causes a massive and selective induction of plant defense genes. Plant Physiol. 31, 205–211.

Xu, S., Li, J., Zhang, X., Wei, H., Cui, L., 2006. Effects of heat acclimation pretreatment on changes of membrane lipid peroxidation, antioxidant metabolites, and ultrastructure of chloroplasts in two cool-season turfgrass species under heat stress. Environ. Exp. Bot. 56, 274–285.

Yang, J., Liu, B., Liang, G., Ning, Z., 2009. Structure–activity relationship of flavonoids active against lard oil oxidation based on quantum chemical analysis. Molecules 4, 46–52.

Yong, Z., Hao-Ru, T., Lou, Y., 2008. Variation in antioxidant enzyme activities of two strawberry cultivars with short-term low temperature stress. World J. Agric. Sci. 4, 458–462.

Young, J., 1991. The photoprotective role of carotenoids in higher plants. Physiol. Plantarum. 83, 702–708.

Zhou, K., Yin, J., Yu, L., 2006. ESR determination of the reactions between selected phenolic acids and free radicals or transition metals. Food Chem. 95, 446–457.

Zhou, Y.H., Yu, J.Q., Mao, W.H., Huang, L.F., Song, X.S., Nogues, S., 2006. Genotypic variation of Rubisco expression, photosynthetic electron flow and antioxidant metabolism in the chloroplasts of chill-exposed cucumber plants. Plant Cell Physiol. 47, 192–199.

Zhu, J.K., 2002. Salt and drought stress signal transduction in plants. Annu. Rev. Plant Biol. 53, 247–273.

Źróbek-Sokolnik, A., Asard, H., Górska-Koplińska, K., Górecki, R.J., 2009. Cadmium and zinc-mediated oxidative burst in tobacco BY-2 cell suspension cultures. Acta. Phys. Plant 31, 43–49.

Plant Polyphenols: Do They Control Freshwater Planktonic Nuisance Phototrophs?

6

Hanno Bährs*, Pauline Laue*†, Shumon Chakrabarti*, Christian E.W. Steinberg*

*Humboldt-Universität zu Berlin, Laboratory of Freshwater & Stress Ecology, Berlin, Germany,
†Lausitz University of Applied Sciences, Senftenberg, Germany

CHAPTER OUTLINE HEAD

6.1 Introduction

According to prevailing paradigms, plant polyphenols, of autochthonous as well as allochthonous origin, excert major ecological control by structuring primary producer communities in freshwater ecosystems. Two issues are relevant:

1. Shallow lakes usually display two strongly contrasting states (Scheffer, 1998): a clear state dominated by aquatic vegetation, and a turbid state characterized by a high phytoplankton biomass. Stabilizing mechanisms tend to keep the system in either the vegetation-dominated or the phytoplankton-dominated state (Gross *et al.*, 2007, and see references therein). One mechanism underlying the macrophyte dominance appears to be the release of allelochemicals, particularly polyphenols, by aquatic macrophytes (Scheffer, 1998; Gross *et al.*, 2007). Although several laboratory studies (Table S1) showed that at least 37 macrophyte species affect phytoplankton (Hilt and Gross, 2008), convincing field studies with documented environmental fate of the chemicals and the physiological state of the target phototrophs, however, are sparse or even lacking. Furthermore, laboratory studies with identical species are often inconsistent, indicating

Polyphenols in Plants. http://dx.doi.org/10.1016/B978-0-12-397934-6.00006-1

that some governing environmental variables might have been disregarded. More than 25 years ago, Lewis (1986) raised concerns regarding allelopathy in lakes by pointing out two central problems: (i) Between producers and receivers, a considerable distance has to be covered and several environmental mechanisms impact on the stability of the released allelochemical which do not apply to small-scale studies in the laboratory. (ii) In the dynamic water-body of a lake, non-producers can also share the benefit from the production of a costly allelopathic compound. If the receiver fails, much energy would be lost for the producer. In addition to Lewis' concerns, there appears to be an ongoing paradigmatic shift of the evolutionary and ecological significance of plant polyphenols (Close and McArthur, 2002). These compounds are not primarily produced to protect from biotic stress such as herbivores or competitors, but to protect from abiotic stresses such as excess light energy and increasing UV irradiation and ozone concentration. Accepting this concept also for freshwater macrophytes would probably diminish an obvious bias concerning the allelochemical role of macrophytes in lakes.

2. The second aspect concerning how plant polyphenols structure phytoplankton biocenoses is an applied issue: massive algal growth in lakes can lead to toxic cyanobacterial blooms and causes risk to wildlife and humans. The reason for the growth of nuisance freshwater phototrophs is the input of nutrients, accelerated by increasing temperature due to global warming. To combat the symptoms, rather than the causes of eutrophication, several allelochemical strategies have been developed, based on the understanding that many plant secondary metabolites are effective against competitors (Whittaker and Feeny, 1971). Consequently, this environmentally friendly application of allochemicals from various sources, such as barley or rice straw, are increasingly applied in water quality management measures with a recent focus on East-Asia. Although first attempts to control, e.g., "nuisance blooms," were already reported some 60 years ago (Fitzgerald et al., 1952), the underlying environmental and physiological mechanisms, however, are still a matter of scientific debate. Hence, it is not surprising that reports exist with contradictory results, even from the same laboratory (e.g., Gibson et al., 1990; Welch et al., 1990). A comprehensive overview of such studies is presented in Table S1, which includes laboratory as well as field studies, trials against planktonic nuisance phototrophs and applications of single phenols, as well as leachates from macrophytes and the Gramineae. Furthermore, it also collates some recent reports that question whether phenols are the only effective compounds.

There is no doubt that leachates from the Gramineae have the potential to reduce algal growth, as demonstrated with an aqueous leachate from the giant reed (*Arundo donax*, Figure 6.1). In this trial, the coccal green alga, *Pseudokirchneriella subcapata*, was cultured as described by Bährs and Steinberg (2012) and challenged by increasing leachate concentrations. After the control culture reached the stationary phase, exposure to 1.67 or 3.33 µM DOC resulted in strongly reduced biomass. However, one question is whether polyphenols are the only allelochemical compounds

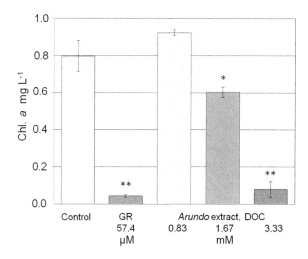

FIGURE 6.1

Chlorophyll *a* content of *Pseudokirchneriella subcapitata* cultures after 4 days under the influence of 0.83–3.33 mM (equals 10, 20, 40 mgl^{-1}) DOC (Dissolved Organic Carbon) of an aqueous extract of the giant reed *Arundo donax* in comparison to 57.4 µM gramine (equals 10 mgl^{-1} DOC).

active against planktonic phototrophs that can be found in the Gramineae. Hong *et al.* (2010) identified several non-phenolic compounds in an algicidal fraction of a methanol-extract of *A. donax*, including indole gramine (GR). The efficacy of GR is also displayed in Figure 6.1.

We assume that the effects of polyphenols in both issue matters are over-rated and believe that the following issues should be addressed, since evidence suggests that they have previously not been considered in depth when polyphenols have been applied in laboratory assays or field trials: (i) role of ambient pH; (ii) physiological state of the phototrophs; (iii) action of allelochemicals other than polyphenols; and (iv) mode of action in *Microcystis aeruginosa*, the most common cyanobacterial bloom-former.

6.2 Role of ambient pH

It is well understood that most phenolics are susceptible to autoxidation at pH values >7 in the presence of oxygen (Cason, 1967). Furthermore, the toxicity of polyphenols is significantly enhanced when oxidized (Lyr, 1965). Interestingly, Lyr performed toxicity tests at pH ~5.5. In the phytoplankton system, this mechanism might apply to the very proximity of phototrophs, since under alkaline pH conditions, autoxidation would not maintain polyphenols as oxidized monomers, but continue to polymerization (Lyr, 1965; Cason, 1967). Polymers, in turn, reduce bioavailability and toxicity.

Overall, the observed action of polyphenolic toxicity can only be very local and short term. Unfortunately, most authors do not provide any kinetic data on the toxic action.

In eutrophicated freshwater bodies with nuisance development of phototrophs, associated CO_2 uptake and subsequent release of OH^-, at pH values significantly > 8.3, prevail (refer to Table S1) with oxidative polymerization of the polyphenols. In fact, a few reports have shown that increasing polymerization results in decreasing algal toxicity (Pillinger *et al.*, 1994).

To demonstrate the pH effect on the efficacy of polyphenolic action, Bährs *et al.* (2013b) used the smallest and most reactive polyphenol, hydroquinone (HQ). HQ was aged through exposure to media at pH 7 and 11 for a given time prior to exposure

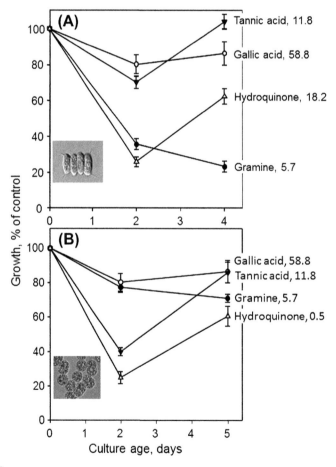

FIGURE 6.2

Growth inhibition of (A) *Desmodesmus armatus* and (B) *Microcystis aeruginosa* by the polyphenols tannic acid, gallic acid, hydroquinone, and the alkaloid gramine. Concentrations in μM.

to various phototrophs. HQ continues to completely inhibit algal growth even after 4 days if aged at pH 7.0. When aged at pH 11, toxicity vanished rapidly within 48 h and the growth potential of the phototrophs recovered. Furthermore, this effect did not significantly differ between coccal greens and coccal cyanobacteria. We therefore assume that if even this highly toxic polyphenol loses its potency above pH 7, other less reactive polyphenols would act similarly, probably with slower kinetics.

The neglect of prevailing pH-conditions in many field experiments and the corresponding neglect of the environmental fate of the polyphenols almost certainly largely explains the high variability of the results and the lack of success of several nuisance algal treatment trials (Table S1). Even laboratory studies that use BG-11 medium with an initial pH of 7.4 do not mirror environmental reality, because this pH is at least one pH unit below that of highly productive lakes (see above); again, under circumneutral pH conditions in the laboratory trials, applied polyphenolics might have sustainably intoxicated the initial phototroph population, an effect that might last until the experiment was terminated. In lakes with a pH > 8.3, however, this intoxication is likely outcompeted by oxidative polymerization. This process can be observed in Figure 6.2, with polyphenols often used in laboratory and field studies (Bauer *et al.*, 2008): the initial pH 8.2 and the polyphenols reduced the growth rates of all phototrophs tested. During the growth of the cultures, the pH rose above 10 and the toxicity of tannic acid, gallic acid, and HQ towards *D. armatus* and *M. aeruginosa* vanished (Figure 6.2). Only the alkaloid GR continued to be toxic to both phototrophs, since its environmental fate differs from that of the polyphenols and appears to be pH-independent.

6.3 Physiological state of the phototrophs

If, in field trials, nuisance developments of phototrophs were successfully combated by the use of barley or rice leachates, the question arises whether this effect is attributable exclusively to polyphenolic action, or whether potential physiological variables of the phototrophs support or accelerate this action. Conversely, if these potentially accelerating conditions are excluded, do the polyphenols really perform their expected function? In the literature, the physiological status of the phototrophs has never been reported.

Mainly following the "Great Phosphorus Controversy" (Likens, 1972) concerning the key factors of eutrophication of freshwaters, the emphasis in scientific and water management has been almost exclusively placed on phosphorus and its reduction. Carbon dioxide (CO_2) has only been assigned a temporary and minor role in this process. Congruent with this assumption is that from studies on nuisance algal development in shallow lakes, it is well understood that cyanobacteria are able to utilize much lower CO_2 concentrations than most eukaryotic phototrophs (Scheffer, 1998). However, during periods of mass development of phototrophs, lakes are usually significantly undersaturated with CO_2 (Gelbrecht *et al.*, 1998) and this undersaturation, in turn, might limit growth. In the same line of evidence, reports by Schippers *et al.* (2004)

and Jansson *et al.* (2012) showed that CO_2 supersaturation promoted primary production even in unproductive lakes, by approximately one order of magnitude. Keeping a potential CO_2-limitation in mind, a paradigm of ecology applies: organisms living in conditions close to their environmental tolerance limits appear to be more vulnerable to additional challenges than organisms close to their optimum (Heugens *et al.*, 2001). Hence, one can hypothesize that improved CO_2 supply might modify or even reduce the toxic action of allelochemicals on phototrophs. Most laboratory and field studies have not considered suboptimal CO_2 supply to be a problem; many of them do not even mention how the CO_2 supply was maintained. Algal inocula in dialysis bags, flasks, tanks, or on agar plates will rapidly suffer from suboptimal CO_2 supply; this fact might also explain the inconstant success of polyphenol applications.

To show this CO_2 effect, a new CO_2 dialysis bag technique developed by Pörs *et al.* (2010) providing almost optimal CO_2 supply was used for the experiment in Figure 6.3.

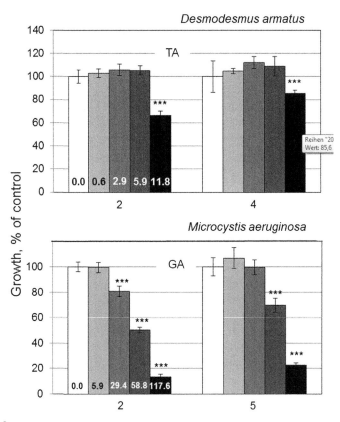

FIGURE 6.3

Growth dependency of *Desmodesmus armatus* on increasing concentrations of tannic acid (TA) after 2 days and 4 days (above) and dependency of *Microcystis aeruginosa* on increasing concentrations of gallic acid (GA) after 2 days and 5 days (below), in μM.

Small bags containing a concentrated carbonate buffer with a CO_2 partial pressure of 32 mbar were inserted as CO_2 reservoirs into the culture vessels. Using this technique, the stationary phase was reached at the latest, after 4 days for coccal greens and 5 days for cyanobacteria. Results with tannic acids (TA, Figure 6.3) show that these had no effect on the growth rate of *D. armatus* with exposure concentrations up to 5.9 μM. In contrast to these results, growth inhibition in the same species were reported by daily additions of 1.2 nM TA over one week (Bauer *et al.*, 2012). Another evidence for inadequate CO_2 supply are the low growth rates for *D. armatus* in the controls: Bauer *et al.* (2010) found growth rates between 0.2 and 0.3 days^{-1} in dialysis bags, instead of 0.74 days^{-1} with optimized CO_2 supply (Heinze *et al.*, 2012).

Also, with *M. aeruginosa* exposed to gallic acid, significant, but less pronounced differences emerge between published values and our findings. For instance, Saito *et al.* (1989) reported an EC50 of 19 μM in closed glass tubes that were agitated daily, and Wang *et al.* (2008) detected an EC50 of 32 μM in axenic cultures with an inital pH of 7.4. Conversely, our experiment with an initial pH of 8.2 (Figure 6.3) revealed an EC50 of 58.8 μM.

Although strain-specific differences in sensivity cannot be excluded, these examples highlight a significant gap in the understanding of polyphenolic action against nuisance phototrophs: their general physiological condition is of paramount importance for a sustainable application of polyphenols and disregarding this might explain the observed failures of polyphenolic applications (Table S1).

6.4 **Action of allelochemicals other than polyphenols**

Provided that polyphenols are active and that the environmental and physiological variables discussed above do not apply, some skepticism still remains concerning the mechanisms behind successful combats of nuisance phototrophs. For instance, Nakai *et al.* (2005) showed that several fatty acids released by *Myriophyllum spicatum* were more toxic to *M. aeruginosa* than polyphenols themselves. Furthermore, the inhibitory effect even of the sum of polyphenols and fatty acids contributed only about 53% to the observed total allelopathic effect of *M. spicatum* crude extracts (Nakai *et al.*, 2012). One question is, what kind of natural xenobiotics underlie the remainder of the effect? Apart from polyphenols, several other chemicals have been, and continue to be, isolated and identified from litter leachates or aquatic macrophytes: alcohols found in East-Asian *Potamogeton* species (Zhang *et al.*, 2011a), dicyclohexanyl orizane and β-sitosterol-β-D-glucoside, recently discovered in rice hulls (Park *et al.*, 2009) and ethyl 2-methylacetoacetate (EMA) from the common reed, *Phragmites australis* (Li and Hu, 2005), amongst others. The EC50 of synthesized EMA on the coccal green *Chlorella pyrenoidosa* and the coccal cyanobacterium *M. aeruginosa* were 4.3 and 5.6 μM, respectively. Recently, Ni *et al.* (2012) isolated the sesquiterpene lactone artemisinin from the riparian sweet sagewort, *Artemisia annua*. However, the interesting compounds are those that occur in high concentrations and are widespread in plants and can therefore easily be applied against nuisance blooms of phototrophs or have an allelochemical potential in riparian plants.

Whittaker and Feeny (1971) stated that "the known agents of allelopathy belong to... phenolic acids, flavonoids, and other aromatic compounds, terpenoid substances, steroids, alkaloids, and organic cyanides" and considered alkaloids as the most toxic ones. Furthermore, alkaloids are prominent in barley, *Hordeum vulgare*, and other Gramineae (e.g., giant reed) as are GR, hordenine, and others. In barley, the maximum reported GR concentrations of GR reached $8\,mg\,g^{-1}$ dry matter (DM) (Hanson *et al.*, 1983). Ostrofsky and Zettler (1986) analyzed the alkaloid content of 15 suberged macrophytes and found a range from $0.13\,mg\,g^{-1}$ DM in *Heteranthera dubia* to $0.56\,mg\,g^{-1}$ DM in *Potamogeton crispus*. Interestingly, Ostrofsky and Zettler (1986) and papers cited therein, discussed alkaloids only as deterrents against herbivores, but not as allelochemicals against phototrophs, however, toxicity is not limited to herbivorous animals.

Although GR was demonstrated to be effective against primary producers and the most common alkaloids, gramine and hordenine, are known to act synergistically (Liu and Lovett, 1993), the idea of applying alkaloids against cyanobacterial blooms entered the water management community only very recently (Hong *et al.*, 2010). In trials with increasing pH and progressive exposure, GR was effective against both, with no obvious dependence on pH (Figure 6.2). Hong *et al.* (2010) reported a 3-day EC50 as low as $2.7\,\mu M$. Hence, it might be hypothesized that in those field trials in which barley straw was successfully and sustainably applied, not polyphenols alone, but alkaloids and probably other natural xenobiotics either alone or in combination with polyphenols were the effective ingredients.

6.5 **Modes of action in *Microcystis aeruginosa***

Understanding the modes of toxic action in *M. aeruginosa* following challenge by polyphenols and other natural xenobiotics is a crucial prerequisite for successful combat, because it provides the platform for improvements of applied xenobiotics alone or in mixtures. However, the available literature shows that there is no unique response of *M. aeruginosa*. Several studies report symptoms of oxidative stress upon exposure to EMA (Hong *et al.*, 2008), anthraquinone (Schrader *et al.*, 2005), or plant extracts (Zhang *et al.*, 2011b). Oxidative stress-induced reduction of photosynthesis is a common avoidance mechanism of excess reactive oxygen in exposed plants (Mittler, 2002). This mechanism also occurs in *M. aeruginosa* exposed to gallic acid, with the cyanobacterium responding only to the highest concentration tested ($117.6\,\mu M$) with oxidative stress symptoms. This suggests that for the observed growth reduction, other mechanisms apply, because it began already at much lower concentrations ($29.4\,\mu M$, Figure 6.3). Ni *et al.* (2012) showed that artemisinin most likely inactivates the PSII reaction centers and electron transport in the acceptor side.

For *p*-benzoquinone (PBQ), Bährs *et al.* (2013a) recently demonstrated that *M. aeruginosa* behaved differently to other cyanobacteria and coccal greens.

Table 6.1 EC50 and EC90 of PBQ after 96 h (green algae) and 120 h (cyanobacteria)

Phototroph		
Green algae	EC50	EC90
P. subcapitata	33.6	89.5
M. braunii	27.5	112
D. armatus	22.5	54.9
Cyanobacteria	EC50	EC90
Synechocystis sp.	8.54	34.4
Nostoc sp.	5.7	15.8
M. aeruginosa	0.49	1.97

Modified from Bährs et al. (2013a) Ann. Environ. Sci. 7, 1–15; courtesy of Northeastern University, Boston.

Whereas all coccal greens tested, *Synechocystis* sp., and *Nostoc* sp. showed PBQ concentration-dependent increases of lipid peroxidation and reductions in photosynthetic oxygen release, *M. aeruginosa* did not respond similarly, although it was a far more sensitive phototroph (Table 6.1), suggesting that other mechanisms must apply. For instance, quinones have long been known for their interference also with the respiratory electron transport chain (Trebst and Draber, 1979). However, benzoquinones can also develop highly diverse cytotoxic effects at different sites of cellular metabolism (Siraki *et al.*, 2004). Hence, in future approaches, the different modes of action of the applied natural xenobiotics and their sustainability are worth deciphering rather than extending the existing catalog of phenotypic effect descriptions.

6.6 Conclusion from an ecological perspective

Many studies of polyphenols and their effect on planktonic phototrophs are purely phytochemical assessments, often with poor environmental realism. They attempt to explain the self-stabilizing mechanisms of the macrophyte-dominated state in shallow lakes or the mechanisms behind the combat of nuisance blooms. Although studies are often performed in enclosures or real lakes, most studies do not consider environmental (prevailing pH, photolysis, hydrolysis) or physiological (mode of action, physiological state) variables. Furthermore, laboratory studies have often used exposure concentrations that were too high and did not take into account the environmental fate of the xenobiotics. Furthermore, many studies are biased towards polyphenols, and seldom consider that these chemicals act in concert with other xenobiotic classes of likely higher efficacy, if applied as barley or rice straw. Due to the high concentrations applied, laboratory studies poorly explain environmental pathways and modes of action for the disappearance of planktonic phototrophs after the application of phenolic materials. In all approaches, it appears that one ecological compartment has been overlooked: the periphyton. The periphyton is a biofilm of primary producers, bacteria, fungi and invertebrates that colonizes macrophytes and competes with them for nutrients and light. As one tool in this competition scenario, macrophytes have developed their xenobiotic strategy. This interaction is not disturbed by scaling problems and long distances between sender and receiver. This means that most of the studies discussed above can explain low densities of periphyton instead of phytoplankton. During eutrophication, however, the increase in the periphyton cannot sufficiently be combated by the xenobiotic strategy of the macrophytes. The periphyton grows too quickly and is most likely the major cause for the decline of macrophytes in shallow lakes (for details, see Jones and Sayer, 2003). Consequently, the increase of phytoplankton in eutrophicated shallow lakes is not the cause, but the consequence of macrophyte declines. Corresponding water management should focus more on this ecological interaction rather than on that of macrophyte-phytoplankton.

Associated content

Supporting information concerning studies with plant polyphenols and related material as tools against nuisance blooms of phototrophs (Table S1) can be found on the website for this book.

Acknowledgments

We gratefully acknowledge the financial support to H.B. by the Deutsche Forschungsgemeinschaft, grant STE-673/17-1. We also thank Cornelia Brumby and Eberhard Spreng, Saint-Quentin-la-Poterie (Uzès), Languedoc-Roussillon, France, for providing giant reeds from their garden. Furthermore, we declare that there is no financial conflict of interests.

References

Bährs, H., Steinberg, C.E.W., 2012. Impact of two different humic substances on selected coccal green algae and cyanobacteria—changes in growth and photosynthetic performance. Environ. Sci. Pollut. Res. 19, 335–346.

Bährs, H., Heinze, T., Gilbert, M., Wilhelm, C., Steinberg, C.E.W., 2013a. How p-benzoquinone inhibits growth of various freshwater phototrophs: different susceptibility and modes of action? Ann. Environ. Sci. 7, 1–15.

Bährs, H., Putschew, A., Steinberg, C.E.W., 2013b. Toxicity of hydroquinone to different freshwater phototrophs is influenced by time of exposure and pH. Environ. Sci. Pollut. Res. 20, 146–154.

Bauer, N., Grossart, H.-P., Hilt, S., 2008. Do bacteria influence the sensitivity of phytoplankton to allelochemicals from submerged macrophytes? Verh. Intern. Verein. Limnol. 30, 307–311.

Bauer, N., Grossart, H.P., Hilt, S., 2010. Effects of bacterial communities on the sensitivity of the phytoplankton *Stephanodiscus minutulus* and *Desmodesmus armatus* to tannic acid. Aquat. Microb. Ecol. 59, 295–306.

Bauer, N., Zwirnmann, E., Grossart, H.-P., Hilt, S., 2012. Transformation and allelopathy of natural dissolved organic carbon and tannic acid are affected by solar radiation and bacteria. J. Phycol. 48, 355–364.

Cason, J., 1967. Synthesis of benzoquinones by oxidation. In: Adams, R., Fieser, L.F., Blatt, A.H., Johnson, J.R., Snyder, H.R. (Eds.), Organic Reactions. John Wiley, New York, p. 305.

Close, D.C., McArthur, C., 2002. Rethinking the role of many plant phenolics—protection from photodamage not herbivores? Oikos 99, 166–172.

Fitzgerald, G.P., Gerloff, G.C., Skoog, F., 1952. Studies on chemicals with selective toxicity to blue-green algae. Sewage Ind. Wast. 24, 888–896.

Gelbrecht, J., Fait, M., Dittrich, M., Steinberg, C.E.W., 1998. Use of GC and equilibrium calculations of CO_2 saturation index to indicate whether freshwater bodies in north-eastern Germany are net sources or sinks for atmospheric CO_2. Fresenius J. Anal. Chem. 361, 47–53.

Gibson, M.T., Welch, I.M., Barrett, P.R.F., Ridge, I., 1990. Barley straw as an inhibitor of algal growth II: laboratory studies. J. Appl. Phycol. 2, 241–248.

Gross, E.M., Hilt, S., Lombardo, P., Mulderij, G., 2007. Searching for allelopathic effects of submerged macrophytes on phytoplankton—state of the art and open questions. Hydrobiologia 584, 77–88.

Hanson, A.D., Ditz, K.M., Singletary, G.W., Leland, T.J., 1983. Gramine accumulation in leaves of barley grown under high-temperature stress. Plant Physiol. Biochem. 71, 896–904.

Heinze, T., Bährs, H., Gilbert, M., Steinberg, C.E.W., Wilhelm, C., 2012. Selected coccal green algae are not affected by the humic substance Huminfeed® in term of growth or photosynthetic performance. Hydrobiologia 684, 215–224.

Heugens, E.H.W., Hendriks, A.J., Dekker, T., van Straalen, N.M., Admiraal, W., 2001. A review of the effects of multiple stressors on aquatic organisms and analysis of uncertainty factors for use in risk assessment. Crit. Rev. Toxicol. 31, 247–284.

Hilt, S., Gross, E.M., 2008. Can allelopathically active submerged macrophytes stabilise clear-water states in shallow lakes? Basic Appl. Ecol. 9, 422–432.

Hong, Y., Hu, H.Y., Li, F.M., 2008. Physiological and biochemical effects of allelochemical ethyl 2-methyl acetoacetate (EMA) on cyanobacterium *Microcystis aeruginosa*. Ecotox. Environ. Saf. 71, 527–534.

Hong, Y., Hu, H.Y., Sakoda, A., Sagehashi, M., 2010. Isolation and characterization of antialgal allelochemicals from *Arundo donax* L. Allelopath. J. 25, 357–368.

Jansson, M., Karlsson, J., Jonsson, A., 2012. Carbon dioxide supersaturation promotes primary production in lakes. Ecol. Lett. 15, 527–532.

Jones, J.I., Sayer, C.D., 2003. Does the fish–invertebrate–periphyton cascade precipitate plant loss in shallow lakes? Ecology 84, 2155–2167.

Lewis, W.M.J., 1986. Evolutionaly interpretion of allelochemicals interactions in phytoplankton algae. Am. Nat. 127, 184–194.

Li, F.M., Hu, H.Y., 2005. Isolation and characterization of a novel antialgal allelochemical from *Phragmites communis*. Appl. Environ. Microbiol. 71, 6545–6553.

Likens, G.E.e., 1972. Nutrients and eutrophication. Special Symposium 1. American Society of Limnology and Oceanography, 328.

Liu, D.L., Lovett, J.V., 1993. Biologically active secondary metabolites of barley. II. Phytotoxicity of barley allelochemicals. J. Chem. Ecol. 19, 2231–2244.

Lyr, H., 1965. On the toxicity of oxidized polyphenols. J. Phytopathol. 52, 229–240.

Mittler, R., 2002. Oxidative stress, antioxidants and stress tolerance. Trends Plant Sci. 7, 405–410.

Nakai, S., Yamada, S., Hosomi, M., 2005. Anti-cyanobacterial fatty acids released from *Myriophyllum spicatum*. Hydrobiologia 543, 71–78.

Nakai, S., Zou, G., Okuda, T., Nishijima, W., Hosomi, M., Okada, M., 2012. Polyphenols and fatty acids responsible for anti-cyanobacterial allelopathic effects of submerged macrophyte *Myriophyllum spicatum*. Water Sci. Technol. 66, 993–999.

Ni, L., Acharya, K., Hao, X., Li, S., 2012. Isolation and identification of an anti-algal compound from *Artemisia annua* and mechanisms of inhibitory effect on algae. Chemosphere 88, 1051–1057.

Ostrofsky, M.L., Zettler, E.R., 1986. Chemical defenses in aquatic plants. J. Ecol. 74, 279–287.

Park, M.H., Chung, I.M., Ahmad, A., Kim, B.H., Hwang, S.J., 2009. Growth inhibition of unicellular and colonial *Microcystis strains* (Cyanophyceae) by compounds isolated from rice (*Oryza sativa*) hulls. Aquat. Bot. 90, 309–314.

Pillinger, J.M., Cooper, J.A., Ridge, I., 1994. Role of phenolic compounds in the antialgal activity of barley straw. J. Chem. Ecol. 20, 1557–1569.

Pörs, Y., Wüstenberg, A., Ehwald, R., 2010. A batch culture method for microalgae and cyanobacteria with CO_2 supply through polyethylene membranes. J. Phycol. 46, 825–830.

Saito, K., Matsumoto, M., Sekine, T., Murakoshi, I., Morisaki, N., Iwasaki, S., 1989. Inhibitory substances from *Myriophyllum brasiliense* on growth of blue-green algae. J. Nat. Prod. 52, 1221–1226.

Scheffer, M., 1998. Ecology of Shallow Lakes. Chapman Hall, London.

Schippers, P., Lürling, M., Scheffer, M., 2004. Increase of atmospheric CO_2 promotes phytoplankton productivity. Ecol. Lett. 7, 446–451.

Schrader, K.K., Dayan, F.E., Nanayakkara, N.P.D., 2005. Generation of reactive oxygen species by a novel anthraquinone derivative in the cyanobacterium *Planktothrix perornata* (Skuja). Pestic. Biochem. Physiol. 81, 198–207.

Siraki, A.G., Chan, T.S., O'Brien, P.J., 2004. Application of quantitative structure-toxicity relationships for the comparison of the cytotoxicity of 14 *p*-benzoquinone congeners in primary cultured rat hepatocytes versus PC12 cells. Toxicol. Sci. 81, 148–159.

Trebst, A., Draber, W., 1979. Structure activity correlations of recent herbicides in photosynthetic reactions. In: Geissbühler, H., Brooks, G.T., Kearney, P.C. (Eds.), Advances in Pesticide Science (Part 2). Pergamon, New York, pp. 222–234.

Wang, H.Q., Wu, Z.B., Zhang, S.H., Cheng, S.P., He, F., Liang, W., 2008. Relationship between the allelopathic activity and molecular structure of hydroxyl derivatives of benzoic acid and their effects on cyanobacterium *Microcystis aeruginosa*. Allelopath. J. 22, 205–211.

Welch, I.M., Barrett, P.R.F., Gibson, M.T., Ridge, I., 1990. Barley straw as an inhibitor of algal growth I: studies in the Chesterfield Canal. J. Appl. Phycol. 2, 231–239.

Whittaker, R.H., Feeny, P.P., 1971. Allelochemics: chemical Interactions between species. Science 171, 757–770.

Zhang, S.H., Sun, P.S., Ge, F.J., Wu, Z.B., 2011a. Different sensitivities of *Selenastrum capricornutum* and toxic strain *Microcystis aeruginosa* to exudates from two *Potamogeton* species. Pol. J. Environ. Stud. 20, 1359–1366.

Zhang, T.T., Wang, L.L., He, Z.X., Zhang, D.A., 2011b. Growth inhibition and biochemical changes of cyanobacteria induced by emergent macrophyte *Thalia dealbata* roots. Biochem. Syst. Ecol. 39, 88–94.

Isolation and Analysis of Polyphenol Structure

Analysis Techniques for Polyphenols

Gas Chromatography–Mass Spectrometry Analysis of Polyphenols in Foods

7

Pilar Viñas, Natalia Campillo

Department of Analytical Chemistry, Faculty of Chemistry, University of Murcia, Spain

CHAPTER OUTLINE HEAD

7.1 Polyphenolic compounds

The beneficial properties of many vegetable foods are related to their secondary metabolites, also called phytochemical substances or phytonutrients, which can be grouped as polyphenols, terpenes, and sulfur compounds (Croteau *et al.*, 2000).

Phenolic compounds or polyphenols constitute a very numerous group comprising at least 8000 different known substances (Gómez-Caravaca *et al.*, 2006), which can be classified according to their structures (Manach *et al.*, 2004; Soleas *et al.*, 1997a). Among the most important for their biological activity are flavonoids, which have a basic C6-C3-C6 structure. Another important subgroup is made up of phenylpropanoids, which include hydroxycinnamic acid derivatives (caffeic, ferulic, sinapic, *p*-coumaric), stilbenoids (resveratrol and piceatannol), and benzoic acid derivatives (gallic and ellagic acids, etc.) (Tomás-Barberán and Espín, 2001). Therefore, polyphenols run from simple molecules, such as phenolic acids, to highly polymerized compounds, such as tannins (Pérez-Jiménez *et al.*, 2010). Polyphenols tend to be water soluble, because they are frequently present as glycosides located in the cell vacuole of plants (Proestos *et al.*, 2006a).

Flavonoids, which constitute the largest and most diverse group of phytochemicals in plants, show a wide spectrum of important functions, including pigmentation,

Polyphenols in Plants. http://dx.doi.org/10.1016/B978-0-12-397934-6.00007-3

103

plant–pathogen interactions, fertility, and protection against UV radiation. The basic skeleton of flavonoids consists of two aromatic rings with six carbon atoms (rings A and B) bound by a heterocycle including three carbon atoms (ring C) (Bovy *et al.*, 2007). Modifications of the central C-ring divide flavonoids into different structural classes: flavones (e.g., luteolin), flavonols (e.g., quercetin), flavanones (e.g., hesperetin), isoflavonoids (e.g., dadzein), anthocyanins (e.g., cyanidin), flavanols, also named flavan-3-ols [e.g., monomeric (catechin and epicatechin), oligomeric (proanthocyanidins) and polymeric compounds also known as condensed tannins], chalconoids (e.g., chalcone), dihydrochalcones (e.g., phloretin), and aurones (e.g., aureusidin). Flavonols and flavanols are the most widespread flavonoid compounds (Valls *et al.*, 2009). The great diversity of flavonoid structures is due to enzymatic modification of the basic skeleton. Some of the hydroxyl groups are often methylated, acetylated, prenylated or sulfated, leading to significant variations in the biological properties of each group of compounds. The flavonoids are predominantly present in plants as glycosides (El Gharras, 2009), although they can also occur as esters. The terms aglycone and glycoside are used for flavonoids when they contain no sugar groups or one or more sugar groups, respectively. The sugar substituents in *O*-glycosides are bound to a hydroxyl group of the aglycone, whereas in C-glycosides they are bound to a carbon of the aglycone. *O*-Glycosides are more frequent than C-glycosides, although some exceptions have been described (Krafczyk and Glomb, 2008). Glucose, rhamnose, galactose, and arabinose are the most common carbohydrates. Flavonoid-diglycosides are also commonly found (de Rijke *et al.*, 2006).

Flavonols are the most ubiquitous flavonoids in foods, the main representative compounds being quercetin glycoside, followed by kaempferol glycoside (El Gharras, 2009; Tomás-Barberán and Espín, 2001). The biological activity of quercetin has been attributed to its ability to neutralize free radicals *in vivo*. Other flavonols, such as myricetin and isorhamnetin, and flavones, such as luteolin and apigenin, may also be relevant. Flavonols and flavones are commonly present in foods as more complex combinations with sugars and aliphatic or aromatic organic acids (Tomás-Barberán and Espín, 2001).

Flavanols can be present in food as monomers (the best known form), as oligomers (with different degrees of polymerization, up to decamers), and even as very large molecules (classified as polymers, e.g., elagitanin or punicalagin) (Tomás-Barberán and Espín, 2001). These flavonoids show different degrees of hydroxylation in the A- and B-rings, and the 3-position on the C-ring is commonly a hydroxyl group or is esterified with gallic acid. The most common flavanol monomers are catechin, epicatechin, catechin gallate, and epigallocatechin gallate (Valls *et al.*, 2009). Proanthocyanidins, oligomers of flavanol monomer units, are classified as procyanidins (derived from catechin, epicatechin, and their gallic esters) or prodelphinidins (derived from gallocatechin, epigallocatechin, and their galloylated derivatives). Highly polymerized proanthocyanidin compounds are known as condensed tannins, which are the second-most abundant family of natural polyphenolic compounds, after lignins (Shadkami *et al.*, 2009).

The antimicrobial, estrogenic, antioxidant and wide spectrum of other pharmacological activities of isoflavonoids make them of great importance in the daily diet (Farag *et al.*, 2007). The benzene ring B position is the difference between isoflavones and other flavonoids. Isoflavonoids from soybeans include the isoflavones genistein and daidzein, which are mainly present as the glycosides genistin and daidzin. Other aglycones such as biochanin A, formonetin, and glycitein are contained in red clover, alfalfa, and pueraria. The positive effect of isoflavones against osteoporosis and menopausal symptoms has led to the commercialization of different soy-enriched products (Valls *et al.*, 2009).

Anthocyanins consist of an aglycone unit (anthocyanidin), which is responsible for the plant color, with one or more *O*-glycosides attached. At pH values lower than 3, the anthocyanidin appears as the red flavylium cation, and at pH values higher than 6 it is present as the blue quinonoidal base form. The main differences between anthocyanins are the number of hydroxylated groups in the anthocyanidin, the nature and the number of bonded sugars in their structure, the aliphatic or aromatic carboxylates bound to the sugar in the molecule, and the position of these bonds. The different anthocyanidins and patterns of sugar substitution provide a large number of structures, with about 500 natural anthocyanins with different chemical properties having been identified in plants (Comeskey *et al.*, 2009). The food industry uses anthocyanins as natural food colorants, although their health properties mean that they are also used as nutritional supplements, in functional food formulations and medicines (Valls *et al.*, 2009).

Phenylpropanoids include phenolic acids, which are aromatic secondary plant metabolites widely spread throughout the plant kingdom (Herrmann, 1989). These naturally occurring organic acids contain two hydroxycinnamic and hydroxybenzoic structures, and several hydroxyl groups on the aromatic ring in different numbers and positions. In many cases, aldehyde analogs are also grouped as phenolic acids (e.g., vanillin). Caffeic, *p*-coumaric, vanillic, ferulic and protocatechuic acids are present in nearly all plants (Shahidi and Wanasundara, 1992). Other acids are found in given foods or plants (e.g., gentisic and syringic).

Stilbenoids are secondary products that can act as phytoalexins. In chemical terms, they are hydroxylated derivatives of stilbene. In biochemical terms, they belong to the family of phenylpropanoids and share most of their biosynthesis pathway with chalcones (Sobolev *et al.*, 2006). An example of a stilbenoid is resveratrol, which is found in grapes and wines (Soleas *et al.*, 1997b) and which has been suggested to have many health-related benefits (Goldberg *et al.*, 1995a; Jang *et al.*, 1997). Other known stilbenoids are aglycones, such as piceatannol, pinosylvin and pterostilbene and their glycosides, astringin, and piceid. In general, the amount of piceatannol in plants is much lower than that of resveratrol (Lin *et al.*, 2007).

Conversely, phenolic compounds in virgin olive oils provide a source of natural antioxidants because of the presence of phenolic and *o*-diphenolic structures (tyrosol and hydroxytyrosol) with free radical scavenging properties. Thus, hydroxytyrosol shows strong antioxidant activity (Papadopoulos and Boskou, 1991) in the oxidation of methyl linolenate. Several nutraceuticals from olive leaves contain significant amounts of hydroxytyrosol as well as other phenols (tyrosol, oleuropein, vanillic acid, etc.).

Moreover, pomegranate contains a complex mixture of anthocyanins and hydrolyzable tannins (HT). The HT structures are classified into gallotannins, ellagitannins (ellagic acid esters of D-glucose with one or more galloyl substitutions), and gallagyl esters such as punicalagin and punicalin (Martin *et al.*, 2009).

7.2 Polyphenolic compounds in foods

Polyphenols show an antioxidant activity that is even higher than that of antioxidant vitamins, such as vitamin C and E, and play an important role in the prevention of chronic diseases such as cardiovascular and neurodegenerative disorders, cancer, type 2 diabetes and osteoporosis. This beneficial activity has been attributed to the phenolic groups in their structure, which are capable of reducing reactive oxygen species and other organic or non-organic substrates (Pérez-Jiménez *et al.*, 2010). Thus, foods rich in polyphenols, all of them of plant origin, such as red wine, green tea, coffee, pomegranate and bilberry juices, nuts, prune, and chocolate are considered as antioxidant foods.

Polyphenols are also responsible for the organoleptic properties of vegetable foods (Tomás-Barberán and Espín, 2001). Thus, anthocyanins are related to the red, blue, or violet color of fruits (strawberry, plum, grape), vegetables (aubergine, red cabbage, radish) and red wine, while flavonols are responsible for the yellow color of the external parts of fruits and vegetables (Tomás-Barberán and Clifford, 2000). Even though flavonols are mainly found in fruits and vegetables, flavones are much less common. Parsley and celery are edible sources of flavones. Large quantities of polymethoxylated flavones (tangeretin, nobiletin and sinensetin), the most hydrophobic flavonoids, have been found in the skin of citrus fruit (El Gharras, 2009; Manach *et al.*, 2004).

Commonly consumed fruits, such as blueberries, cranberries, grapes, apples, kiwis, and pears, and tea are important sources of proanthocyanidins (Valls *et al.*, 2009). While flavonoids are found throughout the plant kingdom, isoflavonoids are more restricted to the Papilionoideae subfamily of the Leguminosae (Farag *et al.*, 2007). Condensed tannins are found in the bark, root, and leaves of most plant species and also in seeds and fruits (Shadkami *et al.*, 2009).

Citrus fruit is the most flavanone-rich food, and the flavanones are generally present in the form of glycosides with a disaccharide. The whole fruit may contain up to five-times more flavanones than the corresponding juice because the solid parts of citrus fruit have the highest concentration of these compounds (El Gharras, 2009). Until 2004, more than 1600 isoflavonoids had been described, soybeans being the principal source in food. In fact, soybeans are widely used in the production of functional foods such as soy-enriched products (Valls *et al.*, 2009).

Conversely, cinnamic and benzoic acid derivatives exist in almost all plant foods (fruits, vegetables, and grains) where they are physically dispersed in seeds, leaves, roots, and stems (Macheix *et al.*, 1990; Shahidi and Nacsk, 1995). Only a minor fraction exist as free acids, while most are linked, through ester, ether or acetal bonds, to structural components of the plant (cellulose, proteins, lignin) (Andreasen *et al.*, 2000), to larger polyphenols (flavonoids), smaller organic molecules (e.g., glucose, quinic, maleic, or tartaric acids) or other natural compounds (e.g., terpenes) (Klick and Herrmann, 1988; Winter and Herrmann, 1986). This diversity is one of the major factors that makes the analysis of phenolic acids so complex. Phenolic acids have been associated with color, sensory qualities, and nutritional and antioxidant properties of foods (Maga, 1978), contributing to organoleptic properties such as flavor, astringency, and hardness (Brenes-Balbuena *et al.*, 1992; Naim *et al.*, 1992; Peleg *et al.*, 1991; Tan, 2000). Phenolic acids also affect fruit maturation and help prevent enzymatic browning (Robards *et al.*, 1999).

Additionally, ellagitannins are widely distributed through vegetables, the main sources in the human diet being pomegranate, strawberry, raspberry, blackberry, nuts, and aged wines. In the case of pomegranate, ellagitannins (especially punicalagin) migrate to the juice when the fruit is pressed directly (Gil *et al.*, 2000).

Fruits, vegetables, nuts, cereals, legumes, olive oil, red wine, cacao, and tea have beneficial properties for health due to the vitamins, minerals, and secondary metabolites they contain. The health action of antioxidant substances present in plants, such as flavonoids and carotenoids, depends on their bioavailability. Functional foods are defined as those which satisfy basic nutritional needs and promote health (Diplock *et al.*, 1999; Juárez *et al.*, 2005; Serrano Ríos *et al.*, 2005). Recently, the application of different types of polyphenols in functional foods, and in the nutraceutical and pharmaceutical industries has given rise to great interest in them. Nevertheless, such applications have come up against some limitations related with the unpleasant taste of most polyphenols, although such problems can be overcome by encapsulation technologies (Fang and Bhandari, 2010). The number of functional products containing bioactive ingredients has increased in recent years in the field of nutrigenomics and foodomics (Herrero *et al.*, 2012; Ibáñez *et al.*, 2012).

7.3 **Methods for determining polyphenols**

The most common colorimetric method employed for the quantification of phenolic compounds is that which uses Folin-Ciocalteu reagent (FCR). FCR involves the reduction of a phosphomolybdic-phosphotungstic acid to a blue colored complex in an alkaline solution (Singleton and Rossi, 1965). However, quantification is difficult because other components in the food extracts behave as reducing agents and may contribute to the absorbance. Identification of individual phenolic compounds is not possible with this method, which is characterized by its poor specificity.

The complexity of the matrices which commonly contain polyphenolic compounds has led to the search for techniques with high separation powers, with chromatographic techniques being the most widely used: thin-layer chromatography (TLC), liquid chromatography (HPLC, LC), and gas chromatography (GC) combined with different detection methods (Gao *et al.*, 2010).

TLC uses different stationary phases, including silica gel, cellulose and polyamide layers, for the determination of phenolic acids and flavonoids in plant material (Bylak and Matlawska, 2001; Ellnain-Wojtaszek, 1997a, b). However, the main disadvantages of TLC include a limited quantitation capacity, although it is useful for rapid screening of samples.

By far the most widely used analytical technique for the separation and characterization of phenolic compounds is LC with reversed phase (RP) coupled to different detection systems (Sun *et al.*, 2006). A photodiode array is frequently used, while other detection methods include electrochemical and fluorimetric systems to increase specificity. Mass spectrometry (MS) coupled to LC (LC-MS) permits the analysis of complex matrices, combining an efficient separation capacity with the structural characterization of MS. In most cases, single-stage MS is combined with photodiode array detection to facilitate confirmation of the identity of analytes in a sample with the aid of standards and reference data. Tandem mass spectrometry (MS/MS) is used for the identification of unknown members of phenolic compounds, this detection system having replaced single-stage MS due to its higher selectivity and because of the greater information it provides (de Rijke *et al.*, 2006).

The speed, resolution, small sample size, low consumption of solvents, simplicity and low operating costs make capillary electrophoresis (CE) an alternative or complement to LC for the analysis of polyphenolic compounds in food analysis. Indeed, CE has been successfully applied for such analyses in a wide variety of foods (Carrasco-Pancorbo *et al.*, 2005). The most commonly used CE modes are capillary zone electrophoresis (CZE) and micellar electrokinetic chromatography (MEKC), usually performed with UV, fluorescence, electrochemical and MS detection (de Rijke *et al.*, 2006). Even though the use of CE for phenolic compounds is increasing, its main drawbacks, such as its generally lower sensitivity and worse reproducibility compared with LC, mean that LC is still the first choice (Valls *et al.*, 2009). Several reviews dedicated to analytical chemistry and separative strategies of polyphenolic compounds have been published (Escarpa and González, 2001; Fan *et al.*, 2011; Robards and Antolovich, 1997; Robards, 2003).

7.4 Determination of polyphenols by gas chromatography– mass spectrometry (GC-MS). Derivatization reactions

One of the most powerful detectors for use in GC is the mass spectrometric detector, and hyphenation of these techniques gives GC-MS (Niessen, 2004; Sparkman *et al.*, 2011). MS measures the mass/charge (*m/z*) ratio of ions produced in the sample. The sample is injected into the GC capillary column and the effluent is transferred to the mass spectrometer. The source fragments and ionizes the molecules, while the formed ions are separated according to their *m/z* value in the analyzer and detected in the electron multiplier. The data system can analyze the data in different ways. First, it can calculate the abundance of ions in each spectrum and represent them in a graph *versus* time to obtain a total ion chromatogram (TIC). It is also possible to select the mass spectra at a given time of the chromatogram to identify the species eluting at that time, or to select an *m/z* value to be monitored during the chromatographic experiment, a technique known as selected ion monitoring (SIM). Thus, the use of GC-MS coupling with associated fragmentation patterns provides important advantages based on the combination of the separation capabilities of GC and the power of MS as an identification and confirmation method.

Few studies have used GC for the direct analysis of polyphenols in foods, because these compounds are not sufficiently volatile, and it is necessary to include a previous chemical derivatization step to obtain their volatile and thermostable derivatives (de Rijke *et al.*, 2006; Rosenfeld, 2003). The chemical characteristics of the hydroxyl group of hydrogen bonding increases the melting point. Analysis in the gas phase requires a chemical derivatization step, in addition to sample isolation, clean up, and preparation. The hydroxyl groups are converted to ethers or esters. However, GC has also been performed on underivatized phenols and acids (Christov and Bankova, 1992) using flame ionization (FID), electron capture (ECD), and MS as detection methods.

Derivatization allows non-volatile polyphenols to be transformed into volatile derivatives, thus improving the separation, selectivity, and sensitivity of GC determination. The derivatization step can be carried out in the sample solution before the extraction, once the analytes have been extracted (post-extraction derivatization) or in the GC injection port. A variety of reagents is used to generate volatile derivatives, the most used of which is the trialkylsilyl group, which is generated by covalently linking the alkyl substituted silicon atom to the oxygen of the hydroxyl groups. The most commonly used reagent, *N,O*-bis-(trimethylsilyl)trifluoroacetamide (BSTFA), reacts with alcohols and other polar organic compounds, substituting the active hydrogen by the trimethylsilyl (TMS) group, $-Si(CH_3)_3$. Other silylating reagents are *N,O*-bis-(trimethylsilyl)acetamide (BSA) and *N*-methyl-*N*-(trimethylsilyl)trifluoroacetamide (MSTFA). The main advantages of using silylated derivatives are the simplicity of the reaction, since the process is almost instantaneous (both functional groups acids and phenols are derivatized in the same step) and the side products of these reactions are extremely volatile and do not interfere with the analysis (Ng *et al.*, 2000). Others reagents used for the formation of alkyl esters are diazomethane (Waksmundzka-Hajnos, 1998), ethyl and methyl chloroformate (Husek, 1992), and dimethyl sulfoxide with methyl iodide in an alkaline medium.

The use of high-temperature gas chromatography (HT-GC) coupled to MS has been proven to act as a powerful analytical tool in natural products, permitting the direct characterization of many compounds without derivatization and, in some cases, without clean-up procedures. The extension of the working range of capillary columns used in GC from 370 to 420–480°C may appear of little practical significance. Nevertheless, it implies the extension of the mass units of the analyzed compounds by more than 400–600 u (dos Santos Pereira *et al.*, 2004). Flavonoids have been analyzed using HT-GC (Branco *et al.*, 2001; dos Santos Pereira *et al.*, 2000; Prytzyk *et al.*, 2003; Shadkami *et al.*, 2009).

Single-column GC offers high separation power along with a number of available configurations and detectors. Nevertheless, when highly complex samples such as food matrices are analyzed, peak overlapping may lead to identification problems. Two-dimensional gas chromatography (GC×GC) provides several advantages over conventional single-column GC in terms of greater separation power (peak positions in two dimensions) and signal-to-noise ratio enhancement. Moreover, it provides unique structured chromatograms when related substances such as analogs, congeners or isomers are analyzed. GC×GC has been successfully applied to identify a high number of flavonoids in different food matrices (Gao *et al.*, 2010).

Several studies have been performed, determining resveratrol and other stilbenoids in a variety of matrices, such as nuts (Medina-Bolivar *et al.*, 2007; Tokusoglu *et al.*, 2005), wines (Antoneli *et al.*, 1996; Flamini and Vedova, 2004; Frankel *et al.*, 1995; Goldberg *et al.*, 1994, 1995b; Jeandet *et al.*, 1993, 1995; Montes *et al.*, 2010; Soleas *et al.*, 1995), olive oils (Ougo *et al.*, 2009), and fruits, aromatic plants, and vegetables (Kalogeropoulos *et al.*, 2010a, 2010b; Ragab *et al.*, 2006; Rimando *et al.*, 2004; Wang and Zuo, 2011).

GC has also been used for the analysis of phenolic acids in different food samples. Earlier works were performed with FID (Hartley, 1971; Leung *et al.*, 1981; Packert and Steinhart, 1995; Schulz and Herrmann, 1980; Vande Casteele *et al.*, 1976) and the most recent using MS detection for the analysis of cereals, fruits, vegetables, and aromatic plants (Ayaz *et al.*, 1997; Ayaz and Kadioglu, 1997; Chu *et al.*, 2001; Heimler and Pieroni, 1994; Krygier *et al.*, 1982; Orr *et al.*, 1993; Smolarz, 2001; Sosulski *et al.*, 1982; Wu *et al.*, 1999, 2000; Zuo *et al.*, 2002), wines (Baranowski and Nagel, 1981; Chu *et al.*, 2001), and beer related products (Mallouchos *et al.*, 2007; Pejin *et al.*, 2009).

Soy-based foods are the samples that have been most studied for their isoflavonoid content (Bluck *et al.*, 2002; Ferrer *et al.*, 2009; Liggins *et al.*, 1998; Mazur *et al.*, 1996; Morton *et al.*, 1999), although GC-MS has also been used for the same purpose in fruit (Liggins *et al.*, 2000), and propolis (Campo Fernández *et al.*, 2008). Several polyphenolic compounds, including flavonoids, phenolic acids, phenolic aldehydes, and/or phenolic alcohols, have also been determined in wines (Anli *et al.*, 2008; Chiou *et al.*, 2007; Di Tommaso *et al.*, 1998; Dugo *et al.*, 2001; Esteruelas *et al.*, 2011; Flamini, 2003; Karagiannis *et al.*, 2000; Minuti *et al.*, 2006; Soleas *et al.*, 1997c, 1997d, 1999; Spranger *et al.*, 2004), fruits and vegetables (Amzad Hossain *et al.*, 2009; Caligiani *et al.*, 2007; Cheng and Zuo, 2007; Damak *et al.*, 2008; Dini *et al.*, 2004; Füzfai *et al.*, 2004; Füzfai and Molnár-Perl, 2007; Marsilio *et al.*, 2001; Owen *et al.*, 2003; Rudell *et al.*, 2008; Sudjaroen *et al.*, 2005), aromatic and medicinal plants (Canini *et al.*, 2007; Deng and William Zito, 2003; Fiamegos *et al.*, 2004; Kakasy *et al.*, 2006; Kalinova *et al.*, 2006; Krafczyk and Glomb, 2008; Leal *et al.*, 2008; Lee *et al.*, 2004; Lin *et al.*, 1999; Obmann *et al.*, 2011; Proestos *et al.*, 2006a, 2006b; Shadkami *et al.*, 2009), honey and related products (dos Santos Pereira *et al.*, 2000; Isidorov *et al.*, 2009; Kalogeropoulos *et al.*, 2009; Lachman *et al.*, 2010; Maciejewicz *et al.*, 2001; Márquez-Hernández *et al.*, 2010; Prytzyk *et al.*, 2003; Sabatier *et al.*, 1992), and distilled alcoholic beverages (Ng *et al.*, 2000).

The most commonly used methods for polyphenolic determination in edible oils are based on HPLC (Brenes *et al.*, 1999) since, in contrast with GC, no previous derivatization is required. Despite this, several studies based on GC have been reported (Angerosa *et al.*, 1995, 1996; Kalogeropoulos *et al.*, 2007; Liberatore *et al.*, 2001; Ríos *et al.*, 2005; Rojas *et al.*, 2005; Romani *et al.*, 2001; Saitta *et al.*, 2002; Salta *et al.*, 2007; Solinas, 1987; Stremple, 1998; Tasioula-Margari and Okogeri, 2001).

GC-based methods developed for polyphenol determination in foods are summarized in Table 7.1, which includes the chromatographic characteristics, such as type of column, temperature program, injection parameters, mobile phase and detection characteristics, as well as the way of extracting analytes from the sample and, if used, the derivatization reagents.

Commonly used columns are fused silica capillary columns with lengths of 25–30 m, and inner dimensions ranging from 0.25 to 0.5 mm. The stationary phase film thickness is typically 0.10–0.25 μm, although thicknesses of between 0.33 and 1.4 μm have been used on a few occasions. The most widely used column coating material is 5% phenyl–95% dimethylpolysiloxane (DB-5, HP-5, SPB-5, XTI-5, BPX5, Rtx-5, ZB-5, TC 5, HP ultra2, VF-5, DB XLB). Other coatings include 100% dimethylpolysiloxane (DB-1, HP-1, BP-1, OV1, HP ultra1) and, to a lesser extent, 5% phenyl–1% vinyl-methylpolysiloxane (SE-54) and 50% phenyl-50% polydimethylsiloxane (CP-SIL 24CB, BP50). Only early GC methods used glass packed columns (Vande Casteele *et al.*, 1976).

The temperature program is generally based on gradients, using initial column temperatures ranging from 40 to 150°C (60–80°C is the most commonly chosen), and final temperatures of between 200 and 320°C, achieved in different steps and with rate increases ranging from 2 to 50°C/min. Liggins *et al.* (1998, 2000) proposed an isothermal GC program for the separation of two isoflavonoid compounds. The application of HT-GC with a final column temperature of 370°C permitted easy separation of six monoisoprenylated flavonoids of high molecular weight without a prior derivatization in spite of their thermolabile character (Branco *et al.*, 2001).

Injection volumes of between 1 and 2 μl are commonly used, although some authors (Amzad Hossain *et al.*, 2009; Kalinova *et al.*, 2006) have injected lower (0.2 μl) or higher (3 μl) volumes (Canini *et al.*, 2007; Ríos *et al.*, 2005) into the capillary columns. Injection is usually carried out via splitless or split (commonly, split ratios of between 1:10 and 1:30) modes maintaining the GC injector at a constant temperature ranging from 220 to 310°C, although some authors (Damak *et al.*, 2008; Kakasy *et al.*, 2006; Sudjaroen *et al.*, 2005) have applied temperature programs using septum-equipped programmable injectors (SPI). The selected injection temperature is a function of the thermal stability of the chromatographed compounds. Some authors have directly injected the sample into the column without heating by using the on-column injection mode (Branco *et al.*, 2001; Füzfai *et al.*, 2004; Liberatore *et al.*, 2001). Helium is commonly used as the carrier gas at a flow-rate of between 0.4 and 3 ml/min, although hydrogen has also been used in capillary GC-MS (Branco *et al.*, 2001).

Table 7.1 Summary of Analytical Methods for Polyphenols in Foods by GC-MS

Compounds	Food	GC Column	Temperature Program	Mobile Phase and Injection	Ionization and Detection	Sample Preparation	Derivatization Method	References
p-Coumaric, ferulic, sinapic, vanillic and syringic acids	Cereals (wheat, oats, corn, brown rice, potatoes)	WCOT OV-101 (24 m × 0.2 mm)	40–150°C at 40°C/min, 150–300°C at 4°C/min, 300°C (10min)	Injection at 300°C	EI-QMF. m/z: 40–800	Extraction with MeOH–acetone–water (7:7:6). LLE extraction. Hydrolysis with NaOH or HCl and extraction with ether-ethyl acetate	BSA	Krygier et al., 1982; Sosulski et al., 1982
Resveratrol	Wine	DB-5 (30 m × 0.25 mm, 0.25 μm)	100°C (1 min), 100–290°C at 10°C/min, 290°C (15 min)	—	EI-QMF. SIM	SPE with diatomaceous earth cartridges, eluting with ethyl acetate	50 μl BSTFA, 70°C for 15min	Frankel et al., 1995
11 Phenolic acids. Resorcin, tyrosol and hydroxytyrosol	Virgin olive oil	SE-54 (25 m × 0.32 mm, 0.1 μm)	70–135°C at 2°C/min, 135°C (10 min), 135–220°C at 4°C/min, 220°C (10 min), 220–270°C at 3.5°C/min, 270°C (20 min)	He, 10kPa	EI-QMF. SIM	Extraction with MeOH. Washing with hexane (to eliminate traces of glyceride), evaporation and reconstitution in acetone	150 μl BSTFA to the acetone solution or 500 μl BSTFA to the evaporated acetone solution	Angerosa et al., 1995

Compounds	Sample	Column	Temperature program	Carrier gas	Detection	Extraction	Derivatization	Reference
Tyrosol, hydroxytyrosol, 3,4-dihydroxyphenylacetic acid and oleuropein aglycone	Virgin olive oil	DB5-MS (30 m × 0.25 mm, 0.25 µm)	40°C (2 min), 40–180°C at 20°C/min, 180–320°C at 4°C/min, 320°C (20 min)	He, 38 cm/s. Injection at 310°C with split 1:30	CI with ammonia. QMF	Extraction with MeOH. Washing with hexane (to eliminate traces of glyceride), evaporation and reconstitution in acetone	120 µl BSTFA to 1 ml of the acetone extract, room temperature for 1h	Angerosa et al., 1996
5 Isoflavonoids (formononetin, biochanin A, daidzein, genistein, coumestrol) and 2 lignans (secoisolariciresinol, matairesinol)	Plant-derived foods (bread, seeds, tea, flour)	BP-1 (12.5 m × 0.22 mm, 0.25 µm)	100°C (1 min), 100–280°C at 30°C/min	He	EI-QMF. SIM. ID	Enzymatic hydrolysis, extraction of isoflavonoids with diethyl ether. Acid hydrolysis and extraction of lignans with diethyl ether. Purification by ion-exchange chromatography	100 µl Py:HMDS:TMCS (9:3:1), room temperature for at least 30 min	Mazur et al., 1996

Continued

Table 7.1 Summary of Analytical Methods for Polyphenols in Foods by GC-MS—cont'd

Compounds	Food	GC Column	Temperature Program	Mobile Phase and Injection	Ionization and Detection	Sample Preparation	Derivatization Method	References
15 Polyphenols (vanillic, gentisic, m- and p-coumaric, gallic, ferulic and caffeic acids. cis- and trans-Resveratrol, epicatechin, catechin, morin, quercetin, cis- and trans-polydatin	Wine	DB-5HT (30m×0.25mm, 0.1µm)	80°C (1min), 80–250°C at 20°C/min, 250°C (1min), 250–300°C at 6°C/min, 300°C (2min), 300–320°C at 20°C/min, 320°C (4min)	He, 60psi. Injection of 1µl at 280°C in splitless mode	EI-QMF. SIM	Wine was diluted 1:1 with water and submitted to SPE C8, elution with ethyl acetate	1ml BSTFA:Py (1:1), 70°C for 30min	Soleas et al., 1997c
Vanillic, protocatechuic, p-hydroxybenzoic, caffeic and p-coumaric acids	Cherry laurel cultivars	HP-1 (25m×0.32mm, 0.17µm)	60°C (2min), 60–250°C at 8°C/min, 250°C (3min)	He, 3ml/min	EI-QMF	Hydrolysis with 2 N NaOH, acidified with HCl and extraction with ethyl acetate	BSTFA:Py (1:1)	Ayaz et al., 1997
8 Phenolic acids	Diospyros lotus L. (persimmon)	HP-1 (25m×0.32mm, 0.17µm)	60°C (2min), 60–250°C at 8°C/min, 250°C (3min)	He, 3ml/min	EI-QMF	Hydrolysis with 2 N NaOH, acidified with HCl and extraction with ethyl acetate	BSTFA:Py (1:1)	Ayaz and Kadioglu, 1997

Compound	Sample	Column	Temperature	Injection	Detection	Extraction	Derivatization	Reference
Tyrosol and hydroxytyrosol	Wine	CP-SIL 24CB (15m×0.25mm, 0.25μm)	150°C (1min), 150–200°C at 10°C/min	He, 40psi. Injection of 1μl at 220°C in pulsed splitless	EI-QMF. SIM	SPE on a C18 column and elution with ethyl acetate	50μl BSTFA, 60°C for 30min and then overnight at room temperature	Di Tommaso et al., 1998
Dadzein and genistein	Fruits and vegetables	DB-1 (15m)	320°C, isothermal	Injection of 3μl with split ratio 25:1	EI-QMF. SIM	Extraction with MeOH, assisted by sonication. Enzymatic hydrolysis and extraction with ethyl acetate	0.6ml Py - 0.4ml MTBSTFA (1% TBDMS-chloride), 25°C for 1h	Liggins et al., 2000
3 Flavones (baicalin, baicalein, wogonin)	Roots of *Scutellaria baicalensis*	HP-5MS (30m×0.25mm, 0.25μm)	100°C (2min), 100–280°C at 10°C/min, 280°C (15min)	He, 1ml/min. Injection of 1μl at 250°C, splitless	EI-QMF	SFE: Packing of the sample into a cartridge, addition of MeOH. Extraction with liquid CO_2, the soluble fraction collected in MeOH	No derivatization	Lin et al., 1999
Dadzein and genistein	Soy-based foodstuffs	SE54 (12m×0.32mm, 1μm)	190–245°C at 49.9°C/min	He, 25kPa. Injection at 230°C in splitless mode	EI-QMF. SIM. ID	Extraction with MeOH. Enzymatic hydrolysis and extraction with ethyl acetate. Isolation by gel permeation chromatography	BSTFA	Morton et al., 1999

Continued

Table 7.1 Summary of Analytical Methods for Polyphenols in Foods by GC-MS—cont'd

Compounds	Food	GC Column	Temperature Program	Mobile Phase and Injection	Ionization and Detection	Sample Preparation	Derivatization Method	References
cis- and trans-p-Coumaric, cis- and trans-ferulic, p-hydroxy-benzoic, vanillic and syringic acids	Wheat and root tissues	DB-5 MS (30 m × 0.25 mm, 0.25 μm)	80°C (1 min), 80–160°C at 10°C/min, 160–235°C at 5°C/min, 235–280°C at 50°C/min, 280°C (5 min)	He, 34 cm/s. Injection of 1 μl at 280°C, splitless	EI-QIT-MS and EI-QIT-MS/MS. m/z: 50–450	Maceration in 0.001 M HCl. Extraction with diethyl ether	1 m MSTFA, 60°C for 30 min	Wu et al., 1999, 2000
Vanillic acid, gallic acid, syringalde-hyde, conifer-aldehyde and vanillin	Distilled alcohol beverages	DB-5 MS (30 m × 0.25 mm, 0.5 μm)	75°C (2 min), 75–100°C at 25°C/min, 100–300°C at 10°C/min, 300°C (5 min)	He, 1.1 ml/min. Injection at 250°C in splitless	EI-QMF. m/z: 39–450. SIM	Sample purification and preconcentration by solid- phase (anion-exchange) disk extraction	MSTFA in acetonitrile, 80°C for 20 min	Ng et al., 2000
p-Coumaric, ferulic, caffeic, vanillic, gentisic and gallic acids	Wine and agricultural products	HP-5MS (30 m × 0.25 mm, 0.25 μm)	80°C (1 min), 80–240°C at 15°C/min, 240°C (3 min)	He, 1 ml/min. Injection at 280°C in splitless mode	EI-QMF. SIM	Wine was diluted 1:1 with water and submitted to SPE C8, elution with ethyl acetate	0.5 ml Py and 0.3 ml BSA, microwave oven assistance for 30 s	Chu et al., 2001

Compounds	Sample	Column	Temperature program	Carrier/injection	Detector	Sample preparation	Derivatization	Reference
Tyrosol and p-hydroxy-benzoic, p-coumaric, ferulic, caffeic, protocatechuic, vanillic and syringic acids	Virgin olive oil	SPB-5 (30 m × 0.32 mm, 0.25 μm)	70–135°C at 2°C/min, 135°C (10 min), 135–220°C at 4°C/min, 220°C (10 min), 220–270°C at 4°C/min, 270°C (20 min)	He, 10 kPa. On-column injection mode	EI-QMF	Sample dissolved in hexane. SPE on a C18 column and elution with hexane and MeOH	150 μl BSTFA	Liberatore et al., 2001
9 Flavonoid aglycones	Propolis	OV-1 (12.5 m × 0.2 mm)	180–280°C at 3°C/min	He, 3 ml/min. Injection of 2 μl	EI-QMF	Extraction with ethanol. Cleaning with hexane	No derivatization	Maciejewicz et al., 2001
Cinnamic, p-coumaric, ferulic, caffeic, benzoic, p-hydroxy-benzoic, vanillic and protocatechuic acids	Taxons of Polygonum L. genus	XTI-5 (30 m × 0.25 mm, 0.25 μm)	140°C (2 min), 140–300°C at 5°C/min	He, 1 ml/min. Injection at 300°C	EI-QMF. m/z: 50–650	Refluxed pulverized plant material in CHCl$_3$ for 30 min, SPE on C18 silica gel	Methylation: dry KOH in DMSO, CH$_3$I. Silylation: BSTFA, 90°C for 20 min	Smolarz, 2001
Hydroxytyrosol, tyrosol, oleuropein, oleuropein aglycone and oleoside-11-methyl ester	Olive fruits	HP1 (30 m × 0.32 mm, 0.1 μm)	70–90°C at 20°C/min, 90–300°C at 4°C/min, 300°C (40 min)	He, 35 kPa. On-column injection of 0.3 μl	EI-QMF	Extraction with ethanol and washing with water:hexane (1:1). Extraction of the aqueous phase with ethyl acetate	Py:HMDS:TMCS (2:1:1), 25°C for 1 h	Marsilio et al., 2001

Continued

Table 7.1 Summary of Analytical Methods for Polyphenols in Foods by GC-MS—cont'd

Compounds	Food	GC Column	Temperature Program	Mobile Phase and Injection	Ionization and Detection	Sample Preparation	Derivatization Method	References
6 Monoiso-prenylated flavonols	*Vellozia graminifolia* (Vellozia-ceae)	Silaren 30 (10 m × 0.3 mm, 0.1 μm)	40–370°C at 12°C/min	H_2, 2 ml/min. On-column injection of 0.2 μl at room temperature	EI-QMF. *m/z:* 40–700	Successive extractions with hexane and ethyl acetate. Evaporation and cleaning of oily residue on a silica gel column, elution with ethyl acetate and MeOH	No derivatization	Branco *et al.*, 2001
o- and *p*-Coumaric, ferulic, sinapic, caffeic, benzoic and vanillic acids	Cranberry fruit	DB-5 (30 m × 0.35 mm, 0.25 μm)	80–120°C at 5°C/min, 120–240°C at 10°C/min, 240–280°C at 20°C/min	He, 13.3 psi. Injection of 1 μl at 280°C, splitless	EI-QMF. *m/z:* 50–550	Sample blending in water. Acidification and extraction with ether. Cleaning of the organic phase	50 μl Py and 50 μl BSTFA:TMCS, 60°C for 30 min	Zuo *et al.*, 2002
23 Polyphenols	Sicilian olive oil	DB-5MS (30 m × 0.25 mm, 0.25 μm)	60–275°C at 15°C/min, 275°C (15 min)	He, 40 cm/s. Injection in splitless	EI-QMF-MS/ MS. *m/z:* 40–600. SIM	Extraction with MeOH:water (80:20) and washing with hexane	200 μl BSTFA:TMCS (99:1), 25°C for 30 min	Saitta *et al.*, 2002

Compounds identified	Sample	Column	Temperature program	Carrier gas	MS detection	Extraction	Derivatization	Reference
Dadzein and genistein	Soy plants and flour and iso-flavone-enriched wheat flour	DB1 (6m × 0.25 mm, 0.25 μm)	70°C (1 min), 70–300°C at 25°C/min, 300°C (5 min)	He	Combustion/ Isotope ratio MS	Extraction with MeOH, assisted by sonication. Enzymatic hydrolysis and extraction with ethyl acetate	0.3 ml Py and 0.2 ml MTBSTFA (1% TBDMS-chloride), 1 h at room temperature	Bluck et al., 2002
24 Polyphenols were identified, the major being: gallic acid, cinnamic acid, myricetin and quercetin	Carob fiber	HP 5MS (30 m × 0.25 mm, 0.25 μm)	100–270°C at 4°C/min, 270°C (20min)	He, 0.9 ml/s. Injection of 1 μl at 250°C	EI-QMF. SIM	Successive Soxhlet extractions with hexane and MeOH. Fractionation by normal-phase chromatography. Alkaline or acid hydrolysis and extraction with diethyl ether	100 μl BSTFA, 37°C for 30 min	Owen et al., 2003
Kaempferol, isorhamnetin and quercetin	Ginkgo biloba L. extract and pharmaceutical products	HP Ultra 1 (25 m × 0.20 mm, 0.33 μm)	80°C (0.1 min), 80–245°C at 25°C/ min, 245°C (25.5 min), 245–270°C at 60°C/min, 270°C (8 min)	He, 0.5 ml/ min. Injection of 2 μl at 275°C in splitless mode	EI-QMF. m/z: 70–650. SIM	Acid hydrolysis and extraction with ethyl acetate. 50 μl of the organic layer was used for derivatization	250 μl DMF - 250 μl BSTFA (1% TMCS), 115°C for 45 min	Deng and William Zito, 2003

Continued

Table 7.1 Summary of Analytical Methods for Polyphenols in Foods by GC-MS—cont'd

Compounds	Food	GC Column	Temperature Program	Mobile Phase and Injection	Ionization and Detection	Sample Preparation	Derivatization Method	References
8 Phenolic acids and 4 flavonoids	Methanolic and water plant extracts	SPBM-5 (30 m × 0.32 mm, 0.25 μm)	50°C (5 min), 50–150°C at 5°C/min, 150–210°C at 10°C/ min, 210°C (11 min)	He. Injection of 1 μl at 260°C with split ratio 50:1	EI-QMF. SIM	In-vial combined derivatization and extraction. Soxhlet extraction with MeOH or hot water. Addition of tri-n-butyl-methylphos-phonium chloride and 1 ml CH_2Cl_2	180 mg CH_3I added simultaneously with in-vial extraction reagents	Fiamegos et al., 2004
4 Flavonoid glycosides and their aglycones (cyanin, naringin, hesperidin, rutin and cyanidin, naringenin, hesperitin and quercetin)	Sour cherries	BPX5 (30 m × 0.25 mm, 0.25 μm)	60°C (2 min), 60–120°C at 20°C/min, 120–155°C at 6°C/ min, 155°C (10 min), 155–250°C at 13°C/ min, 250°C (12 min), 250–330°C at 20°C/ min, 330°C (10 min)	Injection of 1 μl. Program: 60°C (2 min), 60–320°C at 180°C/ min, 320°C (6 min)	EI-QIT. m/z: 40–650	Acid hydrolysis with TFA	500 μl Py (hydroxylamine), 70°C for 30 min. Next, 1 ml HMDS and 100 μl TFA, 100°C for 60 min	Füzfai et al., 2004

17 Polyphenols	Medicinal herbs	Ultra-2 SE-54 (25 m × 0.2 mm, 0.33 μm)	160–270°C at 10°C/min, 270°C (10 min), 270–315°C at 5°C/min, 315°C (2 min)	He, 0.4 ml/min. Injection of 2 μl with split 5:1	EI- QMF. SIM	Succesive extractions with water:MeOH and diethyl ether. Enzymatic and acid hydrolysis. Extraction with diethyl ether. SPE, elution with MeOH	50 μl MSTFA:NH$_4$: Dithioerythritol (1000:4:5), 60°C for 15 min	Lee et al., 2004
Naringenin-chalcone	Surface wax of tomatoes	Rtx-5 (60 m × 0.25 mm, 0.1 μm)	200°C (3 min), 200–300°C at 2°C/min, 300°C (5 min)	He, 2 ml/min. Injection: 1 μl, 350°C, split 1:10	EI-QMF	Extraction with TBME: MeOH (9:1). Purification by mixing with Celite 545 and silica gel column	0.8 ml TBME and 0.2 ml TMSI	Bauer et al., 2004
21 Polyphenols	Virgin olive oil	ZB-5 ms (30 m × 0.25 mm, 0.25 μm)	150°C (5 min), 150–295°C at 3°C/min, 295°C (18 min)	He, 1 ml/min. Injection of 3 μl in split	EI-QIT	Oil dissolved in hexane and submitted to SPE, elution with MeOH	100 μl HMDS: DMCS:Py (3:1:9)	Ríos et al., 2005

Continued

Table 7.1 Summary of Analytical Methods for Polyphenols in Foods by GC-MS—cont'd

Compounds	Food	GC Column	Temperature Program	Mobile Phase and Injection	Ionization and Detection	Sample Preparation	Derivatization Method	References
Catechin, epicatechin, taxifolin, apigenin, luteolin, naringenin and eriodictyol	Tamarind seeds and pericarp	HP 5MS (30 m × 0.25 mm, 0.25 μm)	100–270°C at 4°C/min, 270°C (20 min)	He, 0.9 ml/s. Injection of 1 μl at 250°C	EI-QMF. SIM	Succesive Soxhlet extractions with hexane and MeOH	100 μl BSTFA, 37°C for 30 min	Sudjaroen et al., 2005
22 Polyphenols	Wine	DB-5MS (30 m × 0.25 mm, 0.25 μm)	120°C (3 min), 120–320°C at 5°C/min, 320°C (5 min)	He, 0.8 ml/min. Injection: 1 μl, 300°C, splitless	EI-QMF. SIM	Addition of NaCl and sodium metabisulfite to the sample, extraction with ethyl acetate	50 μl BSTFA in Py, 25°C for 1 h	Minuti et al., 2006
15 Polyphenols	Aromatic plants	CP-Sil 8 CB-MS (30 m × 0.32 mm, 0.25 μm)	70–135°C at 2°C/min, 135°C (10 min), 135–220°C at 4°C/min, 220°C (10 min), 220–270°C at 3.5°C/min, 270°C (20 min)	He, 1.9 ml/min. Injection at 280°C in splitless (1 min)	EI-QMF. m/z: 25–700	Extraction with MeOH containing butylated hydroxytoluene and HCl. Extraction with ethyl acetate	100 μl TMCS and 200 μl BSTFA, 80°C for 45 min	Proestos et al., 2006a, 2006b

Analytes	Sample	Column	Oven program	Injection / carrier	MS detection	Sample preparation	Derivatization	Reference
6 Phenolic acids and 2 flavones	*Dracocephalum* (medicinal plant)	BPX5 (30m×0.25mm, 0.25µm)	150°C (4.6 min), 150-330°C at 8°C/min, 330°C (7 min)	Injection of 1µl. Pogram: to 150°C at 2°C/min, 150°C (2min), 150-330°C at 180°C/min, 330°C (5min)	EI-QIT. m/z: 40–650	Extraction in MeOH:acetone (1:1) acidified mixture. Acid hydrolysis	500µl Py (hydroxylamine). Next, 1ml HMDS and 100µl TFA, 100°C for 60min	Kakasy et al., 2006
9 Phenolic acids	Brewer's spent grains	CP-Sil 8CB (30m×0.32mm, 0.25µm)	80-250°C at 10°C/min, 250–280°C at 20°C/min, 280°C (2 min)	He, 1ml/min. Injection of 1µl at 260°C, splitless	EI-QMF. m/z: 50–550	Addition of 1m NaOH, incubation for 20h. Acidification to pH2 and extraction six times with diethyl ether: ethyl acetate (1:1)	Microwave-assisted BSTFA derivatization	Mallouchos et al., 2007
14 Polyphenols	Enriched vegetable oils with olive leaf extract	HP-5 MS (30m×0.25mm, 0.25µm)	70°C (5 min), 70-130°C at 15°C/min, 130-160°C at 4°C/min, 160°C (15min), 160-300°C at 10°C/min, 300°C (15min)	He, 0.6ml/min. Injection of 1µl with split 1:20	EI-QMF. SIM	Extraction with MeOH. Evaporation, reconstitution in acetonitrile and washing with hexane	250µl BSTFA, 70°C for 20min.	Salta et al., 2007

Continued

Table 7.1 Summary of Analytical Methods for Polyphenols in Foods by GC-MS—cont'd

Compounds	Food	GC Column	Temperature Program	Mobile Phase and Injection	Ionization and Detection	Sample Preparation	Derivatization Method	References
8 Flavonoids (pelargonidin, cyanidin, malvidin, quercetin, apigenin, luteolin, naringenin and hesperetin)	Citrus fruits	BPX5 (30 m × 0.25 mm, 0.25 μm)	150°C (4.5 min), 150–330°C at 10°C/min, 330°C (7 min)	Injection of 1 μl. Program: 150°C (2 min), 150–330°C at 180°C/min, 330°C (5 min)	EI-QIT. m/z: 40–650	Homogenized juice and pulverized albedo were dried and submitted to derivatization. Some polyphenols were quantified including previous acid hydrolysis	500 μl Py (hydroxylamine), 70°C for 30 min. Next, 0.9 ml HMDS and 100 μl TFA, 100°C for 60 min	Füzfai and Molnár-Perl, 2007
(−)-Epicatechin and (+)-catechin	Cocoa beans	DB5 (30 m × 0.25 mm, 0.25 μm)	60°C (3 min), 60–280°C at 20°C/min	He. Injection at 280°C	EI-QMF. m/z: 40–550 SIM.	Extraction with MeOH, evaporation and reconstitution in dry DMF	0.3 ml TMCS and 0.6 ml HMDS, 60°C for few minutes	Caligiani et al., 2007

Analytes	Sample	Column	Temperature program	Carrier gas/Injection	Detection	Sample preparation	Derivatization	Reference
Sugar moieties of 2 flavonol glycosides (quercetin galactoside and quercetin arabinoside)	Cranberry fruit	DB-5ms (30 m × 0.32 mm, 0.25 μm)	100°C (1 min) 100–280°C at 8°C/min, 280°C (5 min)	He. Injection of 1 μl at 260°C	EI-QMF. m/z: 50–800	Maceration in aqueous MeOH. Evaporation and cleaning with petroleum ether and ethyl ether. Concentration and purification by HPLC. Acid hydrolysis	50 μl Py and 50 μl BSTFA containing 1% TMCS, 70°C for 15 min	Cheng and Zuo, 2007
Protocatechuic, p-coumaric, caffeic and chlorogenic acids. Kaempferol, quercetin and 5,7-dimethoxycoumarin	Carica papaya L. leaf	SPB-5 (30 m × 0.25 mm, 1.4 μm)	100°C (3 min), 100–315°C at 20°C/min	He, head pressure 100 kPa. Injection of 3 μl	EI-QMF. SIM	Soxhlet extraction with acidified MeOH. Two aliquots treated separately: a) Free phenolics: Extraction with diethyl ether. b) Total phenolics: Evaporation, alkaline hydrolysis and extraction with diethyl ether	1.5 ml BSTFA (TMCS 1%):Py (1:1), 90°C for 1 h. 5,7-Dimethoxycoumarin is not derivatized	Canini et al., 2007

Continued

Table 7.1 Summary of Analytical Methods for Polyphenols in Foods by GC-MS—cont'd

Compounds	Food	GC Column	Temperature Program	Mobile Phase and Injection	Ionization and Detection	Sample Preparation	Derivatization Method	References
25 Polyphenols	Fresh and fried oils and fish	HP-5 MS (30 m × 0.25 mm, 0.25 μm)	70°C (5 min), 70–130°C at 15°C/min, 130–160°C at 4°C/min, 160°C (15 min), 160–300°C at 10°C/min, 300°C (15 min)	He, 0.6 ml/min. Injection of 1 μl at 280°C with split ratio 1:20	EI-QMF. SIM	Extraction with MeOH and clean-up with hexane	250 μl BSTFA, 70°C for 20 min	Kalogeropoulos et al., 2007
p-Coumaric, gallic, ferulic and caffeic acids. trans-Resveratrol, epicatechin, catechin and quercetin	Varietal Turkish red wines	TC-5 (30 m × 0.25 mm, 0.25 μm)	80–250°C at 20°C/min, 250°C (1 min), 250–300°C at 6°C/min, 300°C (2 min), 300–320°C at 20°C/min, 320°C (4 min)	He, 1.5 ml/min	EI-QMF	Sample pH adjusted to 2 and extraction three times with diethyl ether and then three times with ethyl acetate	Py:TMCS: BSA	Anli et al., 2008
17 Isoflavonoids	Propolis	HP-5 ms (17 m × 0.20 mm, 0.33 μm)	80°C (1 min), 80–310°C at 5°C/min, 310°C (20 min)	He, 0.8 ml/min. Injection of 1 μl at 280°C with split 1:10	EI-QMF. m/z: 40–800. SIM	Extraction with MeOH	75 μl MSTFA, 60°C for 15 min	Campo Fernández et al., 2008

Compounds identified	Sample	Column	Temperature program	Carrier/Injection	Detection	Extraction	Derivatization	Reference
7 Flavonoid glycosides, catechin, epicatechin and dihydrochalcone	Skin of apple fruit	HP-5MS (30 m × 0.25 mm, 0.25 µm)	40°C (2 min), 40–330°C at 18°C/min, 330°C (6 min)	He, 66 cm/s and 40 cm/s. Injection of 0.2 µl at 230°C, splitless	EI-QMF. m/z: 30–500	Extraction with MeOH	125 µl Methoxyamine (in Py), 30°C for 90 min. 125 µl BSTFA, 37°C for 30 min	Rudell et al., 2008
6 Phenolic acids (3,5-dihydroxybenzoic, p-hydroxybenzoic, salicylic, ferulic and sinapinic acids) and 4 flavonoids (catechin, quercetin, luteolin, chrysoeriol)	Rooibois tea	HP-5 (30 m × 0.32 mm, 0.25 µm)	Phenolic acids: 100–200°C at 5°C/min, 200–270°C at 10°C/min, 270°C (25 min). Flavonoids: 200–270 at 2.3°C/min, 270°C (30 min)	He, 27.5 cm/s. Injection at 220°C with split 1:10	EI-QMF. m/z: 50–650	Extraction with acetone:water (7:3). Extraction with diethyl ether. Fractionation by mLCCC. Evaporation and purification of flavonoids by preparative HPLC	100 µl Py and 100 µl BSA, room temperature for 1 h	Krafczyk and Glomb, 2008
p-Coumaric acid, kaempherol and isorhamnetin are quantified. Ferulic and caffeic acids, naringenin and apigenin were identified	Beebread	HP-5ms (30 m × 0.25 mm, 0.25 µm)	40–310°C at 3°C/min, 310°C (15 min)	He, 1 ml/min. Injection at 250°C with split 1:30	EI-QMF. m/z: 41–700	Extraction with diethyl ether	220 µl Py and 80 µl BSTFA (1% TMCS) to 2 ml of reconstituted extract, 60°C for 30 min	Isidorov et al., 2009

Continued

Table 7.1 Summary of Analytical Methods for Polyphenols in Foods by GC–MS—cont'd

Compounds	Food	GC Column	Temperature Program	Mobile Phase and Injection	Ionization and Detection	Sample Preparation	Derivatization Method	References
8 Isoflavonoids (biochanin A, coumestrol, daidzein, equol, formononetin, glycitein, genistein and prunetin)	Soy milk	DB-5MS (25 m × 0.25 mm, 0.25 μm)	80°C (2 min) 80–300°C at 4°C/min, 300°C (3 min), 300–310°C at 10°C/min, 310°C (1 min)	He, 1.2 ml/min. Injection at 280°C in splitless mode	EI-QIT-MS/MS. m/z: 50–500	Extraction with ethyl acetate	200 μl BSTFA (10% TMCS), 60°C for 4 h. Evaporation to dryness and dissolving in 200 μl BSTFA:Py (5:1)	Ferrer et al., 2009
Sinensetin, rutin, 3'-hydroxy-5,6,7,4'-tetramethoxyflavone and rosmarinic acid	Skin of apple fruit	VF-5 (30 m × 0.25 mm, 0.25 μm)	50–200°C at 8°C/min, 200°C (20 min), 200–300°C at 10°C/min	He, 1 ml/min. Injection of 0.2 μl at 250°C in splitless mode	EI-QMF	Extraction with acidified MeOH	No derivatization	Amzad Hossain et al., 2009
Sugar moieties of 5 anthocyanins	Red kiwifruit	ZB-5 ms (30 m × 0.25 mm, 0.25 μm)	40°C (1 min) 40–300°C at 7°C/min, 300°C (4 min)	—	EI-QMF. SIM	Extraction with acidified ethanol. Cleaning with hexane and next with XAD resine, eluting with MeOH. Preparative RP-HPLC	50 μl methoxyamine in Py, 80°C for 30 min. 50 μl MSTFA, 80°C for 30 min	Comeskey et al., 2009

Analyte	Sample	Column	Temperature program	GC conditions	MS	Sample preparation	Derivatization	Reference
Catechins and condensed tannins	Green tea and tea leaves extracts	ZB-5HT Inferno (15 m × 0.32 mm, 0.1 µm)	100°C (5 min), 100–375°C at 20°C/min, 375°C (5 min)	Column head pressure, 5 psi. Injection of 5 µl. Pyrolysis interface and GC injector at 350°C	EI-QMF. m/z: 60–650	20 µl of tea extract placed in a hot metallic container was purged with N_2 and dried. Dissolution in MeOH	Two steps: 30 µl TMS-diazomethane. 30 min of sonication, evaporation. 40 µl of 2.5% TMSH or 0.25% TMAH	Shadkami et al., 2009
31 Polyphenols	Propolis	HP-5 MS (30 m × 0.25 mm, 0.25 µm)	100–310°C at 5°C/min, 310°C (8 min)	He, 0.7 ml/min. Injection of 1 µl at 220°C with split ratio 1 : 20	EI-QMF. m/z: 50–800. 27 analytes quantified in SIM and 4 in SCAN mode	Extraction with ethanol	250 µl BSTFA, 70°C for 20 min	Kalogeropoulos et al., 2009
trans-Resveratrol	Wine	BP-5 (30 m, 0.25 mm, 0.25 µm)	90°C (1 min), 90–280°C at 15°C/min, 280°C (15 min)	He, 1.2 ml/min. Injection of 1–2 µl at 280°C in splitless (1 min)	EI-QMF. m/z: 150–450. SIM	Mixed mode SPE (MAX sorbent), eluting with MeOH	50 µl AA and 10 ml 5% K_2HPO_4 aqueous solution. Extraction with isooctane	Montes et al., 2010

Continued

Table 7.1 Summary of Analytical Methods for Polyphenols in Foods by GC–MS—cont'd

Compounds	Food	GC Column	Temperature Program	Mobile Phase and Injection	Ionization and Detection	Sample Preparation	Derivatization Method	References
14 Phenolic acids, 6 flavonoids and resveratrol	Cooked dry legumes	HP-5 MS (30 m × 0.25 mm, 0.25 μm)	100–310°C at 5°C/min, 310°C (8 min)	He, 0.7 ml/min. Injection of 1 μl with split 1:20	EI-QMF. SIM	Extraction with MeOH	250 μl BSTFA, 70°C for 20 min	Kalogeropoulos et al., 2010a
27 Polyphenols	Hypericum perforatum	HP-5 MS (30 m × 0.25 mm, 0.25 μm)	100–310°C at 5°C/min, 310°C (8 min)	He, 0.7 ml/min. Injection of 1 μl with split 1:20	EI-QMF. SIM	Extraction with MeOH. Purification by successive extractions in CHCl₃ and hexane	250 μl BSTFA, 70°C for 20 min	Kalogeropoulos et al., 2010b
34 Flavonoids (flavones, flavonols, isoflavones, flavanonols, chalcones and flavan-3-ols)	Chocolate and propolis	GCxGC: BPX50 (30 m × 0.25 mm, 0.25 μm) and BPX5 (30 m × 0.1 mm, 0.1 μm)	100°C (1 min), 100–210°C at 20°C/min, 210–320°C at 2°C/min	He, 1.5 ml/min. Injection of 1 μl at 310°C in splitless mode	EI-TOFMS. m/z: 45–760	Ultrasound assisted extraction with aqueous ethanol at pH 2. Extraction with ethyl acetate	50 μl Py and 50 μl BSTFA (1% TMCS), 100°C for 30 min	Gao et al., 2010

Analytes	Sample	Column	Temperature program	Carrier/Injection	Detector	Sample preparation	Derivatization	Reference
Dihydroxycinnamic, ferulic isomers, 3,4-dimethoxycinnamic isomers, salicylic and syringic acids. Vanillin, 4-hydroxybenzaldehyde, syringaldehyde, chrysin, dihydrochrysin, galangin and tectochrysin	Honey	DB-5MS (30 m × 0.25 mm, 0.25 µm)	5°C (3 min), 5–290°C at 10°C/min	He, 1 ml/min. Injection of 1 µl at 250°C with split ratio 1:10	EI-QMF. m/z: 35–450	Sample dissolved in water. Purification by column chromatography (Amberlite XAD resine). Preparative HPLC using a Nucleosil C18 column and H_2O:water gradient elution	No derivatization	Lachman et al., 2010
Sugar moieties of peonidin 3,7-O-β-diglucoside	Grapes	DB-XLB-DG (30 m × 0.25 mm, 0.25 µm)	40°C (1 min), 40–300°C at 7°C/min, 300°C (4 min)	—	EI-QMF	Extraction with MeOH:water:HCOOH. Cleaning and fractionation by column chromatography	50 µl of methoxyamine in Py, 80°C for 30 min. Next, 50 µl MSTFA, 80°C for 30 min	Castillo-Muñoz et al., 2010

Continued

Table 7.1 Summary of Analytical Methods for Polyphenols in Foods by GC-MS—cont'd

Compounds	Food	GC Column	Temperature Program	Mobile Phase and Injection	Ionization and Detection	Sample Preparation	Derivatization Method	References
10 Polyphenols (8 phenolic acids, catechin and tyrosol)	White wine and natural precipitate of white wine	HP-5 (30 m × 0.25 mm, 0.25 μm)	120°C (3 min), 120–320°C at 5°C/min, 320°C (5 min)	He, 0.8 ml/min. Injection of 1 μl at 300°C, splitless	EI-QMF	Concentrated wine and dissolved natural precipitate submitted to SPE C18 cartridges, elution with MeOH	100 μl BSTFA:Py (1:1), 70°C for 30 min	Esteruelas et al., 2011
Free and total forms of 20 polyphenols: 15 phenolic acids, 4 flavonoids and resveratrol	Cranberry products	DB-5 (30 m × 0.32 mm, 0.25 μm)	80°C (1 min), 80–220°C at 10°C/min, 220–310°C at 20°C/min, 310°C (6 min)	He, column head pressure 13 psi. Injection of 1 μl at 280°C, splitless	EI-QMF. m/z: 50–650	Free polyphenols: Extraction with ethyl acetate. Total polyphenols: Ultrasound assisted acid hydrolysis and extraction with ethyl acetate	50 μl Py and 50 μl of BSTFA(1% TMCS), 70°C for 4 h	Wang and Zou, 2011

| Sugar moieties of 8 flavonoid C- and O-glycosides | Mongolian medicinal plant (*Dianthus versicolor*) | ZB-5 (60 m × 0.25 mm, 0.25 μm) | 100–270°C at 3°C/min | He, 2 ml/min Injection at 270°C with split 1:10 | EI-QMF. *m/z*: 40–500 | Extraction with TFA. Fractionation by SPE and CPC. Acid hydrolysis, extraction with ethyl acetate and evaporation of the aqueous phase containing the monosaccharides | BSTFA | Obmann et al., 2011 |

AA, acetic anhydride; BSA, N,O-bis-(trimethylsilyl)acetamide; BSTFA, N,O-bis-(trimethylsilyl)trifluoroacetamide; CI, chemical ionization; CPC, centrifugal partition chromatography; DMF, dimethylformamide; EI, electron impact ionization mode; HMDS, hexamethyldisilazane; ID, isotope dilution; mLCCC, multilayer countercurrent chromatography; MTBSTFA, N-(tert-butyldimethylsilyl)-N-methyltrifluoroacetamide; MSTFA, N-methyl-N-(trimethylsilyl)trifluoro-acetamide; Py, pyridine; QIT, quadrupole ion trap analyzer; QMF, quadrupole mass filter analyzer; TBDMS, N-(tert-butyldimethylsilyl)-N-methyl-trifluoro-acetamide; TBME, tert-butylmethyl ether; TFA, trifluoroacetic acid; TLC, thin layer chromatography; TMAH, tetramethylammonium hydroxide; TMCS, trimethylchlorosilane; TMS, trimethylsilyl; TMSH, trimethylsulfonium hydroxide; TMSI, 1-trimethylsilyl-imidazol; TOF, time-of-flight mass analyzer.

Trialkylsilylation is the preferred derivatization method to increase the volatility of polyphenols and, since it is one of the most reactive silyl donors, BSTFA is the reagent selected by most authors. In fact, BSTFA is more volatile than many other silylating reagents and consequently its chromatographic interference is minimal. The optimized BSTFA reaction times and temperatures are in the 15–60 min and 25–115°C ranges, respectively, these conditions varying depending on the specific compounds being derivatized. Other trialkylsilyl derivatizing reagents employed are MSTFA (Campo Fernández *et al.*, 2008; Castillo-Muñoz *et al.*, 2010; Comeskey *et al.*, 2009; Lee *et al.*, 2004; Ng *et al.*, 2000; Wu *et al.*, 1999, 2000), BSA (Chu *et al.*, 2001; Krafczyk and Glomb, 2008; Krygier *et al.*, 1982; Sosulski *et al.*, 1982), the weak trimethylsilyl donor hexamethyldisilazane (HMDS) (Amzad Hossain *et al.*, 2009; Damak *et al.*, 2008; Füzfai *et al.*, 2004, 2007; Kakasy *et al.*, 2006; Mazur *et al.*, 1996; Ríos *et al.*, 2005; Sudjaroen *et al.*, 2005), and to a lesser extent *N*-methyl-*N*-(*tert*-butyldimethylsilyl)trifluoroacetamide (MTBSTFA) (Bluck *et al.*, 2002; Liggins *et al.*, 1998, 2000), and the selective 1-trimethylsilyl-imidazole (TMSI) (Bauer *et al.*, 2004). The silylating potential of the common reagents used can be increased by using 1 or 10% trimethylchlorosiloxane (TMCS) as catalyst and by adding an appropriate polar solvent such as dimethylformamide (DMF) (Deng and William Zito, 2003), or, more frequently, pyridine (Zafra *et al.*, 2006). The addition of a polar solvent to the reaction medium is also used to favor the dissolution of the analyzed material in the derivatizing reagent. Silylation reactions have been much accelerated by the application of microwaves using microwave ovens, an approach that has been applied in the analysis of phenolic acids (Chu *et al.*, 2001; Mallouchos *et al.*, 2007). Using microwaves, reaction times of 30 s can provide similar results to those obtained in 30 min under conventional conditions.

As previously stated, several authors have determined isoflavones as their *tert*-butyldimethylsilyl (TBDMS) ethers using MTBSTFA as reagent (Bluck *et al.*, 2002; Liggins *et al.*, 1998, 2000). In such cases, *tert*-butyl-chloro-dimethylsilane chloride (TBDMS-chloride) at a concentration of 1% acted as catalyst. The application of MTBSTFA as silylation reagent for polyphenols is infrequent probably due to its reduced ability to silylate tertiary or hindered secondary alcohols.

For polyphenols containing carbonyl groups in their structure, TMS-(oxime) ether/ester derivatives have provided improved chromatographic performance compared with the corresponding TMS-derivatives. For example, hydroxylamine (Damak *et al.*, 2008; Kakasy *et al.*, 2006; Sudjaroen *et al.*, 2005) and methoxyamine (Amzad Hossain *et al.*, 2009) have been used jointly with HMDS and BSTFA, respectively, as trialkylsilyl derivatizing reagents. Castillo-Muñoz *et al.* (2010) and Comeskey *et al.* (2009) used the combination methoxyamine/MSTFA to form (oxime) ether/ester derivatives from anthocyanin compounds.

In spite of the stability of TMS-derivatives, these compounds are easily hydrolyzed, so that when the analytes have been isolated from the sample matrix, the extracts are always evaporated to dryness, in most cases using a nitrogen stream (Gao *et al.*, 2010; Isidorov *et al.*, 2009; Kalogeropoulos *et al.*, 2010b; Shadkami *et al.*, 2009), after the derivatization step.

Etherification of the free hydroxyl groups of some phenolic compounds was carried out by Fiamegos *et al.* (2004) using methyl iodide as reagent. Simultaneous extraction in dichloromethane of the analytes as ion-pairs from the aqueous phase and their derivatization was proposed, based on phase transfer catalysis. Under the finally selected conditions, phenolic acids were derivatized in 30 min, whereas flavonoids needed 90 min at 70°C. Shadkami *et al.* (2009) proposed thermally assisted hydrolysis in the presence of trimethylsulfonium hydroxide (TMSH) as alkylation reagent, to analyze intact methylated flavanols of catechins and a two step alkylation procedure with trimethylsilyl diazomethane (TMS-diazomethane) followed by thermally assisted hydrolysis to analyze condensed tannins.

Only two applications of acetic anhydride (AA) as acylation reagent under basic conditions in aqueous medium can be found in the literature for polyphenol determination by GC-MS, for the quantification of resveratrol in wine (Cacho et al.,2013; Montes *et al.*, 2010). This may be considered surprising because acetylated derivatives are more stable than the corresponding silylated derivatives and they do not suffer hydrolysis. Moreover, the reaction time is shorter than that typically needed in silylation reactions, but, of course, it is necessary to extract the acetylated compound in an organic solvent prior to the chromatographic separation. In spite of the advantages involved in submitting polyphenolic derivatives to GC, some authors have carried out polyphenolic determinations without derivatization of the analytes (Amzad Hossain *et al.*, 2009; Branco *et al.*, 2001; Lachman *et al.*, 2010; Lin *et al.*, 1999; Maciejewicz *et al.*, 2001).

Detection was carried out in most studies using MS, although FID has also been employed. Most GC-MS systems use the electron ionization (EI) mode with an ionizing voltage of 70 eV, although the milder chemical ionization (CI) mode was used by Angerosa *et al.* (1996). The most used analyzer is quadrupole mass filter (QMF), followed by quadrupole ion trap (QIT) and time-of-flight (TOF) analyzers. The spectra are collected from *m/z* 39 to 650 (frequently 50–500) in continuous scanning mode. Sample quantification is generally carried out using SIM mode, which lowers the detection limits compared with the full spectrum acquisition mode. Generally, tandem MS techniques provide higher selectivity than single ion monitoring techniques, because two or more characteristic fragments are used to characterize the analytes. Thus, GC-MS/MS has been used to identify and characterize polyphenols in food samples (Ferrer *et al.*, 2009).

Note that using SIM only makes sense for QMF and double-focusing instruments because the ion current for individual *m/z* values from a continuous beam that contains ions of any *m/z* value is measured. When GC-MS is used with TOF and QIT, all ions of any *m/z* values formed during a cycle are detected, and there is no advantage over SIM mode (Sparkman *et al.*, 2011).

Conversely, for accurate quantitation in GC-MS, the internal standard method is advisable because losses during the separation and concentration steps as well as in the amount of sample injected into the GC are corrected. The best internal standard is one chemically similar to the analytes and eluting in an empty zone in the chromatogram. In MS it is possible to select deuterated, ^{13}C-labeled or ^{15}N-labeled, analogs of the analyte that coelute with it but which can be distinguished from one another by their different *m/z* values (Mazur *et al.*, 1996; Morton *et al.*, 1999).

Most of the published studies using GC-MS for polyphenols in food samples focus on the identification and characterization of these compounds, although some also discuss the analytical characteristics of the methods used. Detection limits of between 0.0002–8 µg/ml and 0.2–30 ng/g have been reported, depending on the compound, the detection conditions, the sample analyzed as well as the sample treatment applied, among other factors.

Table 7.2 summarizes the fragmentation patterns of some polyphenols. As expected, the *m/z* values for both the target ions as well as the fragments formed are directly related with the derivatization reagent selected. Silylation of the hydroxyl groups generates additional distinct fragmentation patterns. In most cases, the molecular ion $[M]^+$ for the TMS derivatives of polyphenols is the main peak in the mass spectrum.

7.5 Sample preparation techniques

The first step of an analytical procedure involves preparation of an appropriate sample (Raynie, 2006). Optimization of sample preparation reduces the total analysis time and avoids potential error sources at the low concentrations required for ultra-trace analysis. The use of MS has permitted analyte detection using specific ions or transitions. However, it is important to develop appropriate extraction and sample preparation procedures for GC-MS and to establish the matrix effect on the detection system, the ionization efficiency, detector noise and detection and quantification limits (Calbiani *et al.*, 2004; Ng *et al.*, 2005).

The sample preparation step depends on the complexity of the food matrix. For liquid foods, such as wines, spirits, beer, and clear juices, only filtration and/or centrifugation of the sample is necessary for most applications, while the analysis of solid samples, such as fruits, requires successive steps that include crushing, pressing, freeze-drying or grinding, and further extraction.

The choice of the polyphenol extraction process is a very important step for achieving good recoveries. Most polyphenols can be extracted using organic solvents (ethanol, methanol, ether, acetone, ethyl acetate, hexane or different solvent mixtures), methanol being the most used. Soxhlet extraction is also used, but to a lesser extent than solvent extraction, to isolate phenolic compounds from solid samples (Canini *et al.*, 2007; de Rijke *et al.*, 2006; Dini *et al.*, 2004; Fiamegos *et al.*, 2004). Conversely, solid-phase extraction (SPE) (Esteruelas *et al.*, 2011; Frankel *et al.*, 1995; Liberatore *et al.*, 2001; Soleas *et al.*, 1997c) and preparative RP-LC have been used to fractionate as well as to remove unwanted matrix components such as sugars or lipids (Branco *et al.*, 2001; Castillo-Muñoz *et al.*, 2010; Lachman *et al.*, 2010). Elution using solvents of different pH values allows the separation of the phenolic compounds. Non-polar adsorbents, such as C8 and C18, or adsorbents based on polystyrene-divinylbenzene copolymers are commonly used because the polyphenol

Table 7.2 Fragmentation Patterns for Several Derivatized Polyphenols

Compound	Target Ion (*m/z*) and Qualifier Ions (*m/z*) in brackets	References
Stilbenoids		
Resveratrol	444 (445, 443), 444.7 (445.6, 446.7), 429 (444, 147), 228 (270, 312)	Kalogeropoulos *et al.*, 2007, 2009; Soleas *et al.*, 1997c; Minuti *et al.*, 2006; Montes *et al.*, 2010
Polydatin	361 (444, 372)	Soleas *et al.*, 1997c
Piceatannol	532 (516, 444), 532 (516, 575, 446)	Asensio-Ramos *et al.*, 2011; Shao *et al.*, 2003
Flavonoids		
Apigenin	486 (471, 73)	Isidorov *et al.*, 2009
Baicalein	326 (270, 168, 140)	Lin *et al.*, 1999
Catechin	368 (355, 369, 267), 368 (355, 474), 368 (355, 267), 369.5 (355.5, 368.5)	Esteruelas *et al.*, 2011; Kalogeropoulos *et al.*, 2007, 2009; Minuti *et al.*, 2006; Soleas *et al.*, 1997c
Cyanidin	356 (382)	Füzfai *et al.*, 2004
Chrysin	383 (384)	Kalogeropoulos *et al.*, 2009
Epicatechin	368 (355, 474), 368 (355, 267), 369.5 (355.5, 368.5), 368 (369, 370, 356, 357)	Kalogeropoulos *et al.*, 2007, 2009; Minuti *et al.*, 2006; Soleas *et al.*, 1997c; Zafra *et al.*, 2006
Fisetin	559 (471, 399), 471 (399, 559.8)	Minuti *et al.*, 2006; Soleas *et al.*, 1997c
Flavonone	224 (223, 147)	Minuti *et al.*, 2006
Genistein	473	Kalogeropoulos *et al.*, 2009
Hesperidin	Two isomers: 298 (517, 606) and 209 (445, 534)	Füzfai *et al.*, 2004
Isorhamnetin	604 (589, 73, 559)	Isidorov *et al.*, 2009
Kaempferol	559 (560)	Kalogeropoulos *et al.*, 2007, 2009
Morin	648 (649, 560)	Soleas *et al.*, 1997c
Myricetin	735 (647, 575), 649 (648, 647)	Kalogeropoulos *et al.*, 2009; Minuti *et al.*, 2006
Naringenin	488 (473, 73), 473 (296)	Isidorov *et al.*, 2009; Kalogeropoulos *et al.*, 2009
Naringin	Two isomers: 268 (486, 577) and 267 (413, 504)	Füzfai *et al.*, 2004

Continued

Table 7.2 Fragmentation Patterns for Several Derivatized Polyphenols—cont'd

Compound	Target Ion (*m/z*) and Qualifier Ions (*m/z*) in brackets	References
Quercetin	647 (559), 647 (559, 575), 648 (649, 559.8)	Minuti *et al.*, 2006; Kalogeropoulos *et al.*, 2007, 2009; Soleas *et al.*, 1997c
Rutin	Two isomers: 649 and 577 (649)	Füzfai *et al.*, 2004
Wogonin	284 (269)	Lin *et al.*, 1999
Phenolic Aldehydes		
Vanillin	224 (209, 194), 194, 194 (209)	Kalogeropoulos *et al.*, 2007, 2009; Minuti *et al.*, 2006; Ng *et al.*, 2000; Salta *et al.*, 2007; Zafra *et al.*, 2006
Coniferalde-hyde	220	Ng *et al.*, 2000
Syringalde-hyde	224	Ng *et al.*, 2000
Phenolic Alcohols		
Homovanillic alcohol	209 (179)	Kalogeropoulos *et al.*, 2007
Hydroxytyrosol	370.1 (267.1, 371.1), 267 (360), 370 (267)	Di Tommaso *et al.*, 1998; Kalogeropoulos *et al.*, 2007; Minuti *et al.*, 2006
Tyrosol	282.2 (179.1, 180.1), 179 (282, 267, 193), 282 (267), 179 (267, 282), 282 (267, 193, 179)	Di Tommaso *et al.*, 1998; Esteruelas *et al.*, 2011; Kalogeropoulos *et al.*, 2007, 2009; Minuti *et al.*, 2006; Zafra *et al.*, 2006
Phenolic Acids		
Caffeic acid	396.5 (381.5, 307.4), 396 (381, 73), 396 (381, 219, 191)	Esteruelas *et al.*, 2011; Isidorov *et al.*, 2009; Kalogeropoulos *et al.*, 2007, 2009; Soleas *et al.*, 1997c; Zafra *et al.*, 2006
Chlorogenic acid	345 (307, 324)	Kalogeropoulos *et al.*, 2007, 2009
Cinnamic acid	162 (131, 103, 77, 51), 205 (220), 220 (205, 161)	Fiamegos *et al.*, 2004; Kalogeropoulos *et al.*, 2007, 2009; Minuti *et al.*, 2006
o-Coumaric acid	308 (293, 219, 73), 293 (308, 147)	Isidorov *et al.*, 2009; Kalogeropoulos *et al.*, 2007, 2009

Table 7.2 Fragmentation Patterns for Several Derivatized Polyphenols—cont'd

Compound	Target Ion (*m/z*) and Qualifier Ions (*m/z*) in brackets	References
p-Coumaric acid	293 (308, 249, 219), 308 (293, 219), 308 (293), 249 (293, 308), 308 (293, 279, 219)	Esteruelas *et al.*, 2011; Kalogeropoulos *et al.*, 2007, 2009; Minuti *et al.*, 2006; Soleas *et al.*, 1997c; Zafra *et al.*, 2006
3,4-Dihy-droxyphenyl acetic acid	384 (267, 179), 384 (369)	Kalogeropoulos *et al.*, 2007, 2009; Minuti *et al.*, 2006;
Ferulic acid	338 (323, 308, 293), 222 (207, 191, 178, 164, 147, 133), 338 (249, 308)	Esteruelas *et al.*, 2011; Fiamegos *et al.*, 2004; Isidorov *et al.*, 2009; Kalogeropoulos *et al.*, 2007, 2009; Minuti *et al.*, 2006; Zafra *et al.*, 2006
Gallic acid	458 (459, 443, 444, 281), 282 (443.6, 460), 281 (458, 443), 281 (443, 458)	Esteruelas *et al.*, 2011; Kalogeropoulos *et al.*, 2007, 2009; Minuti *et al.*, 2006; Soleas *et al.*, 1997c; Zafra *et al.*, 2006
Gentisic acid	355.4 (356.5, 357.4)	Soleas *et al.*, 1997c
Homogentisic acid	384 (341)	Minuti *et al.*, 2006
p-Hydroxy-benzoic acid	166 (135, 107, 92, 77), 267 (223, 193), 282 (267, 223, 193)	Fiamegos *et al.*, 2004; Kalogeropoulos *et al.*, 2007, 2009; Zafra *et al.*, 2006
2-Hydroxycin-namic acid	191 (161, 179, 137, 131, 118)	Fiamegos *et al.*, 2004
4-Hydroxycin-namic acid	192 (161, 178, 133, 118)	Fiamegos *et al.*, 2004
3-Hydroxy-phenyl acetic acid	296 (281, 252, 164, 147)	Zafra *et al.*, 2006
p-Hydroxy-phenyl acetic acid	252 (296, 281)	Kalogeropoulos *et al.*, 2007, 2009
p-Hydroxy-phenyl propa-noic acid	192 (310)	Kalogeropoulos *et al.*, 2007
Homovanillic acid	210 (181, 169, 161), 326 (267, 311), 236 (267)	Fiamegos *et al.*, 2004; Kalogeropoulos *et al.*, 2007, 2009; Minuti *et al.*, 2006

Continued

Table 7.2 Fragmentation Patterns for Several Derivatized Polyphenols—cont'd

Compound	Target Ion (*m/z*) and Qualifier Ions (*m/z*) in brackets	References
Phloretic acid	192 (310)	Kalogeropoulos *et al.*, 2009
Protocat-echuic acid	370 (193, 355, 311), 193 (355, 370)	Esteruelas *et al.*, 2011; Kalogeropoulos *et al.*, 2007, 2009
Shikimic acid	204 (147, 357, 372)	Esteruelas *et al.*, 2011
Sinapinic acid	368 (353, 338), 368 (353, 338, 323)	Kalogeropoulos *et al.*, 2007, 2009; Minuti *et al.*, 2006; Zafra *et al.*, 2006
Syringic acid	342 (327, 312, 297, 253), 327 (342, 312, 297), 226 (195, 154, 125, 77)	Esteruelas *et al.*, 2011; Fiamegos *et al.*, 2004; Kalogeropoulos *et al.*, 2007, 2009; Minuti *et al.*, 2006; Zafra *et al.*, 2006
Vanillic acid	312 (297, 282, 267, 253, 223), 196 (181, 165, 137,125), 297.3 (253, 312.4)	Esteruelas *et al.*, 2011; Fiamegos *et al.*, 2004; Kalogeropoulos *et al.*, 2007; Minuti *et al.*, 2006; Soleas *et al.*, 1997c; Zafra *et al.*, 2006
Veratric acid	254 (239, 195)	Minuti *et al.*, 2006

structures contain hydrophobic groups (Kartsova and Alekseeva, 2008). Counter-current chromatography (CCC) has been applied as an alternative to conventional methods of fractionation and purification (Krafczyk and Glomb, 2008; Valls *et al.*, 2009). However, supercritical fluid extraction (SFE) has demonstrated some advantages over conventional extraction methods when chemically or thermally labile compounds are being analyzed. The control of solvent strength in terms of solvent pressure allows sequential extraction of phenolic compounds of increasing polarity. The use of a co-solvent is necessary (Robards, 2003). In spite of the development of new extraction techniques such as SFE, pressurized liquid extraction (PLE) or solid-phase microextraction (SPME), conventional extraction techniques continue to be the most widely used in the area of polyphenol determinations in food samples.

If intact glycosides are to be analyzed and/or characterized, sample preparation conditions should be mild enough to preserve them, avoiding harsh extraction conditions and heating and hydrolysis should be prevented. Moreover, any hydrolyzing enzymes contained in the plant material which may be released during sample treatment must be inactivated by adding a chemical such as tris(hydroxymethyl)amino-methane. In this case, the number of target analytes is higher and more selective and sensitive analytical methods are required (de Rijke *et al.*, 2006).

However, some polyphenols are bound to carbohydrates and proteins and it is necessary to carry out a hydrolysis step prior to extraction to cleave the ester linkage, especially if appropriate standards are not commercially available or if structural characterization of the phenolic glycoside is required (Robards and Antolovich, 1997). Hydrolysis of the ester to a carboxylic acid can be carried out by acidic hydrolysis, saponification or by enzymatic hydrolysis. To protect labile metabolites, preservatives such as ascorbic acid are sometimes added to the samples before analysis (Mazur *et al.*, 1996).

Acid hydrolysis is the traditional approach for analyzing aglycones from flavonoid glycosides and phenolic acids from phenolic acid esters. Acidic hydrolysis is performed using different conditions, reaction times and temperatures. Frequently, the food sample is treated with hydrochloric acid at reflux in aqueous medium or in the presence of methanol. Concentrations vary from 1 to 2 M HCl, and the reaction times range from 30 min to 1 h. Acid hydrolysis assisted by ultrasounds has been described for polyphenol analysis before extraction, which shortens the hydrolysis step from 16 h in conventional conditions to 1.5 h (Wang and Zuo, 2011).

Alkaline medium is used in the treatment of many plant-derived foods and the stability of phenolic compounds in these conditions needs to be studied. In addition, free phenols are released following alkaline hydrolysis from a wide range of conjugated forms in which many phenols exist in plants (Robards and Antolovich, 1997). Alkaline saponification is frequently applied using 1 to 4 M NaOH at room temperature for times ranging from 15 min to overnight, and the reactions can be performed in the dark, as well as under an inert atmosphere such as argon or nitrogen.

Enzymatic hydrolysis is generally proposed as a milder alternative to acid and alkaline hydrolysis, in which case enzymes such as glucuronidases (Lee *et al.*, 2004; Mazur *et al.*, 1996; Morton *et al.*, 1999), pectinases, cellulases (Bluck *et al.*, 2002), and amylases are used to degrade the carbohydrate linkages.

The thermally assisted hydrolysis of tannins has been proposed by Shadkami *et al.* (2009) who placed the sample in a hot metallic container which was purged and dried. The authors carried out the hydrolysis and derivatization steps simultaneously, so that no sample extraction is needed and any interaction with the solvent is minimal because it is purged in the early steps of the hydrolysis.

7.6 Miniaturized techniques for sample preparation in GC-MS

Environmental analytical studies and the consequent use of toxic reagents and solvents have increased to a point at which they have become unjustifiable if they are continued without taking into account environmentally friendly alternatives. Green Chemistry is the name given to chemical techniques and methodologies that reduce or eliminate the use or generation of products, solvents, reagents, etc., that are hazardous to human health or the environment. Long standing developments in both sample pretreatment and measurement methods have been incorporated into the new integrated approach to analytical chemistry (Armenta *et al.*, 2008).

Miniaturization is applied not only to sample preparation but also to the analytical instruments used in the final determination of analytes. Research into miniaturization is primarily driven by the need to reduce costs, which is achieved by reducing the consumption of reagents, decreasing analysis times and sample volumes, increasing separation efficiency, and enabling automation.

The miniaturization of chromatography started with the introduction of capillary columns into GC. This resulted in the reduced consumption of chemicals, improved separation, better sensitivity and smaller amounts of sample being required (Tobiszewski *et al.*, 2009). GC is an excellent method for the analysis of volatile and semi-volatile and non-polar compounds, and recent developments have led to substantially reduced analysis times and improved sensitivity and selectivity. The problem with GC analysis is the tedious sample pretreatment step involved. On-line coupling of extraction and clean-up steps with the separation system is thus an important goal in the development of sample preparation methods.

Most extraction techniques developed in the last ten years are related to miniaturization (Cuadros-Rodríguez *et al.*, 2005; Ramos *et al.*, 2005). The selective extraction of analytes is based on differences in their physical and chemical properties, such as molecular weight, charge, solubility, polarity or volatility. The usefulness of an extraction technique depends on different analytes and matrices (Mitra, 2003). Classical techniques for sample preparation based on wet digestion are time consuming and require great amounts of reagents, which are expensive, generate numerous residues and contaminate the sample (Buldini *et al.*, 2002).

Miniaturized sample preparation methods have been regarded as the most attractive techniques for the pretreatment of complex sample mixtures prior to the chromatographic process, especially in microscale separation systems. Effective on-line coupling of miniaturized sample preparation and microcolumn separation enables the user to take advantage of the combined system, which include: (i) high speed analysis with great efficiency; (ii) low cost operation due to extremely low or no solvent consumption; (iii) the development of environmentally friendly analytical procedures; and (iv) highly selective analysis made possible by developing tailored systems designed for particular applications (Saito and Jinno, 2003).

Conventional techniques, such as liquid–liquid extraction (LLE), solid–liquid extraction (SLE) and Soxhlet extraction, are still widely accepted and used for routine applications. However, in recent years, some of these techniques have been revisited and upgraded versions, in which their most pressing shortcomings have been solved, are now available. Studies in this field have also led to the development of new faster and more powerful and/or versatile extraction and preconcentration techniques. In many instances, partial and even full hyphenation and automation of the analytical processes, or, at least, of many of the treatment steps involved are now possible. In addition, sample preparation approaches that fulfill the goals of green analytical chemistry are also available. Concepts like miniaturization, integration and simplification became key concepts, which have already proven to effectively contribute to solve some of the drawbacks of conventional sample preparation methods (Ramos, 2012).

Microextraction techniques represent emerging techniques in the miniaturization of the analytical laboratory and can be considered as a challenge in the analytical field (Nerín *et al.*, 2009). These techniques, such as SPME, stir-bar sorptive extraction (SBSE), liquid phase microextraction (LPME), and on-line solid-phase extraction (on-line SPE), have several advantages over the traditional approaches of LLE and conventional SPE. The main advantages are the minimal consumption of harmful solvents and, typically, a high enrichment factor. The improved sensitivity makes it possible to minimize the amount of sample needed in the analysis. Although most microextraction techniques recover only a fraction of the analytes contained in a sample, in contrast to the exhaustive processes of conventional SPE and LLE, they have the advantage of being almost solvent-free and therefore, more sustainable and easily implemented. They also reduce exposure of the analyst to solvents and enable greater selectivity in sample preparation than the above exhaustive extraction approaches. Some of the microextraction techniques can be exploited in on-line combination with GC (Hyötyläinen and Riekkola, 2008).

Sorptive extraction techniques are based on the distribution equilibria between the sample matrix and a non-miscible liquid phase. Matrices are mostly aqueous and the non-miscible phase (e.g., polydimethylsiloxane, PDMS) is often coated onto a solid support. Analytes are extracted from the matrix into the non-miscible extracting phase. Unlike adsorption techniques (such as SPE), where the analytes are bound to active sites on the surface, the total volume of the extraction phase is important. Extraction of analytes depends on the partitioning coefficient of solutes between the phases.

SPME is a solvent-free sample preparation technique that uses a fused silica fiber coated with an appropriate stationary phase attached to a modified microsyringe (Arthur and Pawliszyn, 1990). SPME is essentially a two step process: firstly, the partitioning of analytes between the sample matrix, which can be a liquid sample or headspace (HS) vapor, and the fiber coating, and then the desorption of the (concentrated) extract from the fiber into the analytical instrument, usually a GC, where the sample components are thermally desorbed. The fiber can also be extracted (desorbed) into an LC eluent using a static or dynamic mode. Generally, SPME extraction of the analyte from the matrix is not an exhaustive extraction technique but an equilibrium technique. Although maximum sensitivity is obtained at the equilibrium point, it is not necessary to reach this point and the extractions can, instead, be performed for a defined period of time. The extraction temperature, time and sample stirring must be optimized for each application and operating conditions must be consistent. Another issue with SPME is the limited volume of stationary phase that can be bound to the fiber, which may lead to incomplete extraction and limit the sample enrichment. The main advantages of SPME extraction compared to solvent extraction are the reduction in solvent use, the combination of sampling and extraction into one step and the possibility of examining smaller sample sizes. SPME can be highly sensitive and be used for polar and non-polar analytes in a wide range of matrices with linking to both GC and LC.

SBSE was developed to overcome the limited extraction capacity of SPME fibers (Baltussen *et al.*, 1999). A glass stirrer bar is coated with a thick bonded absorbent layer (PDMS) to give a large stationary phase surface area, leading to a higher phase ratio and hence better recovery and sampling capacity. Transfer of the analyte from the bar is achieved either by GC thermal desorption, or elution with an LC solvent. As with SPME, the stir bar can also be used to sample the volatiles and semi-volatiles in the headspace above the sample. The major advantage of the stir bar technique is the high concentration factors that can be achieved. It can be used for liquid or semi-solid complex matrices and therefore has the potential to be used in many applications in food analysis (Ridgway *et al.*, 2007).

Recent research trends involve miniaturization of the traditional LLE principle by greatly reducing the acceptor-to-donor phase ratio. One of the emerging techniques in this area is LPME, where a microvolume of an organic solvent is used as acceptor solution. This methodology has proved to be an extremely simple, low-cost and virtually solvent-free sample-preparation technique, which provides a high degree of selectivity and enrichment by, additionally, eliminating the possibility of carry-over between runs. This is achieved by using either immiscible liquid phases (solvent microextraction) or a membrane to separate the acceptor-donor phases (membrane extraction) (Psillakis and Kalogerakis, 2003). LPME is normally performed between a small amount of a water-immiscible solvent and an aqueous phase containing the analytes of interest. The acceptor phase can be immersed directly in or suspended above the sample for HS extraction. The volume of the receiving phase is in the microliter or submicroliter range, so that, high enrichment factors can be obtained.

Despite the obvious differences betweeen SPME or SBSE and LPME concerning the nature of the extraction phases, both techniques are based on the equilibrium concept. In all these techniques, high extraction efficiencies are achieved because extraction occurs in a very small drop of liquid phase or a minimal amount of solid phase. Thus, the miniaturization of the analytical process permits high sample treatment efficiency. Additional advantages are portability, sustainability, and its low cost (Nerín *et al.*, 2009). Considering the importance of the development of clean chemical procedures, emerging methods for food matrices will be based on solvent-free procedures (Asensio-Ramos *et al.*, 2011).

Derivatization permits non-volatile polyphenols to be converted into volatile derivatives, thus improving separation, selectivity and sensitivity in the determination by direct GC or GC coupled to SPME (Pan and Pawliszyn, 1997). The derivatization step can be performed in the sample solution before extraction, on the SPME fiber once the analytes have been extracted or during the thermal desorption step in the injection port of the GC. The derivatizing agent can be adsorbed on the fiber coating before or after the extraction. Commonly, the analytes are adsorbed on the fiber and then exposed to the derivatizing agent vapor when a trimethylsilylation agent is used (Dietz *et al.*, 2006; Wang *et al.*, 2006).

Miniaturized techniques (Table 7.3) have been used to determine polyphenols. For example, SPME was used to determine *trans*-resveratrol using silylation with BSTFA and polyacrylate (PA) fibers (Cai *et al.*, 2009; Luan *et al.*, 2000; Shao *et al.*, 2003), phenolic acids with derivatization by ethyl and methyl chloroformate and a PA fiber (Citová *et al.*, 2006), *cis*- and *trans*-resveratrol isomers, piceatannol,

Table 7.3 Miniaturized Sample Preparation Techniques for Polyphenols Analysis by GC-MS

Compounds	Miniaturized Sample Preparation Technique	Characteristics	Food	Reference
trans-Resveratrol	SPME	Post-derivatization by silylation with BSTFA. PA fiber	Wine	Luan *et al.*, 2000
trans-Resveratrol	SPME	On-fiber derivatization. Both BSTFA and AA reagents were tested. PA fiber	Red wine	Shao *et al.*, 2003
trans-Resveratrol	SPME	On-fiber derivatization by silylation with BSTFA. PA fiber	Red wine	Cai *et al.*, 2009
Caffeic, ferulic, gallic, *p*-coumaric, protocatechuic, syringic and vanillic acids	SPME	Derivatization with ethyl and methyl chloroformate. PA fiber, 25°C for 50 min	—	Citová *et al.*, 2006
3 Phenylpropanoids	HS-SPME	DVB/CAR/PDMS fiber, room temperature for 20 min	*Chrysolina herbacea* frass and *Mentha species* leaves	Cordero *et al.*, 2012
Volatile phenols	HS-SPME	DVB/CAR/PDMS fiber	Beer	LeBlanc *et al.*, 2009
21 Polyphenols	LPME	Derivatization with the mixture BSTFA:Py. ME with ethyl acetate (2×0.5 ml)	Wastewater olive oil	Zafra *et al.*, 2006
Cinnamic, *o*-coumaric, caffeic and *p*-hydroxy-benzoic acids	SDME	In-syringe derivatization with BSA and ME with 2.5 µl hexyl acetate	Fruits and fruit juices	Viñas *et al.*, 2011
Cis- and *trans*-Resveratrol, piceatannol, catechin, epi-catechin, fisetin, quercetin	DSDME	In injection-port derivatization with BSTFA and ME with undecanone	Fruits, juices, functional foods, infusions	Viñas *et al.*, 2008

Continued

Table 7.3 Miniaturized Sample Preparation Techniques for Polyphenols Analysis by GC-MS—cont'd

Compounds	Miniaturized Sample Preparation Technique	Characteristics	Food	Reference
Resveratrol isomers	Comparison of SPME and SBSE	DI-SPME: polar CW/TPR 50 µm fiber. DI-SBSE: PDMS twister. Derivatization reaction in the presence of AA and K_2CO_3	Wines and fruit juices	Cacho *et al.*, 2013

CW/TPR, carbowax template resin; DI, direct immersion; DSDME, directly suspended droplet microextraction; HS, headspace; LPME, liquid phase microextraction; ME, microextraction; PA, polyacrylate; PDMS, polydimethylsiloxane; SBSE, stir-bar sorptive extraction; SDME, single-drop microextraction; SPME, solid-phase microextraction.

catechin, epicatechin, quercetin, and fisetin in grapes and wines using direct immersion (DI)-SPME, and then inserting the fiber in the headspace of a BSTFA solution (Viñas *et al.*, 2009), phenylpropanoids in aromatic plants (Cordero *et al.*, 2012), and flavonoids and flavones in bee pollen using a PA fiber (LeBlanc *et al.*, 2009).

Different LPME procedures have also been developed for determining 21 polyphenolic compounds, using derivatization with BSTFA-pyridine and extraction with ethyl acetate (Zafra *et al.*, 2006), phenolic acids by single-drop microextraction (SDME) with in-syringe derivatization with BSA and extraction in hexyl acetate (Saraji and Mousavinia, 2006), polyphenols in fruits, fruit juices, infusions and functional foods, based on the coupling of directly suspended droplet microextraction (DSDME) with GC-MS and derivatization in the injection port with BSTFA (Viñas *et al.*, 2011). The use of SPME and SBSE for determination of resveratrol isomers in wines and fruit juices using LC has been compared (Viñas *et al.*, 2008). Polyphenols have been determined in wines by SBSE-GC-MS (Cacho *et al.*, 2013).

References

Amzad Hossain, M., Salehuddin, S.M., Kabir, M.J., Rahman, S.M.M., Vasantha Rupasinghe, H.P., 2009. Sinensetin, rutin, 3′-hydroxy-5,6,7,4′-tetramethoxyflavone and rosmarinic acid contents and antioxidative effect of the skin of apple fruit. Food Chem. 113, 185–190.

Andreasen, M.F., Christensen, L.P., Meyer, A.S., Hansen, A., 2000. Content of phenolic acids and ferulic acid dehydrodimers in 17 rye (*Secale cereale* L.) varieties. J. Agric. Food Chem. 48, 2837–2842.

Angerosa, F., d'Alessandro, N., Konstantinou, P., Giacinto, L., 1995. GC-MS evaluation of phenolic compounds in virgin olive oil. J. Agric. Food Chem. 43, 1802–1807.

Angerosa, F., d'Alessandro, N., Corana, F., Mellerio, G., 1996. Characterization of phenolic and secoiridoid aglycons present in virgin olive oil by gas chromatography-chemical ionization mass spectrometry. J. Chromatogr. A 736, 195–203.

Anli, R.E., Vural, N., Kizilet, E., 2008. An alternative method for the determination of some of the antioxidant phenolics in varietal Turkish red wines. J. Instute. Brewing 114, 239–245.

Antoneli, A., Fabbri, C., Lercker, G., 1996. Techniques for resveratrol silylation. Chromatographia 42, 469–472.

Armenta, S., Garrigues, S., de la Guardia, M., 2008. Green analytical chemistry. Trends Anal. Chem. 27, 497–511.

Arthur, C.L., Pawliszyn, J., 1990. Solid phase–with thermal desorption using fused silica optical fibers. Anal. Chem. 62, 2145–2148.

Asensio-Ramos, M., Ravelo-Pérez, L.M., González-Curbelo, M.A., Hernández-Borges, J., 2011. Liquid phase microextraction applications in food analysis. J. Chromatogr. A 1218, 7415–7437.

Ayaz, F.A., Kadioglu, A., 1997. Changes in phenolic acid contents of *Diospyros lotus* L. during fruit development. J. Agric. Food Chem. 45, 2539–2541.

Ayaz, F.A., Kadioglu, A., Reunanen, M., Var, M., 1997. Phenolic acid and fatty acid composition in the fruits of *Laurocerasus officinalis* Roem. and its cultivars. J. Food Composition Anal. 10, 350–357.

Baltussen, E., Sandra, P., David, F., Cramers, C., 1999. Stir bar sorptive extraction (SBSE), a novel extraction technique for aqueous samples: Theory and principles. J. Microcol. Sep. 11, 737–747.

Baranowski, J.D., Nagel, C.W., 1981. Isolation and identification of the hydroxycinnamic acid derivatives in white riesling wine. Am. J. Enology Viticulture 32, 5–13.

Bauer, S., Schulte, E., Thier, H.-P., 2004. Composition of the surface wax from tomatoes. Eur. Food Res. Technol. 219, 223–228.

Bluck, L.J.C., Jones, K.S., Thomas, J., Liggins, J., Harding, M., Bingham, S.A., et al., 2002. Quantitative analysis using gas chromatography/combustion/isotope ratio mass spectrometry and standard addition of intrinsically labeled standards (SAIL)-application to isoflavones in foods. Rapid Comm. Mass Spectrom. 16, 2249–2254.

Bovy, A., Schijlen, E., Hall, R.D., 2007. Metabolic engineering of flavonoids in tomato (*Solanum lycopersicum*): the potential for metabolomics. Metabolomics 3, 399–412.

Branco, A., dos Santos Pereira, A., Cardoso, J.N., de Aquino Neto, F.R., Pinto, A.C., Braz-Filho, R., 2001. Further lipophilic flavonols in *Vellozia graminifolia* (Velloziaceae) by high temperature gas chromatography: quick detection of new compounds. Phytochem. Anal. 12, 266–270.

Brenes, M., García, A., García, P., Rios, J.J., Garrido, A., 1999. Phenolic compounds in Spanish olive oils. J. Agric. Food Chem. 47, 3535–3540.

Brenes-Balbuena, M., García-García, P., Garrido-Fernández, A., 1992. Phenolic compounds related to the black color formed during the processing of ripe olives. J. Agric. Food Chem. 40, 1192–1196.

Buldini, P.L., Riccib, L., Sharma, J.L., 2002. Recent applications of sample preparation techniques in food analysis. J. Chromatogr. A 975, 47–70.

Bylak, W., Matlawska, I., 2001. Flavonoids and free phenolic acids from *Phytolacca americana* L. leaves. Acta. Pol. Pharm. 58, 69–72.

Cacho, J.I., Campillo, N., Viñas, P., Hernández-Córdoba, M., 2013. Stir bar sorptive extraction coupled to Gas Chromatography-Mass Spectrometry for the determination of *cis*- and *trans*-resveratrol, oxyresveratrol and piceatannol in wine samples. J. Chromatogr. A. 1315(8), 21–27 http://dx.doi.org/10.1016/j.chroma.2013.09.045.

Cai, L., Koziel, J.A., Dharmadhikari, M., van Leewen, J.H., 2009. Rapid determination of *trans*-resveratrol in red wine by solid-phase microextraction with on-fiber derivatization and multidimensional Gas Chromatography-Mass Spectrometry. J. Chromatogr. A 1216, 281–287.

Calbiani, F., Careri, M., Elviri, L., Mangia, A., Pistara, L., Zagnoni, I., 2004. Development and in-house validation of a liquid chromatography–electrospray–tandem mass spectrometry method for the simultaneous determination of Sudan I, Sudan II, Sudan III and Sudan IV in hot chilli products. J. Chromatogr. A 1042, 123–130.

Caligiani, A., Cirlini, M., Palla, G., Ravaglia, R., Arlorio, M., 2007. GC-MS detection of chiral markers in cocoa beans of different quality and geographic origin. Chirality 19, 329–334.

Campo Fernández, M., Cuesta-Rubio, O., Rosado Pérez, A., Montes De Oca Porto, R., Márquez Hernández, I., Lisa Piccinelli, A., et al., 2008. GC-MS determination of isoflavonoids in seven red cuban propolis samples. J. Agric. Food Chem. 56, 9927–9932.

Canini, A., Alesiani, D., D'Arcangelo, G., Tagliatesta, P., 2007. Gas chromatography-mass spectrometry analysis of phenolic compounds from *Carica papaya* L. leaf. J. Food Composition Anal. 20, 584–590.

Carrasco-Pancorbo, A., Cerretani, L., Bendini, A., Segura-Carretero, A., Gallina-Toschi, T., Fernández-Gutiérrez, A., 2005. Analytical determination of polyphenols in olive oils. J. Sep. Sci. 28, 837–858.

Castillo-Muñoz, N., Winterhalter, P., Weber, F., Gómez, M.V., Gómez-Alonso, S., García-Romero, E., et al., 2010. Structure elucidation of peonidin 3,7-*O*-β-diglucoside isolated from Garnacha Tintorera (*Vitis vinifera* L.) grapes. J. Agric. Food Chem. 58, 11105–11111.

Cheng, H., Zuo, Y., 2007. Identification of flavonol glycosides in American cranberry fruit. Food Chem. 101, 1357–1364.

Chiou, A., Karathanos, V.T., Mylona, A., Salta, F.N., Preventi, F., Andrikopoulos, N.K., 2007. Currants (*Vitis vinifera* L.) content of simple phenolics and antioxidant activity. Food Chem. 102, 516–522.

Christov, R., Bankova, V., 1992. Gas chromatographic analysis of underivatized phenolic constituents from propolis using an electron-capture detector. J. Chromatogr. A 623, 182–185.

Chu, T.Y., Chang, C.H., Liao, Y.C., Chen, Y., 2001. Microwave accelerated derivatization processes for the determination of phenolic acids by gas chromatography-mass spectrometry. Talanta 54, 1163–1171.

Citová, I., Sladkovský, R., Solich, P., 2006. Analysis of phenolic acids as chloroformate derivatives using solid phase microextraction-gas chromatography. Anal. Chim. Acta. 573, 231–241.

Comeskey, D.J., Montefiori, M., Edwards, P.J.B., McGhie, T.K., 2009. Isolation and structural identification of the anthocyanin components of red kiwifruit. J. Agric. Food Chem. 57, 2035–2039.

Cordero, C., Zebelo, S.A., Gnavi, G., Griglione, A., Bicchi, C., Maffei, M.E., et al., 2012. HS-SPME–GC×GC-qMS volatile metabolite profiling of Chrysolina herbacea frass and Mentha spp. Leaves. Anal. Bioanal. Chem. 402, 1941–1952.

Croteau, R., Kutchan, T.M., Lewis, N.G., 2000. Natural products (secondary metabolites). In: Buchanan, B., Gruissem, W., Jones, R. (Eds.), Biochemistry & Molecular Biology of Plants. American Society of Plant Physiologists, Rockville, MD, pp. 1250–1318.

Cuadros-Rodríguez, L., Almansa-López, E.M., García-Campaña, A.M., González-Casado, A., Egea-González, F.J., Garrido-Frenich, A., et al., 2005. Setting up of recovery profiles: A tool to perform the compliance with recovery requirements for residue analysis. Talanta 66, 1063–1072.

Damak, N., Bouaziz, M., Ayadi, M., Sayadi, S., Damak, M., 2008. Effect of the maturation process on the phenolic fractions, fatty acids, and antioxidant activity of the Chetoui olive fruit cultivar. J. Agric. Food Chem. 56, 1560–1566.

de Rijke, E., Out, P., Niessen, W.M.A., Ariese, F., Gooijer, C., Brinkman, U.A.Th., 2006. Analytical separation and detection methods for flavonoids. J. Chromatogr. A 1112, 31–63.

Deng, F., William Zito, S., 2003. Development and validation of a gas chromatographic-mass spectrometric method for simultaneous identification and quantification of marker compounds including bilobalide, ginkgolides and flavonoids in *Ginkgo biloba* L. extract and pharmaceutical preparations. J. Chromatogr. A 986, 121–127.

Di Tommaso, D., Calabrese, R., Rotilio, D., 1998. Identification and quantitation of hydroxytyrosol in Italian wines. J. High Resolut. Chromatogr. A 21, 549–553.

Dietz, C., Sanz, J., Cámara, C., 2006. Recent developments in solid-phase microextraction coatings and related techniques. J. Chromatogr. A 1103, 183–192.

Dini, I., Tenore, G.C., Dini, A., 2004. Phenolic constituents of *Kancolla* seeds. Food Chem. 84, 163–168.

Diplock, A.T., Aggett, P.J., Ashwell, M., Bornet, F.F., Fern, E.B., Roberfroid, M.B., 1999. Scientific concepts of functional foods in Europe consensus document. Br. J. Nutr. 81, S1–S27.

dos Santos Pereira, A., Norsell, M., Cardoso, J.N., de Aquino Neto, F.R., 2000. Rapid screening of polar compounds in Brazilian propolis by high-temperature high-resolution gas chromatography-mass spectrometry. J. Agric. Food Chem. 48, 5226–5230.

dos Santos Pereira, A., Costa Padilha, M., de Aquino Neto, F.R., 2004. Two decades of high temperature gas chromatography (1983–2003): what's next? Microchem. J. 77, 141–149.

Dugo, G., Saitta, M., Salvo, F., Ragusa, M., Manzo, G., Bambara, G., 2001. Determinazione di composti fenolici in vini rossi siciliani mediante un sistema HRGC-MS/MS. Vignevini 28, 81–84.

El Gharras, H., 2009. Polyphenols: food sources, properties and applications—a review. Int. J. Food Sci. Technol. 44, 2512–2518.

Ellnain-Wojtaszek, M., 1997a. Phenolic acids from *Gingko biloba* L. Part I. Qualitative analysis of free and liberated by hydrolysis phenolic acids. Acta. Pol. Pharm. 54, 225–228.

Ellnain-Wojtaszek, M., 1997b. Phenolic acids from *Gingko biloba* L. Part II. Qualitative analysis of free and liberated by hydrolysis of phenolic acids. Acta. Pol. Pharma. 54, 229–232.

Escarpa, A., González, M.C., 2001. An overview of analytical chemistry of phenolic compounds in foods. Crit. Rev. Anal. Chem. 31, 57–139.

Esteruelas, M., Kontoudakis, N., Gil, M., Fort, M.F., Canals, J.M., Zamora, F., 2011. Phenolic compounds present in natural haze protein of Sauvignon white wine. Food Res. Int. 44, 77–83.

Fan, E., Lin, S., Du, D., Jia, Y., Kang, L., Zhang, K., 2011. Current separative strategies used for resveratrol determination from natural sources. Anal. Methods 3, 2454–2462.

Fang, Z., Bhandari, B., 2010. Encapsulation of polyphenols—a review. Trends Food Sci. Technol. 21, 510–523.

Farag, M.A., Huhman, D.V., Lei, Z., Sumner, L.W., 2007. Metabolic profiling and systematic identification of flavonoids and isoflavonoids in roots and cell suspension cultures of *Medicago truncatula* using HPLC-UV-ESI-MS and GC-MS. Phytochemistry 68, 342–354.

Ferrer, I., Barber, L.B., Thurman, E.M., 2009. Gas chromatographic-mass spectrometric fragmentation study of phytoestrogens as their trimethylsilyl derivatives: Identification in soy milk and wastewater samples. J. Chromatogr. A 1216, 6024–6032.

Fiamegos, Y.C., Nanos, C.G., Vervoort, J., Stalikas, C.D., 2004. Analytical procedure for the in-vial derivatization—extraction of phenolic acids and flavonoids in methanolic and aqueous plant extracts followed by gas chromatography with mass-selective detection. J. Chromatogr. A 1041, 11–18.

Flamini, R., 2003. Mass spectrometry in grape and wine chemistry. Part I: Polyphenols. Mass Spectrom. Rev. 22, 218–250.

Flamini, R., Vedova, A.D., 2004. Fast determination of the total free resveratrol content in wine by direct-exposure-probe, positive-ion chemical ionization and collision-induced-dissociation mass spectrometry. Rapid Comm. Mass Spectrom. 18, 1925–1931.

Frankel, E.N., Waterhouse, A.L., Teissedre, P.L., 1995. Principal phenolic phytochemicals in selected California wines and their antioxidant activity in inhibiting oxidation of human low-density lipoproteins. J. Agric. Food Chem. 43, 890–894.

Füzfai, Zs., Molnár-Perl, I., 2007. Gas chromatographic-mass spectrometric fragmentation study of flavonoids as their trimethylsilyl derivatives: analysis of flavonoids, sugars, carboxylic and amino acids in model system and in citrus fruits. J. Chromatogr. A 1149, 88–101.

Füzfai, Zs., Kovács, E., Molnár-Perl, I., 2004. Identification and quantitation of the main constituents of sour cherries: simultaneous, as their trimethylsilyl derivatives, by gas chromatography-mass spectrometry. Chromatographia 60, S143–S151.

Gao, X., Williams, S.J., Woodman, O.L., Marriott, P.J., 2010. Comprehensive two-dimensional gas chromatography, retention indices and time-of-flight mass spectra of flavonoids and chalcones. J. Chromatogr. A 1217, 8317–8326.

Gil, M.I., Tomás-Barberán, F.A., Hess-Pierce, B., Holcroft, D.M., Kader, A.A., 2000. Antioxidant activity of pomegranate juice and its relationship with phenolic composition and processing. J. Agric. Food Chem. 48, 4581–4589.

Goldberg, D.M., Yan, J., Ng, E., Diamandis, E.P., Karumanchiri, A., Soleas, G., et al., 1994. Direct-injection gas-chromatographic mass-spectrometric assay for *trans*-resveratrol. Anal. Chem. 66, 3959–3963.

Goldberg, D.M., Hahn, S.E., Parkes, J.G., 1995a. Beyond alcohol: beverage consumption and cardiovascular mortality. Clin. Chim. Acta. 237, 155–187.

Goldberg, D.M., Karumanchiri, A., Ng, E., Yan, J., Diamandis, E.P., Soleas, G.J., 1995b. Direct gas-chromatographic mass-spectrometric method to assay *cis*-resveratrol in wines—preliminary survey of its concentration in commercial wines. J. Agric. Food Chem. 43, 1245–1250.

Gómez-Caravaca, A.M., Gómez-Romero, M., Arráez-Román, D., Segura-Carretero, A., Fernández-Gutiérrez, A., 2006. Advances in the analysis of phenolic compounds in products derived from bees. J. Pharm. Biomed. Anal. 41, 1220–1234.

Hartley, R.D., 1971. Improved methods for the estimation by gas-liquid chromatography of lignin degradation products from plants. J. Chromatogr. 54, 335–344.

Heimler, D., Pieroni, A., 1994. Capillary gas chromatography of plant tissues and soil phenolic acids. Chromatographia 38, 475–478.

Herrero, M., Simó, C., García-Cañas, V., Ibáñez, E., Cifuentes, A., 2012. Foodomics: MS-based strategies in modern food science and nutrition. Mass. Spectrom. Rev. 31, 49–69.

Herrmann, K., 1989. Occurrence and content of hydroxycinnamic and hydroxybenzoic acid compounds in foods. Crit. Rev. Food Sci. Nutr. 28, 315–347.

Husek, P., 1992. Fast derivatization and GC analysis of phenolic acids. Chromatographia 34, 621–626.

Hyötyläinen, T., Riekkola, M.L., 2008. Sorbent- and liquid-phase microextraction techniques and membrane-assisted extraction in combination with gas chromatographic analysis: A review. Anal. Chim. Acta. 614, 27–37.

Ibáñez, C., Valdés, A., García-Cañas, V., Simó, C., Celebier, M., Rocamora-Reverte, L., et al., 2012. Global Foodomics strategy to investigate the health benefits of dietary constituents. J. Chromatogr. A 1248, 139–153.

Isidorov, V.A., Isidorova, A.G., Sczczepaniak, L., Czyzewska, U., 2009. Gas chromatographic-mass spectrometric investigation of the chemical composition of beebread. Food Chem. 115, 1056–1063.

Jang, M.S., Cai, E.N., Udeani, G.O., 1997. Cancer chemopreventive activity of resveratrol, a natural product derived from grapes. Science 275, 218–220.

Jeandet, P., Bessis, R., Maume, B.F., Sbaghi, M., 1993. Analysis of resveratrol in Burgundy wines. J. Wine Res. 4, 79–85.

Jeandet, P., Bessis, R., Maume, B.F., Meunier, P., Peyron, D., Trollat, P., 1995. Effect of enological practices on the resveratrol isomer content of wine. J. Agric. Food Chem. 43, 316–319.

Juárez, M., Olano, A., Morais, F., 2005. Alimentos Funcionales. Fundación Española para la Ciencia y la Tecnología, Madrid.

Kakasy, A., Füzfai, Zs., Kursinszki, L., Molnár-Perl, I., Lemberkovics, É., 2006. Analysis of non-volatile constituents in *Dracocephalum* species by HPLC and GC-MS. Chromatographia 63, S17–S22.

Kalinova, J., Triska, J., Vrchotova, N., 2006. Distribution of vitamin E, squalene, epicatechin, and rutin in common buckwheat plants (*Fagopyrum esculentum* Moench). J. Agric. Food Chem. 54, 5330–5335.

Kalogeropoulos, N., Chiou, A., Mylona, A., Ioannou, M.S., Andrikopoulos, N.D., 2007. Recovery and distribution of natural antioxidants (alpha-tocopherol, polyphenols and terpenic acids) after pan-frying of Mediterranean finfish in virgin olive oil. Food Chem. 100, 509–517.

Kalogeropoulos, N., Konteles, S.J., Troullidou, E., Mourtzinos, I., Karathanos, V.T., 2009. Chemical composition, antioxidant activity and antimicrobial properties of propolis extracts from Greece and Cyprus. Food Chem. 116, 452–461.

Kalogeropoulos, N., Chiou, A., Ioannou, M., Karathanos, V.T., Hassapidou, M., Andrikopoulos, N.K., 2010a. Nutritional evaluation and bioactive microconstituents (phytosterols, tocophenols, polyphenols, triterpenic acids) in cooked dry legumes usually consumed in the Mediterranean countries. Food Chem. 121, 682–690.

Kalogeropoulos, N., Yannakopoulou, K., Gioxari, A., Chiou, A., Makris, D.P., 2010b. Polyphenol characterization and encapsulation in β-cyclodextrin of a flavonoid-rich *Hypericum perforatum* (St. John's wort) extract. LWT-Food Sci. Technol. 43, 882–889.

Karagiannis, S., Economou, A., Lanaridis, P., 2000. Phenolic and volatile composition of wines made from *Vitis vinifera* Cv. muscat lefko grapes from the Island of Samos. J. Agric. Food Chem. 48, 5369–5375.

Kartsova, L.A., Alekseeva, A.V., 2008. Chromatographic and electrophoretic methods for determining polyphenol compounds. J. Anal. Chem. 63, 1024–1033.

Klick, S., Herrmann, K., 1988. Glucosides and glucose esters of hydroxybenzoic acids in plants. Phytochemistry 27, 2177–2180.

Krafczyk, N., Glomb, M.A., 2008. Characterization of phenolic compounds in rooibos tea. J. Agric. Food Chem. 56, 3368–3376.

Krygier, K., Sosulski, F., Hogge, L., 1982. Free, esterified, and insoluble bound phenolic acids. 1. Extraction and purification procedure. J. Agric. Food Chem. 30, 330–334.

Lachman, J., Hejtmánková, A., Sykora, J., Karban, J., Orsák, M., Rygerová, B., 2010. Contents of major phenolic and flavonoid antioxidants in selected Czech honey. Czech J. Food Sci. 28, 412–426.

Leal, P.F., Maia, N.B., Carmello, Q.A.C., Catharino, R.R., Eberlin, M.N., Meireles, M.A.A., 2008. Sweet Basil (*Ocimum basilicum*) extracts obtained by supercritical fluid extraction (SFE): global yields, chemical composition, antioxidant activity, and estimation of the cost of manufacturing. Food Bioprocess Technol. 1, 326–338.

LeBlanc, B.W., Davis, O.K., Boue, S., DeLucca, A., Deeby, T., 2009. Antioxidant activity of Sonoran Desert bee pollen. Food Chem. 115, 1299–1305.

Lee, S.H., Jung, B.H., Kim, S.Y., Chung, B.C., 2004. Determination of phytoestrogens in traditional medicinal herbs using gas chromatography-mass spectrometry. J. Nutr. Biochem. 15, 452–460.

Leung, J., Fenton, T.W., Clandinin, D.R., 1981. Phenolic components of sunflower flour. J. Food. Sci. 46, 1386–1388.

Liberatore, L., Procida, G., d'Alessandro, N., Cichelli, A., 2001. Solid-phase extraction and gas chromatographic analysis of phenolic compounds in virgin olive oil. Food Chem. 73, 119–124.

Liggins, J., Bluck, L.J.C., Coward, W.A., Bingham, S.A., 1998. Extraction and quantification of daidzein and genistein in food. Anal. Biochem. 264, 1–7.

Liggins, J., Bluck, L.J.C., Runswick, S., Atkinson, C., Coward, W.A. Bingham, S.A., 2000. Daidzein and genistein content of fruits and nuts. J. Nutr. Biochem. 11, 326–331.

Lin, L.L., Lien, C.Y., Cheng, Y.C., Ku, K.L., 2007. An effective sample preparation approach for screening the anticancer compound piceatannol using HPLC coupled with UV and fluorescence detection. J. Chromatogr. A. B 853, 175–182.

Lin, M.C., Tsai, M.J., Wen, K.C., 1999. Supercritical fluid extraction of flavonoids from Scutellariae Radix. J. Chromatogr. A 830, 387–395.

Luan, T., Li, G., Zhang, Z., 2000. Gas-phase postderivatization following solid-phase microextraction for rapid determination of *trans*-resveratrol in wine by gas chromatography-mass spectrometry. Anal. Chim. Acta. 424, 19–25.

Macheix, J.J., Fleuriet, A., Billot, J., 1990. Fruit Phenolics. CRC Press, Florida.

Maciejewicz, W., Daniewski, M., Bal, K., Markowski, W., 2001. GC-MS Identification of the flavonoid aglycones isolated from propolis. Chromatographia 53, 343–346.

Maga, J.A., 1978. Simple phenol and phenolic compounds in food flavor. Crit. Rev. Food Sci. Nutr. 10, 323–372.

Mallouchos, A., Lagos, G., Komaitis, M., 2007. A rapid microwave-assisted derivatization process for the determination of phenolic acids in brewer's spent grains. Food Chem. 102, 606–611.

Manach, C., Scalbert, A., Morand, C., Rémésy, C., Jiménez, L., 2004. Polyphenols: food sources and bioavailability. Am. J. Clin. Nutr. 79, 727–747.

Márquez-Hernández, I., Cuesta-Rubio, O., Campo-Fernández, M., Rosado Pérez, A., Montes de Oca Porto, R., Piccinelli, A.L., et al., 2010. Studies on the constituents of yellow cuban propolis: GC-MS determination of triterpenoids and flavonoids. J. Agric. Food Chem. 58, 4725–4730.

Marsilio, V., Campestre, C., Lanza, B., 2001. Phenolic compounds change during California-style ripe olive processing. Food Chem. 74, 55–60.

Martin, K.R., Krueger, C.G., Rodríguez, G., Dreher, M., Reed, J., 2009. Development of a novel pomegranate standard and new method for the quantitative measurement of pomegranate polyphenols. J. Sci. Food. Agric. 89, 157–162.

Mazur, W., Fotsis, T., Wahala, K., Ojala, S., Salakka, A., Adlercreutz, H., 1996. Isotope dilution gas chromatographic-mass spectrometric method for the determination of isoflavonoids, coumestrol, and lignans in food samples. Anal. Biochem. 233, 169–180.

Medina-Bolivar, F., Condori, J., Rimando, A.M., Hubstenberger, J., Shelton, K., O'Keefe, S.F., et al., 2007. Production and secretion of resveratrol in hairy root cultures of peanut. Phytochemistry 68, 1992–2003.

Minuti, L., Pellegrino, R.M., Tesei, I., 2006. Simple extraction method and gas chromatography-mass spectrometry in the selective ion monitoring mode for the determination of phenols in wine. J. Chromatogr. A 1114, 263–268.

Mitra, S., 2003. Sample Preparation Techniques in Analytical Chemistry. Wiley-Interscience, New Jersey.

Montes, R., García-López, M., Rodríguez, I., Cela, R., 2010. Mixed-mode solid-phase extraction followed by acetylation and gas chromatography mass spectrometry for the reliable determination of *trans*-resveratrol in wine samples. Anal. Chim. Acta. 673, 47–53.

Morton, M., Arisaka, O., Miyake, A., Evans, B., 1999. Analysis of phyto-oestrogens by gas chromatography-mass spectrometry. Environ. Toxicol. Pharmacol. 7, 221–225.

Naim, M., Zehavi, U., Nagy, S., Rouseff, R.L., 1992. Hydroxycinnamic acids as off-flavor precursors in citrus fruits and their products. In: Phenolic Compounds in Food and Their Effects on Health. American Chemical Society, Washington, DC, pp. 180–191.

Nerín, C., Salafranca, J., Aznar, M., Batlle, R., 2009. Critical review on recent developments in solventless techniques for extraction of analytes. Anal. Bioanal. Chem. 393, 809–833.

Ng, L.K., Lafontaine, P., Harnois, J., 2000. Gas chromatographic-mass spectrometric analysis of acids and phenols in distilled alcohol beverages. Application of anion-exchange disk extraction combined with in-vial elution and silylation. J. Chromatogr. A 873, 29–38.

Ng, C.K., Tanaka, N., Kim, M., Yap, S.L., 2005. A rapid method of Sudan red I, II, III & IV in tomato sauce using UPLC MS/MS. Waters application note. Waters, Milford.

Niessen, W.M.A., 2004. Hyphenated techniques, Applications of in Mass Spectrometry. Encyclopedia of Spectroscopy and Spectrometry, Mass Spectrometry Applications, 843–849.

Obmann, A., Werner, I., Presser, A., Zehl, M., Swoboda, Z., Purevsuren, S., et al., 2011. Flavonoid C- and O-glycosides from Mongolian medicinal plant *Dianthus versicolor* Fisch. Carbohydr. Res. 346, 1868–1875.

Orr, J.D., Sumner, L.W., Edwards, R., Dixon, R.A., 1993. Determination of cinnamic acid and 4-coumaric acid in alfalfa (*Medicagosativa L*) cell-suspension cultures by gas-chromatography. Phytochem. Anal. 4, 124–130.

Ougo, G.M., La Pera, L., Di Bella, G., Lo, V., Pollicino, D., La Torre, G.L., et al., 2009. Sicilian virgin olive oils and red wines: a potentially rich source of antioxidant compounds in the Mediterranean diet. Rivista Italiana delle Sostanze Grasse 86, 163–172.

Owen, R.W., Haubner, R., Hull, W.E., Erben, G., Spiegelhalder, B., Bartsch, H., et al., 2003. Isolation and structure elucidation of the major individual polyphenols in carob fibre. Food. Chem. Toxicol. 41, 1727–1738.

Packert, M., Steinhart, H., 1995. Separation and identification of some monomeric and dimeric phenolic acids by a simple gas chromatographic method using a capillary column and FID-MSD. J. Chromatogr. Sci. 33, 631–639.

Pan, L., Pawliszyn, J., 1997. Derivatization/solid-phase microextraction: new approach to polar analytes. Anal. Chem. 69, 196–205.

Papadopoulos, G., Boskou, D., 1991. Antioxidant effect of natural phenols in olive oil. J. Am. Oil. Chem. Soc. 68, 669–671.

Pejin, J., Grujic, O., Canadanovic-Brunet, J., Vujic, D., Tumbas, V., 2009. Investigation of phenolic acids content and antioxidant activity in malt production. J. Am. Soc. Brewing Chemists 67, 81–88.

Peleg, H., Naim, M., Rouseff, R.L., Zehavi, U., 1991. Distribution of bound and free phenolic acids in oranges (*Citrus sinensis*) and grapefruits (*Citrus paradisi*). J. Sci. Food Agric. 57, 417–426.

Pérez-Jiménez, J., Neveu, V., Vos, F., Scalbert, A., 2010. Systematic analysis of the content of 502 polyphenols in 452 foods and beverages: an application of the phenol-explorer database. J. Agric. Food. Chem. 58, 4959–4969.

Proestos, C., Boziaris, I.S., Nychas, G.-J.E., Komaitis, M., 2006a. Analysis of flavonoids and phenolic acids in Greek aromatic plants: Investigation of their antioxidant capacity and antimicrobial activity. Food Chem. 95, 664–671.

Proestos, C., Sereli, D., Komaitis, M., 2006b. Determination of phenolic compounds in aromatic plants by RP-HPLC and GC-MS. Food Chem. 95, 44–52.

Prytzyk, E., Dantas, A.P., Salomão, K., Pereira, A.S., Bankova, V.S., De Castro, S.L., et al., 2003. Flavonoids and trypanocidal activity of Bulgarian propolis. J. Ethnopharmacol. 88, 189–193.

Psillakis, E., Kalogerakis, N., 2003. Developments in liquid-phase microextraction. Trends Anal. Chem. 22, 565–574.

Ragab, A.S., van Fleet, J., Jankowski, B., Park, J.H., Bobzin, S.C., 2006. Detection and quantitation of resveratrol in tomato fruit (*Lycopersicon esculentum* Mill.). J. Agric. Food Chem. 54, 7175–7179.

Ramos, L., 2012. Critical overview of selected contemporary sample preparation techniques. J. Chromatogr. A 1221, 84–98.

Ramos, L., Ramos, J.J., Brinkman, U.A.Th., 2005. Miniaturization in sample treatment for environmental analysis. Anal. Bioanal. Chem. 381, 119–140.

Raynie, D.E., 2006. Modern extraction techniques. Anal. Chem. 78, 3997–4004.

Ridgway, K., Lalljie, S.P.D., Smith, R.M., 2007. Sample preparation techniques for the determination of trace residues and contaminants in foods. J. Chromatogr. A 1153, 36–53.

Rimando, A.M., Kalt, W., Magee, J.B., Dewey, J., Ballington, J.R., 2004. Resveratrol, pterostilbene, and piceatannol in *Vaccinium* berries. J. Agric. Food Chem. 52, 4713–4719.

Ríos, J.J., Gil, M.J., Gutiérrez-Rosales, F., 2005. Solid-phase extraction gas chromatography-ion trap-mass spectrometry qualitative method for evaluation of phenolic compounds in virgin olive oil and structural confirmation of oleuropein and ligstroside aglycons and their oxidation products. J. Chromatogr. A 1093, 167–176.

Robards, K., 2003. Strategies for the determination of bioactive phenols in plants, fruit and vegetables. J. Chromatogr. A 1000, 657–691.

Robards, K., Antolovich, M., 1997. Analytical chemistry of fruit bioflavonoids. A review. Analyst 122, 11R–34R.

Robards, K., Prenzler, P.D., Tucker, G., Swatsitiang, P., Glover, W., 1999. Phenolic compounds and their role in oxidative processes in fruits. Food Chem. 66, 400–436.

Rojas, L.B., Quideau, S., Pardon, P., Charrouf, Z., 2005. Colorimetric evaluation of phenolic content and GC-MS characterization of phenolic composition of alimentary and cosmetic argan oil and press cake. J. Agric. Food Chem. 53, 9122–9127.

Romani, A., Pinelli, P., Mulinacci, N., Galardi, C., Vincieri, F.F., Liberatore, L., et al., 2001. HPLC and HRGC analyses of polyphenols and secoiridoid in olive oil. Chromatographia 53, 279–284.

Rosenfeld, J., 2003. Derivatization in the current practice of analytical chemistry. Trends Anal. Chem. 11, 785–798.

Rudell, D.R., Mattheis, J.P., Curry, E.A., 2008. Prestorage ultraviolet-white light irradiation alters apple peel metabolome. J. Agric. Food Chem. 56, 1138–1147.

Sabatier, S., Amiot, M.J., Tacchini, M., Aubert, S., 1992. Identification of flavonoids in sunflower honey. J. Food Sci. 57, 773–774.

Saito, Y., Jinno, K., 2003. Miniaturized sample preparation combined with liquid phase separations. J. Chromatogr. A 1000, 53–67.

Saitta, M., Lo Curto, S., Salvo, F., Di Bella, G., Dugo, G., 2002. Gas chromatographic–tandem mass spectrometric identification of phenolic compounds in Sicilian olive oils. Anal. Chim. Acta. 466, 335–344.

Salta, F.N., Mylona, A., Chiou, A., Boskou, G., Andrikopoulos, N.K., 2007. Oxidative stability of edible vegetable oils enriched in polyphenols with olive leaf extract. Food Sci. Technol. Int. 13, 413–421.

Saraji, M., Mousavinia, F., 2006. Single-drop microextraction followed by in-syringe derivatization and gas chromatography-mass spectrometric detection for determination of organic acids in fruits and fruit juices. J. Sep. Sci. 29, 1223–1229.

Schulz, J.M., Herrmann, K., 1980. Analysis of hydroxybenzoic and hydroxycinnamic acids in plant material II. Determination by gas-liquid chromatography. J. Chromatogr. 195, 95–104.

Serrano Ríos, M., Sastre Gallego, A., Cobo Sanz, J.M., 2005. Tendencias en Alimentación Funcional. You&Us, Madrid, pp. 1–14.

Shadkami, F., Estevez, S., Helleur, R., 2009. Analysis of catechins and condensed tannins by thermally assisted hydrolysis/methylation-GC/MS and by a novel two step methylation. J. Anal. Appl. Pyrol. 85, 54–65.

Shahidi, F., Wanasundara, P.K., 1992. Phenolic antioxidants. Crit. Rev. Food Sci. Nutr. 32, 67–103.

Shahidi, F., Nacsk, M., 1995. Food Phenolics: Sources, Chemistry, Effects, and Application. Technomic Publishing Company, Inc., Lancaster, PA.

Shao, Y., Marriott, P., Hügel, H., 2003. Solid-phase microextraction—On-fibre derivatization with comprehensive two dimensional gas chromatography analysis of *trans*-resveratrol in wine. Chromatographia 57, S349–S353.

Singleton, V.L., Rossi Jr., J.A., 1965. Colorimetry of total phenolics with phosphomolybdic-phosphotungstic acid reagents. Am. J. Enol. Viticulture 16, 144–158.

Smolarz, H.D., 2001. Application of GC-MS method for analysis of phenolic acids and their esters in chloroformic extracts from some taxons of *Polygonum* L. genus. Chemia Analityczna (Warsaw) 46, 439–444.

Sobolev, V.S., Horn, B.W., Potter, T.L., Deyrup, S.T., Gloer, J.B., 2006. Production of stilbenoids and phenolic acids by the peanut plant at early stages of growth. J. Agric. Food Chem. 54, 3505–3511.

Soleas, G.J., Goldberg, D.M., Diamandis, E.P., Karumanchiri, A., Yan, J., Ng, E., 1995. A derivatized gas-chromatographic mass-spectrometric method for the analysis of both isomers of resveratrol in juice and wine. Am. J. Enol. Viticulture 46, 346–352.

Soleas, G.J., Diamandis, E.P., Goldberg, D.M., 1997a. Wine as a biological fluid: history, production, and role in disease prevention. J. Clin. Lab. Anal. 11, 287–313.

Soleas, G.J., Diamandis, E.P., Goldberg, D.M., 1997b. Resveratrol: A molecule whose time has come? And gone? Clin. Biochem. 30, 91–113.

Soleas, G.J., Diamandis, E.P., Karumanchiri, A., Goldberg, D.M., 1997c. A multiresidue derivatization gas chromatographic assay for fifteen phenolic constituents with mass selective detection. Anal. Chem. 69, 4405–4409.

Soleas, G.J., Goldberg, D.M., Ng, E., Karumanchiri, A., Tsang, E., Diamandis, E.P., 1997d. Comparative evaluation of four methods for assay of *cis* and *trans*-resveratrol. Am. J. Enol. Viticulture 48, 169–176.

Soleas, G.J., Goldberg, D.M., David, M., 1999. Analysis of antioxidant wine polyphenols by gas chromatography-mass spectrometry. Methods Enzymol. 299, 137–151.

Solinas, M., 1987. HRGC analysis of phenolic components in virgin olive oils in relation to the ripening and the variety of olives. Rivista Italiana delle Sostanze Grasse 64, 255–262.

Sosulski, F., Krygier, K., Hogge, L., 1982. Free, esterified, and insoluble bound phenolic acids. 3. Composition of phenolic acids in cereal and potato flours. J. Agric. Food Chem. 30, 337–340.

Sparkman, O.D., Penton, Z.E., Kitson, F.G., 2011. Gas chromatography and mass spectrometry. A practical guide. Elsevier, Oxford.

Spranger, M.I., Climaco, M.C., Sun, B., Eiriz, N., Fortunato, C., Nunes, A., et al., 2004. Differentiation of red winemaking technologies by phenolic and volatile composition. Anal. Chim. Acta. 513, 151–161.

Stremple, P., 1998. GC/MS Analysis of polymethoxyflavones in citrus oil. J. High. Resolut. Chromatogr. 21, 587–591.

Sudjaroen, Y., Haubner, R., Würtele, G., Hull, W.E., Erben, G., Spiegelhalder, B., et al., 2005. Isolation and structure elucidation of phenolic antioxidants from Tamarind (*Tamarindus indica* L.) seeds and pericarp. Food Chem. Toxicol. 43, 1673–1682.

Sun, B., Leandro, M.C., de Freitas, V., Spranger, M.I., 2006. Fractionation of red wine polyphenols by solid-phase extraction and liquid chromatography. J. Chromatogr. A. 1128, 27–38.

Tan, S.C., 2000. Determinants of eating quality in fruits and vegetables. Proc. Nutr. Soc. Aust. 24, 183–190.

Tasioula-Margari, M., Okogeri, O., 2001. Isolation and characterization of virgin olive oil phenolic compounds by HPLC/UV and GC-MS. J. Food Sci. 66, 530–534.

Tobiszewski, M., Mechlińska, A., Zygmunt, B., Namieśnik, J., 2009. Green analytical chemistry in sample preparation for determination of trace organic pollutants. Trends Anal. Chem. 28, 943–951.

Tokusoglu, O., Unal, M.K., Yemis, F., 2005. Determination of the phytoalexin resveratrol (3,5,4′-trihydroxystilbene) in peanuts and pistachios by high-performance liquid chromatographic diode array (HPLC-DAD) and gas chromatography–mass spectrometry (GC-MS). J. Agric. Food Chem. 53, 5003–5009.

Tomás-Barberán, F.A., Clifford, M.N., 2000. Flavanones, chalcones and dihydrochalcones – nature, occurrence and dietary burden. J. Sci. Food. Agric. 80, 1073–1080.

Tomás-Barberán, F.A., Espín, J.C., 2001. Phenolic compounds and related enzymes as determinants of quality in fruits and vegetables. J. Sci. Food Agriculture 81, 853–876.

Valls, J., Millán, S., Martí, M.P., Borràs, E., Arola, L., 2009. Advanced separation methods of food anthocyanins, isoflavones and flavanols. J. Chromatogr. A 1216, 7143–7172.

Vande Casteele, K., De Pooter, H., Van Sumere, C.F., 1976. Gas chromatographic separation and analysis of trimethylsilyl derivatives of some naturally occurring nonvolatile phenolic compounds and related substances. J. Chrom. 121, 49–63.

Viñas, P., Campillo, N., Hernández-Pérez, M., Hernández-Córdoba, M., 2008. A comparison of solid-phase microextraction and stir bar sorptive extraction coupled to liquid chromatography for the rapid analysis of resveratrol isomers in wines, musts and fruit juices. Anal. Chim. Acta. 611, 119–125.

Viñas, P., Campillo, N., Martínez-Castillo, N., Hernández-Córdoba, M., 2009. Solid-phase microextraction on-fiber derivatization for the analysis of some polyphenols in wine and grapes using gas chromatography-mass spectrometry. J. Chromatogr. A 1216, 1279–1284.

Viñas, P., Martínez-Castillo, N., Campillo, N., Hernández-Córdoba, M., 2011. Directly suspended droplet microextraction with in injection-port derivatization coupled to gas chromatography-mass spectrometry for the analysis of polyphenols in herbal infusions, fruits and functional foods. J. Chromatogr. A 1218, 639–646.

Waksmundzka-Hajnos, M., 1998. Chromatographic separations of aromatic carboxylic acids. J. Chromatogr. A B 717, 93–118.

Wang, C., Zuo, Y., 2011. Ultrasound-assisted hydrolysis and gas chromatography-mass spectrometric determination of phenolic compounds in cranberry products. Food Chem. 128, 562–568.

Wang, Q., Chong, J.M., Pawliszyn, J., 2006. Determination of thiol compounds by automated headspace solid-phase microextraction with in-fiber derivatization. Flavour Fragance J. 21, 385–394.

Winter, M., Herrmann, K., 1986. Esters and glucosides of hydroxycinnamic acids in vegetables. J. Agric. Food Chem. 34, 616–620.

Wu, H., Haig, T., Prately, J., Lemerle, D., An, M., 1999. Simultaneous determination of phenolic acids and 2,4-dihydroxy-7-methoxy-1,4-benzoxazin-3-one in wheat (*Triticum aestivum* L.) by gas chromatography-tandem mass spectrometry. J. Chromatogr. A 864, 315–321.

Wu, H., Haig, T., Prately, J., Lemerle, D., An, M., 2000. Allelochemicals in wheat (*Triticum aestivum* L.): Variation of phenolic acids in root tissues. J. Agric. Food Chem. 48, 5321–5325.

Zafra, A., Juárez, M.J.B., Blanc, R., Navalón, A., González, J., Vílchez, J.L., 2006. Determination of polyphenolic compounds in wastewater olive oil by gchromatography-mass spectrometry. Talanta 70, 213–218.

Zuo, Y., Wang, C., Zhan, J., 2002. Separation, characterization, and quantitation of benzoic and phenolic antioxidants in American cranberry fruit by GC-MS. J. Agric. Food Chem. 50, 3789–3794.

Novel Techniques Towards the Identification of Different Classes of Polyphenols

Vassiliki G. Kontogianni

Section of Organic Chemistry and Biochemistry, Department of Chemistry,
University of Ioannina, Ioannina, Greece

CHAPTER OUTLINE HEAD

8.1 Introduction

Phytochemicals comprise a wide variety of organic molecules synthesized by plants are also referred to as "secondary metabolites." Although they are not considered as essential nutrients, since plant based foods are complex mixtures of bioactive compounds, information on the potential health effects of individual phyochemicals is linked to information on the health effects of foods that contain them. Polyphenols are secondary metabolites ubiquitously distributed among all higher plants, which are important determinants for the sensory and nutritional quality of fruits, vegetables, and other plants (Tomas-Barberan *et al.*, 2000). They comprise a diverse range of molecules, more than 8000 polyphenolics have been identified so far, which have in common an aromatic ring bearing at least one hydroxyl substituent (a phenol). Polyphenols are generally divided into flavonoids, which account for approximately two-thirds of dietary phenols, and nonflavonoids (Harborne *et al.*, 1999).

Flavonoids share a common carbon skeleton of two benzene rings (ring A and B) joined by a linear three-carbon chain, which forms a closed pyran ring (ring C) with

Polyphenols in Plants. http://dx.doi.org/10.1016/B978-0-12-397934-6.00008-5

the A benzene ring. They can be divided into many subclasses: flavonols, flavones, flavanones, anthocyanidins, flavanols (catechins), and also isoflavones (Figure 8.1). Flavonoids are found in native ("aglycon") form but most commonly exist in plant materials as in flavonoid O-glycosides, where one or more of the aglycone hydroxyl groups are bound to a sugar with formation of an O–C acid-labile acetal bond. Flavonoid C-glycosides are also found, where glycosylation takes place via an acid-resistant C–C bond, by direct linkage of the sugar to the flavonoid basic nucleus (Cuyckens *et al.*, 2003).

Flavonols are the most commonly found flavonoids in foods, and the main representatives are kaempferol and quercetin. These are mostly found in glycosylated forms. The glycosidic sugars are generally glucose, frequently galactose and rhamnose, and sometimes xylose, arabinose and glucuronic acid. Flavones, which are much less common than flavonols in fruit and vegetables, chiefly consist of glycosides of luteolin and apigenin. Flavanones are present in high concentrations only in citrus fruit, naringenin in grapefruit, hesperetin in oranges, and eriodictyol in lemons, but they are also found in tomatoes and certain aromatic plants. Flavanols exist in both the monomer form (catechins) and the polymer form (proanthocyanidins). The flavan-3-ols (+)-catechin, (−)-epicatechin, (−)-epigallocatechin and their gallate esters are found in many types of fruit and in red wine. Tea leaves and chocolate are by far the richest sources. Proanthocyanidins, which are also known as condensed tannins, are dimers, oligomers, and polymers of catechins and they are the major polyphenols in grapes. Anthocyanidins and their glycosides (anthocyanins) are water-soluble vacuolar pigments that occur in all plant tissues. Isoflavones are found almost exclusively in leguminous plants and can be present as aglycones (Andersen and Markham, 2006).

The main group of nonflavonoids is first phenolic acids, which constitute about one-third of the dietary phenols, and may be present in plants in free and bound forms. They can be subdivided into derivatives of hydroxybenzoic acid and derivatives of hydroxycinnamic acid. Hydroxybenzoic acids include gallic, p-hydroxybenzoic, protocatechuic, vanillic and syringic acids, which have in common the C6–C1 structure. The hydroxycinnamic acids are more common than the hydroxybenzoic acids and consist chiefly of p-coumaric, caffeic, ferulic, and sinapic acids. These acids are rarely found in the free form, their bound forms are either glycosylated derivatives or esters of quinic acid, shikimic acid, and tartaric acid (Bravo, 1998). Among the groups of nonflavonoids we find stilbenes, whose main representative is resveratrol, that exists in both *cis* and *trans* isomeric forms, mostly in glycosylated forms. Resveratrol has been extensively studied and its anticarcinogenic effects have been shown during screening of medicinal plants. It is found in low quantities in wine (Delmas *et al.*, 2006). Finally the last group of nonflavonoids is lignans, which are produced by oxidative dimerization of two phenylpropane units; they are mostly present in nature in free form. The interest in lignans and their synthetic derivatives is growing because of potential applications in cancer chemotherapy and various other pharmacological effects (Saleem *et al.*, 2005).

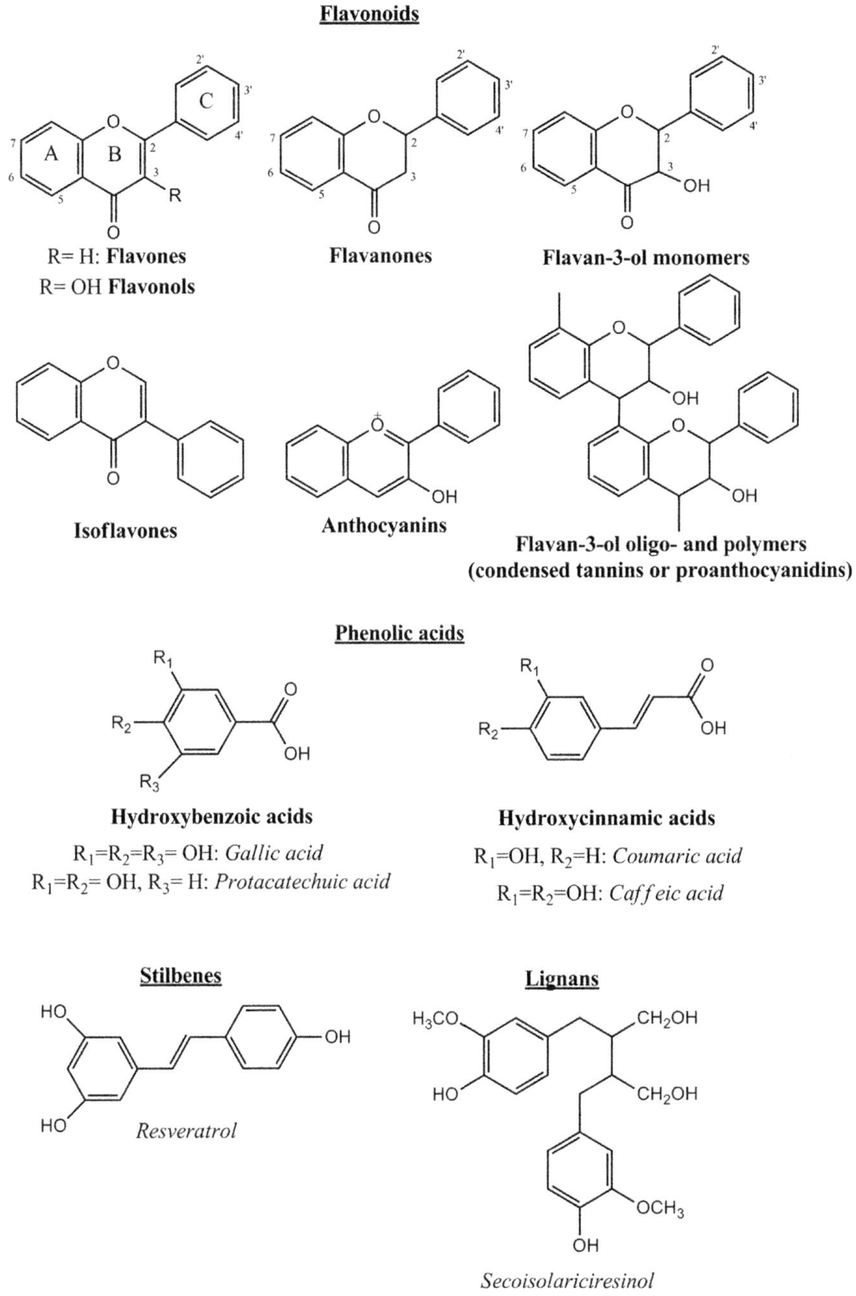

Flavonoids

R= H: **Flavones**
R= OH **Flavonols**

Flavanones

Flavan-3-ol monomers

Isoflavones

Anthocyanins

Flavan-3-ol oligo- and polymers
(condensed tannins or proanthocyanidins)

Phenolic acids

Hydroxybenzoic acids

$R_1=R_2=R_3=$ OH: *Gallic acid*
$R_1=R_2=$ OH, $R_3=$ H: *Protacatechuic acid*

Hydroxycinnamic acids

$R_1=$OH, $R_2=$H: *Coumaric acid*
$R_1=R_2=$OH: *Caffeic acid*

Stilbenes

Resveratrol

Lignans

Secoisolariciresinol

FIGURE 8.1

Chemical structures of the main classes of polyphenols.

The rapid and accurate identification of polyphenols in natural product matrices is an emerging research field for analytical chemists, food chemists, phytochemists, and biochemists because of the structural diversity of these compounds, their potential health effects and their impact on the organoleptic and nutraceutical properties of fruits, vegetables, and other plants. This chapter focuses on the most interesting and novel applications of "state of the art," hyphenated techniques for profiling and structure identification of polyphenols in natural product matrices. Most emphasis is placed on recent LC-MS, NMR, and LC-NMR innovative strategies developed in this direction. Methods for the sample treatment and extraction of different matrices along with HPLC and other methods for the analysis of polyphenols are briefly discussed.

8.2 Sample treatment and extraction

The sample treatment procedure can be considered as a step with a predominant impact on any analytical study since it determines the final result. Various sample preparation methods have been developed to determine polyphenolics and simple phenolics in the wide diversity of matrices which have been analyzed. This process aims for multiple targets, such as to improve sample stability, to increase the efficiency of the extraction method or to transform the analytes into a more appropriate form for detection, separation or quantification. In any case, its final objective is the sample to be uniformly enriched in all the components of interest and ideally free of impurities, in order to spend as little time and energy consumption as possible achieving the highest efficiency and reproducibility of analysis. There are three main types of phenolic-containing matrices, i.e., plants, foods, and liquid samples (including biological fluids and beverages).

Usually, the solid samples are first subjected to milling, grinding, and homogenization, which may be preceded by air-drying, freeze-drying, or freezing with liquid nitrogen.

The chosen sample has to be the most representative and must be conserved and manipulated in appropriated conditions. Conversely, some liquid samples are amenable to direct analysis requiring no treatment other than centrifugation, filtration and/or dilution, like fruit juices and wines. After this, they are either directly injected into the separation system or, in most cases, analytes are first isolated using various extraction techniques. An important aspect of flavonoid analysis is to determine whether the target analytes exist in their various conjugated forms or as aglycones. With a view to the analysis of conjugated aglycones, preliminary hydrolysis must be performed, where enzymatic or chemical treatments (acidic and alkaline) are used. Also, in some cases, a hydrolysis step is included to release compounds from matrix structures.

Extraction is a very important step for the isolation, identification and use of phenolic compounds and there is no single and standardized extraction method. Solvent extraction (SE), that can be performed either by hand shaking, or through mechanical means such as homogenization or sonication, is the most commonly used technique for the isolation of phenolic compounds. Solvent extraction may be liquid–liquid extraction or solid–liquid extraction. SE can be followed by solid phase extraction (SPE), especially when the analyte concentrations are low or as a purification step toward the removal of unwanted phenolics and non-phenolic substances such as waxes, fats, terpenes, and chlorophylls. The most commonly used material for SPE is chemically bonded silica, usually with a C18 or C8 organic group, for the extraction, purification and/or enrichment of flavonoids from different matrices (Wallace and Giusti, 2010; Zgorka and Hajnos, 2003). The most common solvents for extraction are acidified methanol, aqueous methanol, acetonitrile, or ethanol. Furthermore, with dried materials, low-polarity solvents and ethyl acetate will simply leach the sample. Conventional extraction such as heating, boiling, or Soxhlet extraction can be used to extract natural phenolic compounds. However, the disadvantages are the loss of polyphenols as well as the long extraction time.

Recently, various novel extraction techniques have been developed for the extraction of polyphenols, including ultrasound-assisted extraction, microwave-assisted extraction, supercritical fluid extraction, and high-hydrostatic-pressure extraction (Wang and Weller, 2006). Ionic liquid-based ultrasonic assisted extraction has been reported in the literature for the extraction of lignans (Ma *et al.*, 2011, 2012). Furthermore, pressurized liquid extraction has been successfully used for the extraction of flavonoids (Pineiro *et al.*, 2004; Monrad *et al.*, 2010; Wijngaard, and Brunton, 2009), high hydrostatic pressure (Shouqin *et al.*, 2005; Corrales *et al.*, 2009), and matrix solid-phase dispersion. Also, high-speed centrifugal counter-current chromatography has been explored for the fractionation of red wine phenolics (Vitrac *et al.*, 2001). Methods employed for polyphenol extraction from plants and foods are not usually directly applicable to biological fluids and organs. For biological sample preparation, protein precipitation, solvent extraction, and SPE are procedures commonly used. However, new attempts have been made to enhance selectivity (immunoaffinity, molecularly imprinted polymers or aptamers), in combination with approaches that reduce solvent consumption (microextraction approaches) (Novakova and Vlckova, 2009).

8.3 High-performance liquid chromatography (HPLC)

Amongst the different methods available, HPLC has dominated the separation, characterization, and quantification of polyphenols in the last twenty years. Several supports and mobile phases are reported for the analysis of polyphenols including anthocyanins, proanthocyanidins, hydrolysable tannins, flavonols, flavan-3-ols, flavanones, flavones, and phenolic acids in different matrices (Merken and Beecher, 2000; Robbins, 2003; Welch *et al.*, 2008; Valls *et al.*, 2009; Bueno *et al.*, 2012a). Furthermore, the usage of HPLC techniques bears inherent advantages, such as the capacity to analyze simultaneously all components of interest together with their possible derivatives or degradation products (Sakakibara *et al.*, 2003; Downey and Rochfort, 2008). The chromatographic conditions of the HPLC methods include the use of, almost exclusively, a reversed-phase (RP) C18 or C8 column; UV–Vis diode array detector, and a binary solvent system containing acidified water (solvent A) and a polar organic solvent (solvent B).

Acetonitrile and methanol are the most commonly used organic modifiers. Acidification of the solvents, using acetic, formic, and phosphoric acid, is a common strategy in order to suppress the ionization of phenolic hydroxyl groups, giving sharper peak shapes; keeping a low pH (in the range 2–4) helps prevent peak tailing and improves the resolution and reproducibility of the retention characteristics. Acidification of the organic solvent so that the percentage of acid remains constant during elution also gives sharper peak shapes. Acetonitrile usually leads to a better resolution in a shorter analysis time than methanol and usually gives sharper peak shapes. Nevertheless, methanol is more frequently preferred in comparison to acetonitrile because of its nontoxic properties and due to the fact that it can be used in higher percentages in the mobile phase which could protect the HPLC column.

Both isocratic and gradient elution are applied to separate polyphenols. The choice depends on the number and type of the analyte and the nature of the matrix. Under RP conditions, the more polar compounds are generally eluted first, hence, flavonoid diglycosides precede monoglycosides, which precede aglycones. The elution pattern for the different subclasses of flavonoids, is flavanone followed by flavonol, isoflavones and flavones, for both aglycones and glycosides of them. For anthocyanins, molecular structure-retention characteristic relationships are noticed. HPLC analysis of polyphenols has mainly been based on the use of diode-array detectors (DAD) that allow obtaining the UV-visible spectra of the peaks, constituting a useful tool for compound classification (and even identification) and peak purity assessment.

LC with multiple-wavelength or diode-array UV detection is still a satisfactory tool in studies dealing with, for example screening, quantification of the main flavonoid aglycones and/or a provisional sub-group classification (de Rijke *et al.*, 2006). Several flavonoid sub-classes can be initially distinguished from each other using a limited number of monitoring wavelengths sufficient for a general flavonoid screening: flavonoid detection is usually carried out at 250, 265, 290, 350, 370, and/or 400 nm, with an added wavelength in the 500–525 nm range if anthocyanins are included. For the identification the peak spectra are compared with the spectra of reference compounds by evaluating the degree of overlapping (Liu *et al.*, 2008). Detection at 280 nm is most commonly used for the simultaneous separation of mixtures of phenolic acids, although for dual monitoring 254 and 280 nm, or 280 and 320 nm, can be the ideal wavelengths.

Merken and Beecher (2000) determined 17 flavonoids and their aglycones, that are considered to be most prominent in common foods, simultaneously with HPLC-DAD. In the same way, Sakakibara *et al.* (2003) determined a large number of polyphenols in vegetables, fruits, and teas (they reported results for 63 vegetables, fruits, and teas) using HPLC-DAD and constructed a library of retention times, spectra of aglycones, and respective calibration curves for 100 standard chemicals. Nevertheless, UV spectra of polyphenols are often very similar to each other. Using traditional approaches based on HPLC-DAD is not a sufficient tool for the accurate identification of them, especially in complex matrices, such as crude plant extracts or environmental samples and when dealing with the analysis of low levels of fla-vonoid metabolites in plasma and other bio-samples. Therefore, preconcentration and purification procedures of the polyphenols from complex matrices are usually required prior to the instrumental analysis by HPLC, in order to simplify the chro-matograms. Modern high-performance chromatographic techniques combined with instrumental analysis, modern spectroscopic techniques like NMR (Christophori-dou and Dais, 2009; Savage *et al.*, 2011) and MS (Cavaliere *et al.*, 2008; Steinmann and Ganzera, 2011; De la Cruz *et al.*, 2012; Sun *et al.*, 2012), are the "state of the art" techniques for the profiling and quantification of phenolic compounds. MS and NMR detections are more commonly used for structure identification rather than for quantification.

8.4 Liquid chromatography coupled to mass spectrometry (LC-MS) and direct flow injection mass spectrometry

Liquid Chromatography-Mass Spectrometry (LC-MS) techniques are nowadays the "state of the art" analytical approach in polyphenol analysis. Single-stage MS or LC-MS/MS are commonly used in combination with DAD detection to facilitate the confirmation of the identity of polyphenols in a sample with the help of standards and reference data. For the identification of unknowns, tandem mass spectrometry (MS/MS or MS[n]-multi-stage) is used, whereas the same technique has rarely been employed for quantitation of selected compounds. Atmospheric pressure ionization interfaces, such as atmospheric pressure chemical ionization (APCI) and electro-spray ionization (ESI) are used almost exclusively today. Although both positive (PI) and negative ionization (NI) are applied, according to most of the studies, the NI mode provides best sensitivity concerning both APCI and ESI. However, with PI mode, useful complementary information is often obtained in studies dealing with the identification of unknowns. ESI is more frequently used in polyphenol and especially flavonoid analysis, but APCI is gaining ground and, in some cases, better responses are obtained in that mode (de Rijke *et al.*, 2003; Tong *et al.*, 2008). Also, a method based on LC/APCI-MS has been developed for the direct determination of catechins in green and black tea infusions (Zeeb *et al.*, 2000). Table 8.1 lists some selected application of LC-MS including experimental details.

HPLC-DAD-ESI-MS[n] techniques were applied for the identification of poly-phenols, mainly flavonoids and phenolic acids, in plant materials (Lin and Harnly, 2007), vegetables (Olsen *et al.*, 2009), rosemary and sage extracts (Kontogianni *et al.*, 2013), and in "mountain tea" infusion (Vasilopoulou *et al.*, 2013). They were also applied for the characterization of simple phenols, flavonoids, and secoiri-doids in olive products, olive mill waste (Obied *et al.*, 2007), and olive leaf extracts (Kontogianni and Gerothanassis, 2012). Stilbenes and flavonoids were simultane-ously identified by ESI-MS in negative ionization mode and were quantified by PDA detection in tomato fruits of plants genetically modified to synthesize resve-ratrol (Nicoletti *et al.*, 2007). The composition of polyphenols in propolis extracts

Table 8.1 Selected LC-MS Assays for the Analysis of Polyphenols in Different Matrices

Analytes	Matrix	LC-Conditions	Ionization Source	Mass Analyzer	Quantification	Application	References
Stilbenes and Flavonoids	Tomato fruits (plants genetically modified to synthesize resveratrol)	Polaris C18A (5 µm), 0.5% formic acid in water, acetonitrile	ESI	Ion-trap	PDA	S, SA	Nicoletti et al., 2007
Glycosylated, aglycones and polymeric flavonoids, phenolic acids	5 plant materials (cranberry, elderflower, Fuji apple peel, and soybean seed)	Symmetry C18A (5 µm), 0.1% formic acid in water and acetonitrile	ESI	Ion-trap	—	S, SA	Lin and Harnly, 2007
Anthocyanins, flavonol glycosides, hydroxycinnamic acid derivatives	Red mustard greens	UHPLC system, Hypersil Gold AQ RP-C18 (1.9 µm), 0.1% formic acid in water and acetonitrile	ESI	LTQ-Orbitrap	—	S, SA	Lin et al., 2011
Isoflavonoids	Iris germanica extract	Sypelco Discovery C18 (5 µm), 0.1% formic acid in water, acetonitrile	ESI	Triple quadrupole	—	S, SA	Maul et al., 2008
Simple phenols, flavonoids and secoiridoids	Olive products and olive mill waste	C18 Phenomenex column (5 µm), 1% formic acid in water and a mixture of methanol/acetonitrile 90/10%	ESI	Ion-trap	—	S, SA	Obied et al., 2007

Analytes	Sample	Column/conditions	Ionization	Mass analyzer	Detector	S/SA	Reference
Flavonoids and hydroxycinnammic acids	Curly kale (*Brassica oleracea* L.)	Betasil RP-C18 column (5 µm), 2% acetic acid in water and in a mixture of acetonitrile/water (50/48%)	ESI	Ion-trap	PDA	S, SA	Olsen et al., 2009
Phenolic acids and flavonoids	Propolis extracts	Ascentis C18 column (5 µm), 0.1% formic acid in water and acetonitrile	ESI	Ion-trap and triple quadrupole	PDA	S, SA	Pellati et al., 2011
Catechins	Green and black tea	Zorbax Eclipse XDB-C18 column, 0.05% TFA in water and acetonitrile	APCI	Ion-trap	—	S, SA	Zeeb et al., 2000
Anthocyanins and flavonol glycosides	Blueberry, red radish and Hongcaitai	UHPLC system, Hypersil Gold AQ RP-C18 (1.9 µm), 0.1% formic acid in water and acetonitrile	ESI	LTQ-Orbitrap	—	S,SA	Sun et al., 2012
Phenolic acids	*Eucommia ulmodies* Oliver	ODS-3column (5 µm), 0.5% acetic acid in water and methanol	APCI	Ion-trap	—	S, SA	Tong et al., 2008
Flavonoids and phenolic acids	Honey samples	UHPLC system, Hypersil gold C18 (1.9 µm), 0.1% formic acid in water and acetonitrile	HESI (Heated)	LTQ Orbitrap	MS	S, SA	Kečkeš et al., 2013

Continued

Table 8.1 Selected LC-MS Assays for the Analysis of Polyphenols in Different Matrices—cont'd

Analytes	Matrix	LC-conditions	Ionization Source	Mass Analyzer	Quantification	Application	References
Flavonoids and phenolic acids	Soybeans	Phenomenex Sinergy Max-RP (4 μm), 0.1% formic acid in water and acetonitrile	ESI	LTQ Orbitrap	(ECD)	SA	Correa et al., 2010
Flavonoids, oligomeric flavonoids and phenolic acids	Apple fruit	UHPLC system, Waters Acquity UPLC BEH SHIELD RP 18 (1.9 μm), 0.1% formic acid in water and acetonitrile	ESI-am-MS	LTQ Orbitrap	MS	S, SA	De Paepe et al., 2013

S, standard; SA, sample.

was investigated by HPLC-DAD and HPLC-ESI-MS/MS by comparing the performance of ion trap and triple quadrupole mass analyzers (Pellati *et al.*, 2011). Metal complexation is an alternative ionization mode which has been explored extensively for analysis of flavonoids (Davis and Brodbelt, 2005; March and Brodbelt, 2008). Implementation of metal complexation is straightforward: a metal salt is added to a flavonoid solution prior to ESI. Metal complexation results typically in larger ion abundances than those obtained upon protonation or deprotonation of flavonoids, thus enhancing detection sensitivity.

Matrix-assisted laser desorption ionization (MALDI) is another soft ionization technique, MALDI- time of flight (TOF) MS has been applied to the analysis of plant proanthocyanidins and proved to be a powerful tool for the structural elucidation of these complex polymers (Monagas *et al.*, 2010). For anthocyanin analysis, MS techniques such as fast atom bombardment (FAB), MALDI and TOF-TOF have proven to be very useful for structural elucidation (Bueno *et al.*, 2012b). Many combinations of tandem MS have been tried; triple quadrupole (TQ), hybrid quadrupole-time of flight (Q-TOF), and quadrupole ion trap (QIT) mass spectrometers are the most common used (Liu *et al.*, 2008). The mass spectra of polyphenols obtained with quadrupole and ion-trap instruments typically are quite similar, even though relative abundances of fragment ions and adducts do show differences. Therefore, spectra obtained with these instruments can be directly compared. Ion-trap instruments have the advantage to perform MS^n experiments, which enables the confirmation of proposed reaction pathways for fragment ions (de Rijke *et al.*, 2003).

Regarding the mass fragmentations of flavonoids, they originate from two main fragmentation events. One is the ring-opening (RO), which is characterized by the (subsequent) loss of small neutral molecules like H_2O (18 Da), CO (28 Da) and CO_2 (44 Da). The other main fragmentation event results in two broken bonds, the so-called cross ring cleavage (CRC) and for polyphenols Retro-Diels–Alder Reaction (RDA) (Ma *et al.*, 1997; Fabre *et al.*, 2001; Kuhn *et al.*, 2003; Liu *et al.*, 2005). Other, generally less characteristic, fragments common to most flavonoids are those arising from the loss of C_2H_2O (42 Da) and the successive loss of H_2O and CO (46 Da). MS/MS of flavonoid-(di)glycosides is a useful tool to differentiate and also to distinguish the $1\rightarrow2$ and $1\rightarrow6$ glycose linking types of diglycosides, and to differentiate (i) the *O*-glycosidic (3-*O*- and 7-*O*-) and (ii) the *C*-glycosidic (6-*C*- and 8-*C*-) flavonoids (Cuyckens and Claeys, 2004). Papers on utilization of mass spectrometry for the analysis of flavonoids and their structural elucidation were reviewed by Cuyckens *et al.* (2004), March *et al.* (2006) and de Rijke *et al.* (2006). Vukics and Guttman (2010) reviewed the identification options of unknown flavonoid glycosides in complex samples focusing on the differentiation of isomeric compounds. Recently, Gomez-Romero *et al.* (2011) constructed an HPLC library of phenolic compounds identified according to their retention time, MS and MS/MS spectra obtained using LC interfaced to an ESI Qq-TOF mass spectrometer.

Recently, ultra-high-performance liquid chromatography (UHPLC) coupled with hybrid mass spectrometer which combines the Linear Trap Quadrupole (LTQ) and OrbiTrap mass analyzer (LTQ OrbiTrap MS) has been used for the identification of polyphenols. LTQ OrbiTrap MS has several advantages compared to triple quadrupole MS and quadrupole/time-of-flight MS. A mass detector with a triple quadrupole mass analyzer is not a suitable detector for full scan measurements, and it is not possible to analyze components for which we have not predefined their ion transition using this detector. However, the application of this hybrid technique enables a simultaneous determination of qualitative content, quantification, MS^n analysis based on high resolution accurate mass measurement and data dependent experiment. UHPLC-ESI/HRMS/MS^n with LTQ OrbiTrap mass analyzer was used for the identification of anthocyanins, flavonol glycosides, and hydroxycinnamic acid derivatives in red mustard greens (Lin *et al.*, 2011) and for the differentiation of anthocyanins and non-anthocyanin phenolic compounds in botanicals and foods (Sun *et al.*, 2012). Also, the qualitative and quantitative content of polyphenols in honeys was determined using UHPLC with LTQ OrbiTrap MS (Kečkeš *et al.*, 2013) and polyphenols were analyzed in soybeans (Correa *et al.*, 2010) and apple extracts (De Paepe *et al.*, 2013). An example of chromatograms along with MS/MS spectra obtained for polyphenols in honeys with an LTQ OrbiTrap mass analyzer is presented in Figure 8.2, demonstrating its potentials in differentiating flavonoid isomers.

Direct flow injection ESI mass spectrometry analysis can be used to establish polyphenol fingerprints of complex extracts, like wine, discriminating samples on

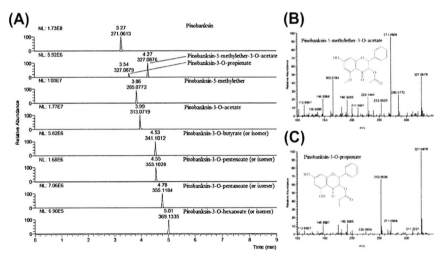

FIGURE 8.2

Chromatograms and MS/MS spectra of flavonoids identified in Sergian unifloral honeys using UHPLC with LTQ OrbiTrap MS analyzer. (A) Chromatograms of pinobanksin and its derivatives with the retention time and accurate mass; (B) MS/MS spectra of pinobanksin-5-methylether-3-*O*-acetate; (C) MS/MS spectra of pinobanksin-3-*O*-propionate. Reprinted with permission from Kečkeš *et al.* (2013) *Food Chem.* 138, 32–40, American Chemical Society.

the basis of their phenolic (i.e., anthocyanin, phenolic acid, and flavan-3-ol) compositions. The MALDI-TOF technique is suitable to determine the presence of molecules of higher molecular weight with high accuracy, and it has been applied with success to study procyanidin oligomers into protein complexes (Fulcrand et al., 2008). Also, polyphenolics from cranberries, grape seed extracts, sorghum, and pomegranate were isolated by liquid chromatography and subjected to MALDI-TOF MS using trans-3-indoleacrylic acid as matrix (Reed et al., 2005). The direct infusion-atmospheric pressure photoionization (APPI) coupled to QqTOF-MS mass analyzer, without prior chromatographic separation, has been used to obtain a classification rule for unknown samples of wine according to their anthocyanins profile. Results obtained were satisfactorily compared with the ESI-MS that has also been used (Gómez-Ariza et al., 2006). Recently, a high-mass resolution multi-stage mass spectrometry (MS^n) fragmentation was tested for differentiation and identification of 121 polyphenolic molecules. A nano-ESI source was used combined with an ion-trap in an Orbitrap Fourier transform (FT) MS mass analyzer. The MS^n spectral tree data obtained consist of a powerful tool to distinguish metabolites with similar elemental formula, thereby assisting compound identification in complex biological samples such as crude plant extracts (Van Der Hooft et al., 2011b).

8.5 Nuclear magnetic resonance spectroscopy (NMR)

Rarely does MS alone, even MS^n, provide an unambiguous structure and it becomes necessary to combine MS with spectroscopic techniques such as UV and nuclear magnetic resonance (NMR) spectroscopy for structure determination. NMR spectroscopy is, unquestionably, the technique that generates more information about a molecule and consequently is the most important tool for complete structure elucidation of polyphenols. Numerous flavonoids have either been examined directly or have been isolated from plants and foods with column chromatography or preparative HPLC and subsequently characterized by ^{13}C and ^1H NMR (Aksnes et al., 1996; Budzianowski et al., 2005; Kim et al., 2006). The recent technique of two-dimensional (2D) NMR has become important for the determination of many anthocyanin linkage positions and aliphatic acyl groups (Andersen and Fossen, 2005; Andersen and Jordheim, 2010). Wawer and Zielinska (2001) have reported the ^1H and ^{13}C NMR spectra for a number of flavonoids in dimethyl sulphoxide (DMSO) in addition to the cross-polarization magic angle spinning (CP/MAS) solid-state ^{13}C NMR spectra. In addition, ^{13}C and ^1H NMR spectra have been reported for acylated derivatives of apigenin-7-O-glucoside (Švehlíková et al., 2004), for flavone and substituted flavones (Park et al., 2007). Burns et al. (2007) described an approach, in which the monohydroxyflavone ^{13}C chemical shifts have been assigned and then compared to those of flavone for the development of a predictive tool for the ^{13}C NMR of polyhydroxylated flavonols and their glycosylated analogs. The ^{13}C chemical shifts for 8- hydroxyflavone, which does not occur in nature, have been predicted using this model (March et al., 2008).

Various NMR techniques have been employed for the structural elucidation of complex polyphenols extracted from complex mixtures, such as plant extracts and foods without previous separation into individual components. Advantages such as simplicity of the sample preparation and measurement procedures, the instrumental stability and the ease with which spectra can be interpreted have contributed to the growing popularity of the technique. Standard ^{1}H, ^{13}C and now high resolution magic angle spinning (HR/MAS) NMR spectra can give a wealth of chemical information on liquid and even semi-solid samples. NMR techniques that have been employed in this direction include, among others: ^{1}H and ^{13}C NMR, 2D ^{1}H-^{1}H correlated NMR spectroscopy (COSY), totally correlated spectroscopy (TOCSY), nuclear Overhauser effect in both the laboratory frame (NOESY) and rotating frame of reference (ROESY) (Gerothanassis *et al.*, 1998), [1]H-^{13}C heteronuclear multiple quantum coherence ([1]H-^{13}C HMQC) and [1]H-^{13}C heteronuclear multiple-bond correlation (HMBC) gradient NMR spectroscopy (Exarchou *et al.*, 2001, 2002; Charisiadis *et al.*, 2010), diffusion-ordered spectroscopy (DOSY) (Rodrigues *et al.*, 2009).

Gerothanassis *et al.* (1998) used a combined NMR methodology consisting of 2D ^{1}H-^{1}H COSY, TOCSY, NOESY, and ROESY techniques for the identification of phenolic acids in oregano extracts. The same team introduced the combination of variable-temperature 2D [1]H-^{1}H double quantum filter DQF-COSY, [1]H-^{13}C HMQC, and ^{1}H-^{13}C HMBC gradient NMR spectroscopy for the identification and quantification of caffeic acid and rosmarinic acid in extracts from plants of the Lamiaceae family (Exarchou *et al.*, 2001). Variable temperature gradient ^{1}H, [1]H–^{13}C Gradient Enhanced GE-HSQC and GE-HMBC NMR studies were used for the discrimination of flavonoids in mixtures of them (Exarchou *et al.*, 2002). Extending this methodology, focusing on –OH NMR region of ^{1}H NMR spectra in dilute DMSO-d_6 solutions with pH value adjusted for a minimum –OH proton exchange rate and/or dilute solutions with 2D ^{1}H-^{13}C HMBC allowed the complete assignment of several flavonoids and phenolic acids in complex natural extracts (Charisiadis *et al.*, 2010, 2011). In Figure 8.3, a ^{1}H-^{13}C HMBC spectrum of an olive leaf extract is illustrated, where the diagnostic connectivities for the identification of flavonoids and some of its other components are indicated. Also, for the simultaneous identification and quantification of hydrogen peroxide, flavonoids and phenolic acids in plant extracts, a rapid and direct low micromolar ^{1}H NMR method was developed (Charisiadis *et al.*, 2012).

^{13}C NMR spectroscopy was used as a complement to HPLC or spectrophotometry to analyze stilbene and anthocyanin metabolism in grape cell cultures (Saigne-Soulard *et al.*, 2006). ^{31}P NMR spectroscopy has been employed to detect and quantify phenolic compounds in the polar fraction of virgin olive oil (Christophoridou and Dais, 2006) and a large number of polyphenols (phenolic acids and flavonoids) and two triterpenic acids in oregano extracts (Agiomyrgianaki and Dais, 2012). DOSY NMR methodologies have been applied for the identification of flavonoid glycosides in a plant extract presenting various biological activities (*Bidens sulphurea*) (Rodrigues *et al.*, 2009) and for the identification of their various components directly

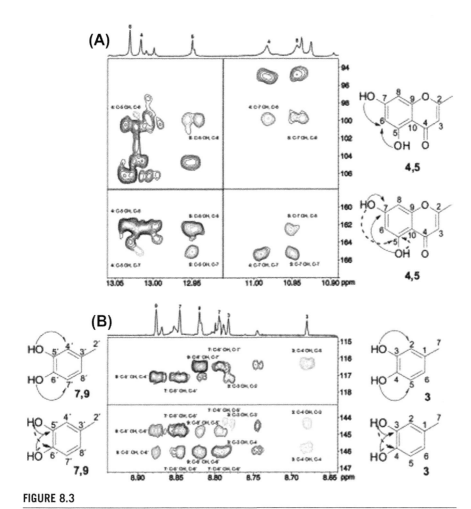

FIGURE 8.3

500 MHz 2D 1H–13C HMBC NMR spectrum of 10 mg of an olive leaf methanol extract in 0.6 mL of DMSO-d$_6$. (A) Cross-peaks of luteolin-4′-O-β-D-glucopyranoside (5) and cross-peaks of luteolin (4) are illustrated in red and blue, respectively. (B) Cross-peaks are illustrated in blue for oleuropein 6-O-β-D-glucopyranoside (7), green for hydroxytyrosol (3), and red for oleuropein (aldehyde form) (9). From Charisiadis et al. (2011) J. Nat. Prod. 74, 2462–2466.

in crude reaction products or mixtures containing polyphenol organic compounds, without any prior separation or isolation (Primikyri et al., 2012). Also, high resolution ^{1}H NMR spectroscopy has been employed as a versatile and rapid method to analyze the polar fraction of extra virgin olive oils containing simple phenols, flavonoids, lignans and secoiridoids, overall 19 constituents were detected and quantified using this methodology. For identification of phenolic compounds,

2D NMR spectroscopy was applied to model standards and phenolic extracts (Christophoridou *et al.*, 2009). Several classes of molecules (simple phenols, flavonols, secoiridoids and lignans) were unambiguously characterized in one-pot analysis in an extract of extra virgin olive oil by means of multiple-quantum (MQ) NMR (Reddy and Caldarelli, 2011). Finally, Nerantzaki *et al.* (2011) introduced a novel method for the determination of the total phenolic content using ^{1}H NMR spectroscopy in the –OH spectral region, with comparable results to the Folin–Ciocalteu (FC) reagent method. The method was applied to model compounds, a mixture of them and several extracts of natural products.

NMR spectra of complex mixtures can act as "fingerprints" that can be used to compare, discriminate, or classify samples. Selected variables (NMR peak heights or integrals) that characterize the samples in a specific way are also used instead of the whole spectra. Chemometric techniques are often employed to analyze the data as the information contained in the spectra is of a high degree of complexity (Le Gall and Colquhoun, 2003). A fast and effective analytical method for metabolomic fingerprinting of plant extracts has been developed, ESI-MS and ^{1}H NMR techniques together with statistical analyses of the acquired data were employed (Mattoli *et al.*, 2006). ^{1}H NMR spectroscopy was used for the characterization and differentiation of apple varieties, on the basis of their catechins, phenolic acids and acid composition, using multivariate statistical analysis (Del Campo *et al.*, 2006). Likewise, wines can be classified according to variety, region, and year of production on the basis of their phenolic profile monitored by NMR in combination with multivariate analysis chemometric methods (Anastasiadi *et al.*, 2009). In the same direction, 1D and 2D ^{1}H and ^{13}C homo- and heteronuclear NMR experiments were examined for the characterization of amino acids and anthocyanins, the "minor" but important compounds of wine (Košir and Kidrič 2002).

^{1}H NMR spectroscopy was used to identify signals belonging to several compound groups (e.g., amino acids, carbohydrates, etc.) along with seven isoflavones, that were assigned and quantified, in soybean extracts (Caligiani *et al.*, 2010). It was also used to detect polyphenols in grape juices (Savage *et al.*, 2011). Samples of extra virgin olive oils were chemically analyzed by means of ^{1}H and ^{31}P NMR spectroscopy and characterized according to their content in fatty acids, phenolic compounds, diacylglycerols, etc. Major phenolic compounds with the highest discriminatory power, relating the harvest year, cultivar, and geographical origin, were the flavonoids, luteolin and apigenin, the lignans pinoresinol, 1-acetoxypinoresinol and syringaresinol, the phenolic acid homovanillic acid, and two simple phenols (Agiomyrgianaki *et al.*, 2012).

8.6 Liquid chromatography coupled to nuclear magnetic resonance spectroscopy (LC-NMR)

LC-NMR is another powerful technique for food and natural products analysis. Its potentials in the discovery of novel biologically active structures and new sources of rare natural products have been amply demonstrated. The main advantages of this technique are its high power of information, the ability of differentiation of isomers and substitution patterns. However, it also has some disadvantages, such as low sensitivity, expensive instrumentation and long run times. The direct, on-line coupling of an NMR spectrometer to HPLC has required the development of special interfaces called flow probes. These systems can work in either on-flow mode, where the solute passes through the system as it passes out of the column and is being analyzed, or in stop-flow mode, where a valve stops the flow of the effluent when a compound is detected by a UV detector, which is connected to the system. NMR spectroscopy has been combined with HPLC/MS, using stop-flow mode, for the identification of quercitin and quercetin glycosides in *Hypericum perforatum* L (Hansen *et al.*, 1999) and for the identification of quercetin glycosides in apple peel (Lommen *et al.*, 2000) and with HPLC-UV-MS for the on-line structural investigation of isoflavones and isoflavanones (Wolfender *et al.*, 2003). de Rijke *et al.* (2004) used stopped-flow reversed-phase LC-NMR for the identification of flavonoid constituents in a red clover extract. Another mode in performing LC-NMR is to use a loop collector, which automatically collects the peaks of interest without stopping the LC flow. An off-line post-chromatographic analysis of the content of the loops is then automatically performed. In a study by Gil *et al.* (2003) on the aromatic composition of beer, grape juice, and a wine phenolic extract, NMR spectroscopy and LC-NMR/MS were performed using the on-flow mode, the time-sliced stop flow method and, in some cases, the loop-sampling method.

Inserting a Solid Phase Extraction (SPE) unit between a HPLC unit and NMR spectrometer (LC-SPE-NMR), in order to trap and accumulate the compounds onto SPE cartridges and hereafter transport them into the NMR probe with deuterated solvent, increases the concentration of the analytes and consequently the sensitivity of the technique. Using an LC-UV-SPE-NMR-MS set-up combined with a cryoflow NMR probe, which exhibits a detectability four-fold better than with conventional probes, five flavonoids were identified in an oregano extract (Exarchou *et al.*, 2003). An LC-SPE-NMR technique using postcolumn solid-phase extraction was introduced by Christophoridou *et al.* (2005) for the direct analysis of phenolic compounds in the polar part of olive oil. Using this technique, 27 constituents (simple phenols, lignans, flavonoids, and a large number of secoiridoid derivatives), which were difficult or impossible to identify by NMR alone, mainly because of their low concentration and severe overlap with other signals in the same region, were assigned. Tatsis *et al.* (2007) applied LC/DAD/SPE/NMR and LC/UV/(ESI)MS techniques for separation and structure verification of flavonoids and phenolic acids in Greek *Hypericum perforatum* extracts. Recent successful applications of HPLC-SPE-NMR and LC/MS in the identification of polyphenols have been reported, such as lignans from *Phyllanthus myrtifolius* and stilbenoids from *Syagrus romanzoffiana* (Wang *et al.*, 2011) and polyphenolic compounds from *Origanum vulgare* L. (Liu *et al.*, 2012).

For the analysis of anthocyanins efficient trapping on to SPE cartridges is not possible for all of them, due to the hydrophilic properties of these compounds. To avoid these problems an alternative technique, the Isco Foxy fraction collector was inserted after the NMR spectrometer, in order to collect the eluting compounds (LC-NMR-Foxy). Each sample was collected after on-flow ^{1}H-LC-NMR analysis into the fraction collector, lyophilized and subsequently analyzed by classical NMR. De la Cruz *et al.* (2012) succeeded in identifying 33 anthocyanins in grape berry skins, using the combination of LC-ESI-MS^2 and LC-NMR-Foxy. Recently, an automated MS-guided HPLC-MS-SPE-NMR approach was used to fully characterize flavonoid structures that are present in crude tomato plant extracts. NMR spectra of plant metabolites, automatically trapped and purified from LC-MS traces, were successfully obtained, leading to the structural elucidation of them (Van Der Hooft *et al.*, 2011a). Last but not least, Van Der Hooft *et al.* (2012) introduced the combined use of LC-LTQ-Orbitrap Fourier transformed (FT)-MS and LC-TOFMS-SPE-NMR for the annotation of 177 polyphenols in tea. Currently, much effort is devoted to the development of micro- or even nano-LC-NMR.

8.7 Other methods

Many other methods have been introduced for the analysis of polyphenols. The electromigration modes primarily used are capillary electrophoresis (CE), capillary zone electrophoresis (CZE) and micellar electrokinetic chromatography (MEKC). Detection is usually performed with UV, but electrochemical and MS detectors are also used. Most studies that use capillary electrophoretic methods for the analysis of phenolics fall in the field of natural product research, including the analysis of plants, vegetables, herbs, and other plant or fruit-derived products (Herrero *et al.*, 2005; Wang and Huang, 2004; Caridi *et al.*, 2007). CE has been indicated to be a suitable technique for the separation, identification and quantification of anthocyanins (Bridle *et al.*, 1996) and CE with UV detection was proven to be an useful tool for the characterization of the polyphenolic fraction of extra-virgin olive oil (Carrasco-Pancorbo *et al.*, 2006). CE coupled to ESI-MS has been used for monitoring anthocyanins and flavonoids in wine (Bednář *et al.*, 2005). Carrasco-Pancorbo *et al.* (2007) utilized CZE coupled with ESI-TOF-MS for the analysis of phenolic compounds in extra-virgin olive oil. Recently, Yang *et al.* (2010) reviewed the uses of CE in the analysis of phytochemical bioactive compounds, including flavonoids and phenolic acids.

Near infrared reflectance (NIR) spectroscopy is another powerful, fast, accurate and non-destructive analytical tool that can be considered as a replacement of the older chemical analysis. Chen *et al.* (2009) reported the results of simultaneous analysis of main catechin contents in green tea by the Fourier transform FT-NIR spectroscopy. Traditional spectroscopic assays may lead to overestimation of polyphenol contents of crude extracts from plant materials due to the overlapping of spectral responses. These problems can be overcome by using a chemometric technique to analyze the spectra such as partial least squares (PLS) or principal component analysis (PCA). Edelmann *et al.* (2001) developed a rapid method for the discrimination of Austrian red wines based on mid-infrared spectroscopy of phenolic extracts of wine. Schulz *et al.* (1999) used a NIR spectroscopic method for prediction of polyphenols in the leaves of green tea. The PLS method was used to calibrate NIR spectra with the contents of gallic acid and catechins in tea.

Fluorescence detection is used only occasionally, because the number of polyphenols that exhibit native fluorescence is limited. Complexation with metal ions can be employed to enhance flavonoid fluorescence and increase sensitivity. HPLC with fluorescence detection using derivatization with aluminium was applied by Paulke *et al.* (2006) to the quantification of quercetin and its methylated metabolites in rat brain. On the other side, most flavonoids are electroactive due to the presence of phenolic groups, which makes them suitable for electrochemical detection (ECD). Thus, HPLC-ECD has also been used for the analysis of flavonoids in plants and food and biological matrices. HPLC-ECD including CoulArray detection has been successfully employed by several authors for the analysis of quercetin and catechins and their metabolites in plasma and other biomatrices (Lee *et al.*, 2000; Bolarinwa and Linseisen, 2005). The multi-channel coulometric detection system may offer a highly sensitive method for the overall characterization of antioxidants (Jandera *et al.*, 2005).

Gas chromatographic (GC) techniques have been widely used especially for separation and quantification of phenolic acids and flavonoids. The major concern with this technique is the low volatility of phenolic compounds. Prior to chromatography, phenolics are usually transformed into more volatile derivatives by methylation, conversion into trimethylsilyl derivatives, etc. Early work with derivatized phenolics was typically performed with flame ionization detection (FID). Mass spectrometry later became widespread, in most cases using electron impact ionization mode. A detailed discussion on application of GC on analysis of phenolic acids and flavonoids was provided by Stalikas (2007). Finally, GC-MS has successfully applied for the characterization of isoflavones; Maul *et al.* (2008) reviewed plenty of its applications. According to their findings, the combination of GC-MS and LC-MS offers reliable structural information without the need to carry out time-consuming cleanup or to reach concentration levels sufficient for NMR analysis.

References

Agiomyrgianaki, A., Dais, P., 2012. Simultaneous determination of phenolic compounds and triterpenic acids in oregano growing wild in Greece by ^{31}P NMR spectroscopy. Magn. Reson. Chem. 50, 739–748.

Agiomyrgianaki, A., Petrakis, P.V., Dais, P., 2012. Influence of harvest year, cultivar and geographical origin on Greek extra virgin olive oils composition: A study by NMR spectroscopy and biometric analysis. Food Chem. 135, 2561–2568.

Aksnes, D.W., Standnes, A., Andersen, Ø.M., 1996. Complete assignment of the ^{1}H and ^{13}C NMR spectra of flavone and its A-ring hydroxyl derivatives. Magn. Reson. Chem. 34, 820–823.

Anastasiadi, M., Zira, A., Magiatis, P., Haroutounian, S.A., Skaltsounis, A.L., Mikros, E., 2009. H NMR-based metabonomics for the classification of Greek wines according to variety, region, and vintage. comparison with HPLC data. J. Agric. Food Chem. 57, 11067–11074.

Andersen, O.M., Fossen, T., 2005. Characterization of Anthocyanins by NMR. In: Wrolstad, R.E. (Ed.), Handbook of Food Analytical Chemistry, Vol. 2. Wiley, New York, pp. 47–69.

Andersen, O.M., Jordheim, M., 2010. Anthocyanins. In: Encyclopedia of Life Sciences. MacMillan, New York, pp. 1–10.

Andersen, Q.M., Markham, K.R., 2006. Flavonoids, Chemistry, Biochemistry and Applications. CRS Press Taylor & Francis Group, London.

Bednář, P., Papoušková, B., Müller, L., Barták, P., Stávek, J., Pavloušek, P., et al., 2005. Utilization of capillary electrophoresis/mass spectrometry (CE/MS^n) for the study of anthocyanin dyes. J. Sep. Sci. 28, 1291–1299.

Bolarinwa, A., Linseisen, J., 2005. Validated application of a new high-performance liquid chromatographic method for the determination of selected flavonoids and phenolic acids in human plasma using electrochemical detection. J. Chromatog. B: Anal. Technol. Biomed. Life Sci. 823, 143–151.

Bravo, L., 1998. Polyphenols: Chemistry, dietary sources, metabolism, and nutritional significance. Nut. Rev. 56, 317–333.

Bridle, P., García-Viguera, C., Tomás-Barberán, F.A., 1996. Analysis of anthocyanins by capillary zone electrophoresis. J. Liq. Chromatogr. Relat. Technol. 19, 537–545.

Budzianowski, J., Morozowska, M, Wesołowska, M., 2005. Lipophilic flavones of Primula veris L. from field cultivation and in vitro cultures. Phytochemistry 66, 1033–1039.

Bueno, J.M., Ramos-Escudero, F., Saez-Plaza, P., Munoz, A.M., Navas, M.J., Asuero, A.G., 2012a. Analysis and Antioxidant Capacity of Anthocyanin Pigments. Part I: General Considerations Concerning Polyphenols and Flavonoids. Crit. Rev. Anal. Chem. 42, 102–125.

Bueno, J.M., Sáez-Plaza, P., Ramos-Escudero, F., Jiménez, A.M., Fett, R., Asuero, A.G., 2012b. Analysis and Antioxidant Capacity of Anthocyanin Pigments. Part II: Chemical Structure, Color, and Intake of Anthocyanins. Crit. Rev. Anal. Chem. 42, 126–151.

Burns, D.C., Ellis, D.A., March, R.E., 2007. A predictive tool for assessing ^{13}C NMR chemical shifts of flavonoids. Magn. Reson. Chem. 45, 835–845.

Caligiani, A., Palla, G., Maietti, A., Cirlini, M., Brandolini, V., 2010. ^{1}H NMR fingerprinting of soybean extracts, with emphasis on identification and quantification of isoflavones. Nutrients 2, 280–289.

Caridi, D., Trenerry, V.C., Rochfort, S., Duong, S., Laugher, D., Jones, R., 2007. Profiling and quantifying quercetin glucosides in onion (Allium cepa L.) varieties using capillary zone electrophoresis and high performance liquid chromatography. Food Chem. 105, 691–699.

Carrasco-Pancorbo, A., Gómez-Caravaca, A.M., Cerretani, L., Bendini, A., Segura-Carretero, A., Fernández-Gutiérrez, A., 2006. A simple and rapid electrophoretic method to characterize simple phenols, lignans, complex phenols, phenolic acids, and flavonoids in extra-virgin olive oil. J. Sep. Sci. 29, 2221–2233.

Carrasco-Pancorbo, A., Neusüß, C., Pelsing, M., Segura-Carretero, A., Fernandez, Gutiérrez, A., 2007. CE- and HPLC-TOF-MS for the characterization of phenolic compounds in olive oil. Electrophoresis 28, 806–821.

Cavaliere, C., Foglia, P., Gubbiotti, R., Sacchetti, P., Samperi, R., Lagana, A., 2008. Rapid-resolution liquid chromatography/mass spectrometry for determination and quantitation of polyphenols in grape berries. Rapid Commun. Mass Spectrom. 22, 3089–3099.

Charisiadis, P., Exarchou, V., Troganis, A.N., Gerothanassis, I.P., 2010. Exploring the "forgotten" –OH NMR spectral region in natural products. Chem. Commun. 46, 3589–3591.

Charisiadis, P., Primikyri, A., Exarchou, V., Tzakos, A., Gerothanassis, I.P., 2011. Unprecedented ultra-high-resolution hydroxy group ^{1}H NMR spectroscopic analysis of plant extracts. J. Nat. Prod. 74, 2462–2466.

Charisiadis, P., Tsiafoulis, C.G., Exarchou, V., Tzakos, A.G., Gerothanassis, I.P., 2012. Rapid and direct low micromolar NMR method for the simultaneous detection of hydrogen peroxide and phenolics in plant extracts. J. Agric. Food Chem. 60, 4508–4513.

Chen, Q., Zhao, J., Chaitep, S., Guo, Z., 2009. Simultaneous analysis of main catechins contents in green tea (*Camellia sinensis* (L.)) by Fourier transform near infrared reflectance (FT-NIR) spectroscopy. Food Chem. 113, 1272–1277.

Christophoridou, S., Dais, P., 2006. Novel approach to the detection and quantification of phenolic compounds in olive oil based on ^{31}P nuclear magnetic resonance spectroscopy. J. Agric. Food Chem. 54, 656–664.

Christophoridou, S., Dais, P., 2009. Detection and quantification of phenolic compounds in olive oil by high resolution ^{1}H-nuclear magnetic resonance spectroscopy. Anal. Chim. Acta. 633, 283–292.

Christophoridou, S., Dais, P., Tseng, L.I.H., Spraul, M., 2005. Separation and identification of phenolic compounds in olive oil by coupling high-performance Liquid Chromatography with Postcolumn Solid-Phase Extraction to Nuclear Magnetic Resonance Spectroscopy (LC-SPE-NMR). J. Agric. Food Chem. 53, 4667–4679.

Corrales, M., Garcia, A.F., Butz, P., Tauscher, B., 2009. Extraction of anthocyanins from grape skins assisted by high hydrostatic pressure. J. Food Eng. 90, 415–421.

Correa, C.R., Li, L., Aldini, G., Carini, M., Oliver Chen, C.Y., Chun, H.K., et al., 2010. Composition and stability of phytochemicals in five varieties of black soybeans (*Glycine max*). Food Chem. 123, 1176–1184.

Cuyckens, F., Claeys, M., 2004. Mass spectrometry in the structural analysis of flavonoids. J. Mass Spectrom. 39, 1–15.

Cuyckens, F., Shahat, A.A., Van den Heuvel, H., Abdel-Shafeek, K.A., El-Messiry, M.M., Seif-El Nasr, M.M., et al., 2003. The application of liquid chromatography-electrospray ionization mass spectrometry and collision-induced dissociation in the structural characterization of acylated flavonol *O*-glycosides from the seeds of *Carrichtera annua*. Eur. J. Mass Spectrom. 9, 409–420.

Davis, B.D., Brodbelt, J.S., 2005. LC-MS^n methods for saccharide characterization of monoglycosyl flavonoids using postcolumn manganese complexation. Anal. Chem. 77, 1883–1890.

De la Cruz, A.A., Hilbert, G., Riviere, C., Mengin, V., Ollat, N., Bordenave, L., et al., 2012. Anthocyanin identification and composition of wild *Vitis* spp. accessions by using LC-MS and LC-NMR. Anal. Chim. Acta 732, 145–152.

De Paepe, D., Servaes, K., Noten, B., Diels, L., De Loose, M., Van Droogenbroeck, B., et al., 2013. An improved mass spectrometric method for identification and quantification of phenolic compounds in apple fruits. Food Chem. 136, 368–375.

de Rijke, E., de Kanter, F., Ariese, F., Brinkman, U.A.T., Gooijer, C., 2004. Liquid chromatography coupled to nuclear magnetic resonance spectroscopy for the identification of isoflavone glucoside malonates in *T. pratense* L. leaves. J. Sep. Sci. 27, 1061–1070.

de Rijke, E., Out, P., Niessen, W.M.A., Ariese, F., Gooijer, C., Brinkman, U.A.T., 2006. Analytical separation and detection methods for flavonoids. J. Chromatogr. A 1112, 31–63.

de Rijke, E., Zappey, H., Ariese, F., Gooijer, C., Brinkman, U.A.T., 2003. Liquid chromatography with atmospheric pressure chemical ionization and electrospray ionization mass spectrometry of flavonoids with triple-quadrupole and ion-trap instruments. J. Chromatogr. A 984, 45–58.

Del Campo, G., Santos, J.I., Iturriza, N., Berregi, I., Munduate, A., 2006. Use of the ^{1}H nuclear magnetic resonance spectra signals from polyphenols and acids for chemometric characterization of cider apple juices. J. Agric. Food Chem. 54, 3095–3100.

Delmas, D., Lancon, A., Colin, D., Jannin, B., Latruffe, N., 2006. Resveratrol as a chemo-preventive agent: A promising molecule for fighting cancer. Curr. Drug Targets 7, 423–442.

Downey, M., Rochfort, S., 2008. Simultaneous separation by reversed-phase high-performance liquid chromatography and mass spectral identification of anthocyanins and flavonols in Shiraz grape skin. J. Chromatogr. A 1201, 43–47.

Edelmann, A., Diewok, J., Schuster, K.C., Lendl, B., 2001. Rapid method for the discrimination of red wine cultivars based on mid-infrared spectroscopy of phenolic wine extracts. J. Agric. Food Chem. 49, 1139–1145.

Exarchou, V., Godejohann, M., Van Beek, T.A., Gerothanassis, I.P., Vervoort, J., 2003. LC-UV-Solid-Phase Extraction-NMR-MS Combined with a Cryogenic Flow Probe and Its Application to the Identification of Compounds Present in Greek Oregano. Anal. Chem. 75, 6288–6294.

Exarchou, V., Troganis, A., Gerothanassis, I.P., Tsimidou, M., Boskou, D., 2001. Identification and quantification of caffeic and rosmarinic acid in complex plant extracts by the use of variable-temperature two-dimensional nuclear magnetic resonance spectroscopy. J. Agric. Food Chem. 49, 2–8.

Exarchou, V., Troganis, A., Gerothanassis, I.P., Tsimidou, M., Boskou, D., 2002. Do strong intramolecular hydrogen bonds persist in aqueous solution? Variable temperature gradient ^{1}H, ^{1}H-^{13}C GE-HSQC and GE-HMBC NMR studies of flavonols and flavones in organic and aqueous mixtures. Tetrahedron 58, 7423–7429.

Fabre, N., Rustan, I., De Hoffmann, E., Quetin-Leclercq, J., 2001. Determination of flavone, flavonol, and flavanone aglycones by negative ion liquid chromatography electrospray ion trap mass spectrometry. J. Am. Soc. Mass Spectrom. 12, 707–715.

Fulcrand, H., Mané, C., Preys, S., Mazerolles, G., Bouchut, C., Mazauric, J.P., et al., 2008. Direct mass spectrometry approaches to characterize polyphenol composition of complex samples. Phytochemistry 69, 3131–3138.

Gerothanassis, I.P., Exarchou, V., Lagouri, V., Troganis, A., Tsimidou, M., Boskou, D., 1998. Methodology for Identification of Phenolic Acids in Complex Phenolic Mixtures by High-Resolution Two-Dimensional Nuclear Magnetic Resonance. Application to Methanolic Extracts of Two Oregano Species. J. Agric. Food Chem. 46, 4185–4192.

Gil, A.M., Duarte, I.F., Godejohann, M., Braumann, U., Maraschin, M., Spraul, M., 2003. Characterization of the aromatic composition of some liquid foods by nuclear magnetic resonance spectrometry and liquid chromatography with nuclear magnetic resonance and mass spectrometric detection. Anal. Chim. Acta 488, 35–51.

Gómez-Ariza, J.L., García-Barrera, T., Lorenzo, F., 2006. Anthocyanins profile as fingerprint of wines using atmospheric pressure photoionisation coupled to quadrupole time-of-flight mass spectrometry. Anal. Chim. Acta. 570, 101–108.

Gomez-Romero, M., Zurek, G., Schneider, B., Baessmann, C., Segura-Carretero, A., Fernandez-Gutierrez, A., 2011. Automated identification of phenolics in plant-derived foods by using library search approach. Food Chem. 124, 379–386.

Hansen, S.H., Jensen, A.G., Cornett, C., Bjørnsdottir, I., Taylor, S., Wright, B., et al., 1999. High-performance liquid chromatography on-line coupled to high-field NMR and mass spectrometry for structure elucidation of constituents of Hypericum perforatum L. Anal. Chem. 71, 5235–5241.

Harborne, J.B., Baxter, H., Moss, G.P., 1999. Phytochemical dictionary: Handbook of bioactive compounds from plants, second ed. Taylor and Francis, London.

Herrero, M., Ibáñez, E., Cifuentes, A., 2005. Analysis of natural antioxidants by capillary eletromigration methods. J. Sep. Sci. 28, 883–897.

Jandera, P., Škeříková, V., Řehová, L., Hájek, T., Baldriánová, L., Škopová, G., et al., 2005. RP-HPLC analysis of phenolic compounds and flavonoids in beverages and plant extracts using a CoulArray detector. J. Sep. Sci. 28, 1005–1022.

Kečkeš, S., Gašić, U., Veličković, T.Ć., Milojković-Opsenica, D., Natić, M., Tešić, Ž., 2013. The determination of phenolic profiles of Serbian unifloral honeys using ultra-high-performance liquid chromatography/high resolution accurate mass spectrometry. Food Chem. 138, 32–40.

Kim, H., Moon, B.H., Ahn, J.H., Lim, Y., 2006. Complete NMR signal assignments of flavonol derivatives. Magn. Reson. Chem. 44, 188–190.

Kontogianni, V.G., Gerothanassis, I.P., 2012. Phenolic compounds and antioxidant activity of olive leaf extracts. Nat. Prod. Res. 26, 186–189.

Kontogianni, V.G., Tomic, G., Nikolic, I., Nerantzaki, A.A., Sayyad, N., Stosic-Grujicic, S., et al., 2013. Phytochemical profile of Rosmarinus officinalis and Salvia officinalis extracts and correlation to their antioxidant and anti-proliferative activity. Food Chem. 136, 120–129.

Košir, I.J., Kidrič, J., 2002. Use of modern nuclear magnetic resonance spectroscopy in wine analysis: Determination of minor compounds. Anal. Chim. Acta. 458, 77–84.

Kuhn, F., Oehme, M., Romero, F., Abou-Mansour, E., Tabacchi, R., 2003. Differentiation of isomeric flavone/isoflavone aglycones by MS^2 ion trap mass spectrometry and a double neutral loss of CO. Rapid Commun. Mass Spectrom. 17, 1941–1949.

Le Gall, G., Colquhoun, I.J., 2003. NMR spectroscopy in food authentication in food authenticity and traceability. In: Lees, Michele (Ed.), Food Science and echnology. Woodhead Publishing, North America, pp. 131–156.

Lee, M.J., Prabhu, S., Meng, X., Li, C., Yang, C.S., 2000. An improved method for the determination of green and black tea polyphenols in biomatrices by high-performance liquid chromatography with coulometric array detection. Anal. Biochem. 279, 164–169.

Lin, L.Z., Harnly, J.M., 2007. A screening method for the identification of glycosylated flavonoids and other phenolic compounds using a standard analytical approach for all plant materials. J. Agric. Food Chem. 55, 1084–1096.

Lin, L.Z., Sun, J.H., Chen, P., Harnly, J., 2011. UHPLC-PDA-ESI/HRMS/MS[n] Analysis of Anthocyanins, Flavonol Glycosides, and Hydroxycinnamic Acid Derivatives in Red Mustard Greens (*Brassica juncea* Coss Variety). J. Agric. Food Chem. 59, 12059–12072.

Liu, E.H., Qi, L.W., Cao, J., Li, P., Li, C.Y., Peng, Y.B., 2008. Advances of Modern Chromatographic and Electrophoretic Methods in Separation and Analysis of Flavonoids. Molecules 13, 2521–2544.

Liu, H., Zheng, A., Yu, H., Wu, X., Xiao, C., Dai, H., et al., 2012. Identification of three novel polyphenolic compounds, origanine A-C, with unique skeleton from *Origanum vulgare* L. using the hyphenated LC-DAD-SPE-NMR/MS methods. J. Agric. Food Chem. 60, 129–135.

Liu, R., Ye, M., Guo, H., Bi, K., Guo, D.A., 2005. Liquid chromatography/electrospray ionization mass spectrometry for the characterization of twenty-three flavonoids in the extract of *Dalbergia odorifera*. Rapid Commun. Mass Spectrom. 19, 1557–1565.

Lommen, A., Godejohann, M., Venema, D.P., Hollman, P.C.H., Spraul, M., 2000. Application of directly coupled HPLC-NMR-MS to the identification and confirmation of quercetin glycosides and phloretin glycosides in apple peel. Anal. Chem. 72, 1793–1797.

Ma, C.H., Liu, T.T., Yang, L., Zu, Y.G., Wang, S.Y., Zhang, R.R., 2011. Study on ionic liquid-based ultrasonic-assisted extraction of biphenyl cyclooctene lignans from the fruit of *Schisandra chinensis* Baill. Anal. Chim. Acta. 689, 110–116.

Ma, C.H., Wang, S.Y., Yang, L., Zu, Y.G., Yang, F.J., Zhao, C.J., et al., 2012. Ionic liquid-aqueous solution ultrasonic-assisted extraction of camptothecin and 10-hydroxycamptothecin from *Camptotheca acuminata* samara. Chem. Eng. Proces. 57–58, 59–64.

Ma, Y.L., Li, Q.M., Van den Heuvel, H., Claeys, M., 1997. Characterization of Flavone and Flavonol Aglycones by Collision-induced Dissociation Tandem Mass Spectrometry. Rapid Commun Mass Spectrom. 11, 1357–1364.

March, R., Brodbelt, J., 2008. Analysis of flavonoids: Tandem mass spectrometry, computational methods, and NMR. J. Mass Spectrom. 43, 1581–1617.

March, R.E., Burns, D.C., Ellis, D.A., 2008. Empirically predicted ^{13}C NMR chemical shifts for 8-hydroxyflavone starting from 7,8,4′-trihydroxyflavone and from 7,8-dihydroxyflavone. Magn. Reson. Chem. 46, 680–682.

March, R.E., Lewars, E.G., Stadey, C.J., Miao, X.S., Zhao, X.M., Metcalfe, C.D., 2006. A comparison of flavonoid glycosides by electrospray tandem mass spectrometry. Int. J. Mass Spectrom. 248, 61–85.

Mattoli, L., Cangi, F., Maidecchi, A., Ghiara, C., Ragazzi, E., Tubaro, M., et al., 2006. Metabolomic fingerprinting of plant extracts. J. Mass Spectrom. 41, 1534–1545.

Maul, R., Schebb, N.H., Kulling, S.E., 2008. Application of LC and GC hyphenated with mass spectrometry as tool for characterization of unknown derivatives of isoflavonoids. Anal. Bioanal. Chem. 391, 239–250.

Merken, H.M., Beecher, G.R., 2000. Measurement of food flavonoids by high-performance liquid chromatography: A review. J. Agric. Food Chem. 48, 577–599.

Monagas, M., Quintanilla-Lopez, J.E., Gomez-Cordoves, C., Bartolome, B., Lebron-Aguilar, R., 2010. MALDI-TOF MS analysis of plant proanthocyanidins. J. Pharm. Biomed. Anal. 51, 358–372.

Monrad, J.K., Howard, L.R., King, J.W., Srinivas, K., Mauromoustakos, A., 2010. Subcritical Solvent Extraction of Procyanidins from Dried Red Grape Pomace. J. Agric. Food Chem. 58, 4014–4021.

Nerantzaki, A.A., Tsiafoulis, C.G., Charisiadis, P., Kontogianni, V.G., Gerothanassis, I.P., 2011. Novel determination of the total phenolic content in crude plant extracts by the use of ^{1}H NMR of the -OH spectral region. Anal. Chim. Acta. 688, 54–60.

Nicoletti, I., De Rossi, A., Giovinazzo, G., Corradini, D., 2007. Identification and quantification of stilbenes in fruits of transgenic tomato plants (Lycopersicon esculentum Mill.) by reversed phase HPLC with photodiode array and mass spectrometry detection. J. Agric. Food Chem. 55, 3304–3311.

Novakova, L., Vlckova, H., 2009. A review of current trends and advances in modern bio-analytical methods: Chromatography and sample preparation. Anal. Chim. Acta 656, 8–35.

Obied, H.K., Bedgood, D.R., Prenzler, P.D., Robards, K., 2007. Chemical screening of olive biophenol extracts by hyphenated liquid chromatography. Anal. Chim. Acta. 603, 176–189.

Olsen, H., Aaby, K., Borge, G.I.A., 2009. Characterization and Quantification of Flavonoids and Hydroxycinnamic Acids in Curly Kale (*Brassica oleracea* L. Convar. acephala Var. sabellica) by HPLC-DAD-ESI-MS^n. J. Agric. Food Chem. 57, 2816–2825.

Park, Y., Moon, B.H., Lee, E., Lee, Y., Yoon, Y., Ahn, J.H., et al., 2007. ^{1}H and ^{13}C-NMR data of hydroxyflavone derivatives. Magn. Reson. Chem. 45, 674–679.

Paulke, A., Schubert-Zsilavecz, M., Wurglics, M., 2006. Determination of St. John's wort flavonoid-metabolites in rat brain through high performance liquid chromatography coupled with fluorescence detection. J. Chromatogr. B: Anal. Technol. Biomed. Life Sci. 832, 109–113.

Pellati, F., Orlandini, G., Pinetti, D., Benvenuti, S., 2011. HPLC-DAD and HPLC-ESI-MS/MS methods for metabolite profiling of propolis extracts. J. Pharmaceut. Biomedical Anal. 55, 934–948.

Pineiro, Z., Palma, M., Barroso, C.G., 2004. Determination of catechins by means of extraction with pressurized liquids. J. Chromatogr. A. 1026, 19–23.

Primikyri, A., Kyriakou, E., Charisiadis, P., Tsiafoulis, C., Stamatis, H., Tzakos, A.G., et al., 2012. Fine-tuning of the diffusion dimension of -OH groups for high resolution DOSY NMR applications in crude enzymatic transformations and mixtures of organic compounds. Tetrahedron 68, 6887–6891.

Reddy, G.N.M., Caldarelli, S., 2011. Maximum-quantum (MaxQ) NMR for the speciation of mixtures of phenolic molecules. Chem. Commun. 47, 4297–4299.

Reed, J.D., Krueger, C.G., Vestling, M.M., 2005. MALDI-TOF mass spectrometry of oligomeric food polyphenols. Phytochemistry 66, 2248–2263.

Robbins, R.J., 2003. Phenolic acids in foods: An overview of analytical methodology. J. Agric. Food Chem. 51, 2866–2887.

Rodrigues, E.D., Da Silva, D.B., De Oliveira, D.C.R., Da Silva, G.V.J., 2009. DOSY NMR applied to analysis of flavonoid glycosides from Bidens sulphurea. Magn. Reson. Chem. 47, 1095–1100.

Saigne-Soulard, C., Richard, T., Mérillon, J.M., Monti, J.P., 2006. ^{13}C NMR analysis of polyphenol biosynthesis in grape cells: Impact of various inducing factors. Anal. Chim. Acta. 563, 137–144.

Sakakibara, H., Honda, Y., Nakagawa, S., Ashida, H., Kanazawa, K., 2003. Simultaneous determination of all polyphenols in vegetables, fruits, and teas. J. Agric. Food Chem. 51, 571–581.

Saleem, M., Kim, H.J., Ali, M.S., Lee, Y.S., 2005. An update on bioactive plant lignans. Nat. Prod. Rep. 22, 696–716.

Savage, A.K., van Duynhoven, J.P.M., Tucker, G., Daykin, C.A., 2011. Enhanced NMR-based profiling of polyphenols in commercially available grape juices using solid-phase extraction. Magn. Reson. Chem. 49, S27–S36.

Schulz, H., Engelhardt, U.H., Wegent, A., Drews, H.H., Lapczynski, S., 1999. Application of near-infrared reflectance spectroscopy to the simultaneous prediction of alkaloids and phenolic substances in green tea leaves. J. Agric. Food Chem. 47, 5064–5067.

Shouqin, Z., Jun, X., Changzheng, W., 2005. Note: Effect of high hydrostatic pressure on extraction of flavonoids in propolis. Food Sci. Technol. Int. 11, 213–216.

Stalikas, C.D., 2007. Extraction, separation, and detection methods for phenolic acids and flavonoids. J. Sep. Sci. 30, 3268–3295.

Steinmann, D., Ganzera, M., 2011. Recent advances on HPLC/MS in medicinal plant analysis. J. Pharmaceut. Biomed. Anal. 55, 744–757.

Sun, J.H., Lin, L.Z., Chen, P., 2012. Study of the mass spectrometric behaviors of anthocyanins in negative ionization mode and its applications for characterization of anthocyanins and non-anthocyanin polyphenols. Rapid Commun. Mass Spectrom. 26, 1123–1133.

Švehlíková, V., Bennett, R.N., Mellon, F.A., Needs, P.W., Piacente, S., Kroon, P.A., et al., 2004. Isolation, identification and stability of acylated derivatives of apigenin 7-O-glucoside from chamomile (*Chamomilla recutita* [L.] Rauschert). Phytochemistry 65, 2323–2332.

Tatsis, E.C., Boeren, S., Exarchou, V., Troganis, A.N., Vervoort, J., Gerothanassis, I.P., 2007. Identification of the major constituents of *Hypericum perforatum* by LC/SPE/NMR and/or LC/MS. Phytochemistry 68, 383–393.

Tomas-Barberan, F.A., Ferreres, F., Gil, M.I., 2000. Antioxidant phenolic metabolites from fruit and vegetables and changes during postharvest storage and processing. In: Rahman, A. (Ed.), Studies in Natural Product Chemistry. Elsevier Science, pp. 739–795. Part D.

Tong, L., Wang, Y.Z., Xiong, J.F., Cui, Y., Zhou, Y.G., Yi, L.C., 2008. Selection and fingerprints of the control substances for plant drug Eucommia ulmodies Oliver by HPLC and LC-MS. Talanta 76, 80–84.

Valls, J., Millan, S., Marti, M.P., Borras, E., Arola, L., 2009. Advanced separation methods of food anthocyanins, isoflavones and flavanols. J. Chromatogr. A 1216, 7143–7172.

Van Der Hooft, J.J.J., Mihaleva, V., De Vos, R.C.H., Bino, R.J., Vervoort, J., 2011a. A strategy for fast structural elucidation of metabolites in small volume plant extracts using automated MS-guided LC-MS-SPE-NMR. Magn. Reson. Chem. 49, S55–S60.

Van Der Hooft, J.J.J., Vervoort, J., Bino, R.J., Beekwilder, J., De Vos, R.C.H., 2011b. Polyphenol identification based on systematic and robust high-resolution accurate mass spectrometry fragmentation. Anal. Chem. 83, 409–416.

Van Der Hooft, J.J.J., Akermi, M., Ünlü, F.Y., Mihaleva, V., Roldan, V.G., Bino, R.J., et al., 2012. Structural annotation and elucidation of conjugated phenolic compounds in black, green, and white tea extracts. J. Agric. Food Chem. 60, 8841–8850.

Vasilopoulou, C.G., Kontogianni, V.G., Linardaki, Z.I., Iatrou, G., Lamari, F.N., Nerantzaki, A.A., et al., 2013. Phytochemical composition of "mountain tea" from *Sideritis clandestina* subsp. *clandestina* and evaluation of its behavioral and oxidant/antioxidant effects on adult mice. Eur. J. Nut. 52, 107–116.

Vitrac, X., Castagnino, C., Waffo-Teguo, P., Delaunay, J.C., Vercauteren, J., Monti, J.P., et al., 2001. Polyphenols newly extracted in red wine from southwestern France by centrifugal partition chromatography. J. Agric. Food Chem. 49, 5934–5938.

Vukics, V., Guttman, A., 2010. Structural characterization of flavonoid glycosides by multi-stage mass spectrometry. Mass Spectrom. Rev. 29, 1–16.

Wallace, T.C., Giusti, M.M., 2010. Extraction and Normal-Phase HPLC-Fluorescence-Electrospray MS Characterization and Quantification of Procyanidins in Cranberry Extracts. J. Food Sci. 75, C690–C696.

Wang, C.Y., Lam, S.H., Tseng, L.H., Lee, S.S., 2011. Rapid screening of lignans from Phyllanthus myrtifolius and stilbenoids from *Syagrus romanzoffiana* by HPLC-SPE-NMR. Phytochem. Anal. 22, 352–360.

Wang, L.J., Weller, C.L., 2006. Recent advances in extraction of nutraceuticals from plants. Trends Food Sci. Technol. 17, 300–312.

Wang, S.P., Huang, K.J., 2004. Determination of flavonoids by high-performance liquid chromatography and capillary electrophoresis. J. Chromatogr. A 1032, 273–279.

Wawer, I., Zielinska, A., 2001. ^{13}C CP/MAS NMR studies of flavonoids. Magn. Reson. Chem. 39, 374–380.

Welch, C.R., Wu, Q.L., Simon, J.E., 2008. Recent advances in anthocyanin analysis and characterization. Curr. Anal. Chem. 4, 75–101.

Wijngaard, H., Brunton, N., 2009. The Optimization of Extraction of Antioxidants from Apple Pomace by Pressurized Liquids. J. Agric. Food Chem. 57, 10625–10631.

Wolfender, J.L., Ndjoko, K., Hostettmann, K., 2003. Liquid chromatography with ultraviolet absorbance-mass spectrometric detection and with nuclear magnetic resonance spectroscopy: A powerful combination for the on-line structural investigation of plant metabolites. J. Chromatogr. A 1000, 437–455.

Yang, F.Q., Zhao, J., Li, S.P., 2010. CEC of phytochemical bioactive compounds. Electrophoresis 31, 260–277.

Zeeb, D.J., Nelson, B.C., Albert, K., Dalluge, J.J., 2000. Separation and identification of twelve catechins in tea using liquid chromatography/atmospheric pressure chemical ionization-mass spectrometry. Anal. Chem. 72, 5020–5026.

Zgorka, G., Hajnos, A., 2003. The application of solid-phase extraction and reversed phase high-performance liquid chromatography for simultaneous isolation and determination of plant flavonoids and phenolic acids. Chromatographia 57, S77–S80.

Characterization of Polyphenolic Profile of *Citrus* Fruit by HPLC/ PDA/ESI/MS-MS

Monica Scordino, Leonardo Sabatino

University of Catania, Central Inspectorate Department of Protection and Prevention of Fraud quality of food products (ICQRF), Catania, Italy

CHAPTER OUTLINE HEAD

9.1 Introduction

The genus *Citrus*, belonging to the rue family (*Rutaceae*) yields pulpy fruits covered with fairly thick skins. Plants in this group include the lemon (*C. limon*), lime (*C. aurantifolia*), sweet orange (*C. sinensis*), sour orange (*C. aurantium*), tangerine (*C. reticulata*), grapefruit (*C. paradisi*), citron (*C. medica*), chinotto (*C. myrtifolia*), bergamot (*C. bergamia*), and shaddock (*C. maxima*, or *C. grandis*; pomelo). *Citrus* is believed to have originated 4000 years ago in the part of Southeast Asia bordered by Northeastern India, Myanmar, and the Yunnan province of China (Dugo and Di Giacomo, 2002). *Citrus* fruits are the most important fruit tree crop in the world, with an annual production of over 120 million tons, of which oranges constitute about 60% of the total production, followed by tangerines with about 20%. Lemons and limes are the third most important *Citrus* species, with an annual production of about 14 million tons (FAOSTAT, 2010). *Citrus* fruits are an important source of nutrients and health-promoting molecules, including vitamins, dietary fiber, carotenoids, and phenolic compounds (Horowitz, 1961; Peterson *et al.*, 2006; Tripoli *et al.*, 2007; González-Molina *et al.*, 2010; Hwang *et al.*, 2012).

This chapter examines literature that investigates the phenolics evaluation of *Citrus* fruits, focusing on HPLC/PDA/ESI/MS-MS profiles.

9.2 Polyphenolic compounds in citrus cultivars

Flavonoids are the most abundant phenolic compounds present in *Citrus* fruit. Flavonoids may be further divided into subclasses: anthocyanidins, flavanols, flavanones,

Polyphenols in Plants. http://dx.doi.org/10.1016/B978-0-12-397934-6.00009-7

187

flavonols, flavonones, and isoflavones. Flavonoids connected to one or more sugar molecules are known as flavonoid glycosides, while those that are not connected to a sugar molecule are called aglycones. Flavonoids occur in plants and most foods as glycosides. Flavonoid glycosides are distributed in different parts of a *Citrus* fruit, but the greatest amount is found in the solid parts: flavedo, albedo, and membranes. The content of flavonoids in peel is 10-times greater than the content of the juice. Among them, flavanone glycosides predominate, together with other small but distinctive flavonoids such as methoxylated flavones, anthocyanins, and flavone glycosides (Horowitz, 1961; Dugo and Di Giacomo, 2002). The literature is interested in *Citrus* flavonoids because of their potential health benefits and their use as a tool for chemotaxonomic markers (Maccarone *et al.*, 1998; Gattuso *et al.*, 2007; González-Molina *et al.*, 2010; Scordino *et al.*, 2011c; Hwang *et al.*, 2012).

Flavonoid is the general name of the compounds based upon a 15-carbon skeleton. The skeleton consists of two phenyl rings (A- and B-rings) connected by a three-carbon bridge (C-ring), which is a pyrone ring in the case of flavones and a dihydropyrone ring in the case of flavanones; anthocyanins are based on the flavylium salt structure (Tables 9.1–9.3). Numerous flavonoids occur in *Citrus* by additional hydroxyl, methoxyl, methyl, and/or glycosyl substituents. Occasionally, aromatic and aliphatic acids, methylenedioxyl or isoprenyl groups also attach to the flavonoid nucleus and their glycosides.

The subclass of flavanones is characterized by the absence of the double bond between the C-2 and C-3 positions, the presence of a chiral center at the C-2 position, and it contains a carbonyl group in the C-4 position of the C-ring (Table 9.1). The majority of *Citrus* flavanone glycosides are *O*-glycosides, with the sugar moiety linked generally to the C-7 hydroxyl group of the aglycone. Sweet orange is characterized by a high content of hesperidin together with small amounts of narirutin and didimin (Mouly *et al.*, 1998; Gattuso *et al.*, 2007; Barreca *et al.*, 2011). Lemon is characterized by the presence of eriocitrin, together with hesperidin, narirutin, and didimin. These last two flavonoids are present in trace amounts (Gattuso *et al.*, 2007; González-Molina *et al.*, 2010). Bergamot and grapefruit have the highest contents of flavonoid glycosides and contain both rutinoside flavonoids (hesperidin and narirutin) and neohesperidoside flavonoids (naringin and neohesperidin) (Gattuso *et al.*, 2006, 2007). The sugar neohesperidoside (2-*O*-α-L-rhamnosyl-β-D-glucose) imparts a bitter taste to the glycosides neohesperidin and naringin, compared to rutinose (6-*O*-α-L-rhamnosyl-β-D-glucose) which is responsible for a neutral taste. Naringin and neohesperidin are the flavanone-*O*-glycosides found in the highest amounts in chinotto fruit, together with small amount of neoriocitrin (Barreca *et al.*, 2010; Scordino *et al.*, 2011a, b). Because of their taxonomic proximity, neoeriocitrin was also found in bergamot (Gattuso *et al.*, 2006). Two 3-hydroxy-3-methylglutaryl flavanone glycosides, melitidin and brutieridin, were found to be present in chinotto and bergamot *Citrus* fruits (Di Donna *et al.*, 2009; Barreca *et al.*, 2011; Scordino *et al.*, 2011a, b).

Flavones (flavus = yellow) have substitutions on the A- and B-rings, a carbonyl group in the C-4 and a double bond between C-2 and C-3 of C-ring (Table 9.2).

Table 9.1 Flavanone-*O*-Glycosides of *Citrus* Cultivars

Name	R1	R2	R3
Eriodictyol 7-*O*-neohesperidoside (Neoeriocitrin)	*O*-Neohesperidose	OH	OH
Eriodictyol 7-*O*-rutinoside (Eriocitrin)	*O*-Rutinose	OH	OH
Hesperetin 7-(2″-α-rhamnosyl-6″-(3‴-hydroxy-3‴-methylglutaryl)-β-glucoside (Brutieridin)	3-Hydroxy-3-Methylglutaryl	H	OH
Hesperetin 7-*O*-neohesperidoside (Neoesperidin)	*O*-Neohesperidose	OH	OCH$_3$
Hesperetin 7-*O*-rutinoside (Hesperidin)	*O*-Rutinose	OH	OCH$_3$
Isosakuranetin 7-*O*-neohesperidoside (Poncirin)	*O*-Neohesperidose	H	OCH$_3$
Isosakuranetin 7-*O*-rutinoside (Didymin)	*O*-Rutinose	H	OCH$_3$
Naringenin 7-(2″-α-rhamnosyl-6″-(3‴-hydroxy-3‴-methylglutaryl)-β-glucoside (Melitidin)	3-Hydroxy-3-Methylglutaryl	OH	OCH$_3$
Naringenin 7-*O*-neohesperidoside (Naringin)	*O*-Neohesperidose	H	OH
Naringenin 7-*O*-rutinoside (Narirutin)	*O*-Rutinose	H	OH

Though flavone glycosides are commonly present in *Citrus* peel as *C*- and *O*-glycosides, other non-glycosilated flavones (polymethylated flavones) occur as components of essential oil fractions of *Citrus* peel (Dugo and Di Giacomo, 2002; Scordino *et al.*, 2011c).

Among flavone glycosides, orange is rich in 6,8-*C*-glucopyranosylapigenin with minor amounts of 6,8-di-*C*-glucopyranosyldiosmetin (Caristi *et al.*, 2006; Gattuso *et al.*, 2007). Lemon shows a large amount of 6,8-di-*C*-glucopyranosyldiosmetin compared to 6,8-di-*C*-glucopyranosylapigenin (Gattuso *et al.*, 2007; González-Molina *et al.*, 2010). Small amounts of these two *C*-glycosides were found in tangerine, while in clementine their relative amounts were negligible (Gattuso *et al.*, 2007). Bergamot and chinotto showed comparable levels of the two *C*-glycosides. Flavones *O*-glycosides (mainly rhoifolin and neodiosmin) were characteristic of chinotto and bergamot fruits (Gattuso *et al.*, 2006; Barreca *et al.*, 2010; Scordino *et al.*, 2011a, b).

The knowledge of the polymethoxylated flavone composition and their relative concentration in various essential oils has an important taxonomic significance for their differentiation. Sweet orange and tangerine are both characterized by tangeretin, 3,5,6,7,8,3′,4′-heptamethoxyflavone, nobiletin, and sinensetin, while 3,5,6,7,3′,4′-hexamethoxyflavone was only found in sweet orange (Mouly *et al.*, 1998; Wu *et al.*, 2007; Zhang *et al.*, 2011, 2012). Considering the high content

Table 9.2 Flavone Glycosides and Polymethoxylated Flavones of *Citrus* Cultivars

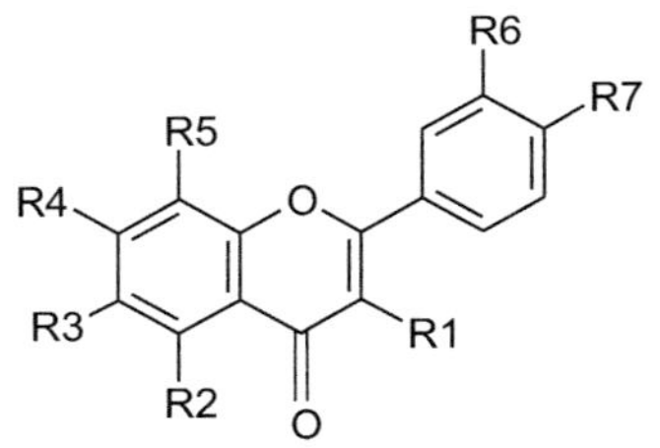

Name	R1	R2	R3	R4	R5	R6	R7
Apigenin 7-O-neohesperidoside (Rhoifolin)	H	OH	H	O-Nh[a]	H	H	OH
Diosmetin 7-O-neohesperidoside (Neodiosmin)	H	OH	H	O-Nh[a]	H	OH	OCH$_3$
Chrysoeriol 7-O-neohesperidoside	H	OH	H	O-Nh[a]	H	OCH$_3$	OH
Diosmetin 7-O-rutinoside (Diosmin)	H	OH	H	O-Ru[b]	H	OH	OCH$_3$
Luteolin 7-O-rutinoside	H	OH	H	O-Ru[b]	H	OH	OH
Diosmetin 6,8-di-C-glucoside (Lucenin-2 4′-Me)	H	OH	Glu[c]	OH	Glu[c]	OH	OCH$_3$
Luteolin 6,8-di-C-glucoside (Lucenin-2)	H	OH	Glu[c]	OH	Glu[c]	OH	OH
Apigenin 6,8-di-C-glucoside (Vicenin-2)	H	OH	Glu[c]	OH	Glu[c]	H	OH
Chrysoeriol 6,8-di-C-glucoside (Stellarin-2)	H	OH	Glu[c]	OH	Glu[c]	OCH$_3$	OH
3′,4′,5,6,7-pentamethoxyflavone (Sinensetin)	H	OCH$_3$	OCH$_3$	OCH$_3$	H	OCH$_3$	OCH$_3$
5,6,7,4′-tetramethoxyflavone (Scutellarein tetramethylether)	H	OCH$_3$	OCH$_3$	OCH$_3$	H	H	OCH$_3$
3′,4′,5,6,7,8-hexamethoxyflavone (Nobiletin)	H	OCH$_3$	OCH$_3$	OCH$_3$	OCH$_3$	OCH$_3$	OCH$_3$
3,5,6,7,8,3′,4′-heptamethoxyflavone	OCH$_3$	OCH$_3$	OCH$_3$	OCH$_3$	OCH$_3$	OCH$_3$	OCH$_3$
4′,5,6,7,8,-pentamethoxyflavone (Tangeretin)	H	OCH$_3$	OCH$_3$	OCH$_3$	OCH$_3$	H	OCH$_3$

[a]*Neohesperidose.*
[b]*Rutinose.*
[c]*Glucose.*

Table 9.3 Anthocyanins of Pigmented *Citrus* Cultivars

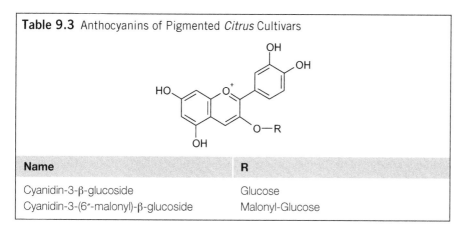

Name	R
Cyanidin-3-β-glucoside	Glucose
Cyanidin-3-(6″-malonyl)-β-glucoside	Malonyl-Glucose

of these compounds in the flavedo, it follows that their concentration in the juice depends on the extraction technology used (Scordino *et al.*, 2011c).

Citrus sinensis varieties Moro, Tarocco, and Sanguinello, also known as red oranges and typically grown in the Etna volcano region of Sicily (Italy) as well as in Florida (USA), are the unique *Citrus* fruit containing anthocyanins responsible for their brilliant red color (Dugo *et al.*, 2003). Anthocyanins are based on the anthocyanidin skeleton, which has extra unsaturation in the C ring forming the pyrylium system (Table 9.3). Anthocyanidins are not stable due to the –OH group at the C-3 position and are found in nature with a sugar linkage at this position. Cyanidin 3-glucoside and cyanidin 3-(6″-malonyl)-β-glucoside (Maccarone *et al.*, 1998) are the main components of the red oranges, together with a dozen other minor constituents (Dugo *et al.*, 2003).

9.3 HPLC/PDA/ESI/MS-MS identification of citrus flavonoids

Various analytical approaches have succeeded over the years in the determination of the content of polyphenolic compounds in fruits. Among the various techniques are gas chromatography and high-performance liquid chromatography (HPLC). The approach with gas chromatography is not used any more because the polyphenols are poorly volatile, especially the glycosylated ones. Therefore, this method would require laborious procedures of extraction and subsequent derivatization. The HPLC technique is the most suitable method of analysis because no complex extraction or derivatization processes are required. Although many authors have extensively studied the identification of flavonoids from *Citrus* (Mouly *et al.*, 1998; Gattuso *et al.*, 2007; Wu *et al.*, 2007; Barreca *et al.*, 2011; Scordino *et al.*, 2011a, b), a well-defined analytical procedure is not standardized any more for either extraction or chromatographic separation. *Citrus* juices can be directly analyzed after filtration without any pretreatment. For the extraction of flavonoids from the whole fruit or any part thereof, the sample may be extracted with acidified aqueous methanol, the mixture centrifuged, and the supernatant decanted. For quantitative purposes, the pellets could be re-extracted under identical conditions. A fast clean-up of the raw extracts using a C-18 Sep-Pak cartridge could be performed before analysis to eliminate carbohydrates and other polar substances or to obtain a concentrated extract. A further separation of anthocyanins from other flavonoids can be achieved by washing the C-18 cartridge with ethyl acetate before elution with methanol.

The use of chromatographic columns in reverse phase (predominantly C18 and C8) and a gradient of aqueous mobile phases containing methanol or acetonitrile, slightly acidified with short-chain organic acids (formic acid and acetic acid) promotes optimum separation of flavonoids, due to their chemical polar structure. Under these chromatographic conditions, more polar compounds elute earlier; consequently, the aglycones will be eluted subsequently to mono- and polyglycosides. Generally, analytical columns used for this approach are thermostated C18 250×4.6 mm, 5 μm i.d., gradient flow rate of about 1 ml/min and an injection volume of 20 μl. Alternatively, fast liquid chromatography systems with reduced column dimensions can be used, lowering flow rate and injection volumes. A binary gradient of formic acid in water and formic acid in acetonitrile or methanol are generally employed for flavonoids separations.

The hyphenated technique of HPLC coupled with photo diode array (PDA) spectrophotometers and soft electrospray ionization (ESI) with tandem mass spectrometry is the approach most recently used for the characterization of polyphenolic profiles of the fruit (Fabre *et al.*, 2001; Dugo *et al.*, 2003; Gattuso *et al.*, 2007; Abad-Garcìa *et al.*, 2009; Mencherini *et al.*, 2013; Simirgiotis *et al.*, 2012). The PDA chromatograms at different wavelengths allow for the discrimination of the various classes of *Citrus* phenolics because of the characteristic absorption spectra due to aromatic molecular structure. A range of photodiode-array detector wavelengths from 200 to 700 nm cover all flavonoids adsorption range, including the red-colored anthocyanins. The flavonoid PDA spectra are characterized by two bands, which can be attributed to bands II (due to the A ring benzoyl system) and I (associated with absorption due to the B ring cinnamoyl system) of a phenolic structure, respectively. The PDA spectra could give information about the rings' substituents, showing adsorption ranges from 240 to 280 nm for band II and from 300 to 380 nm for band I. For example, the flavanone nature of the aglycone has absorptions centered at 285 and 330 nm (Figure 9.1), while λ_{max} at 335 nm suggests the presence of flavones (Figure 9.2). Because of pH-dependent conformational rearrangement of the anthocyan molecule, under analytical acidic conditions, the C-ring acquires aromaticity involving a flavylium cation, which imparts intense red color with increasing absorption at about 520 nm (Figure 9.3). For quantitative determinations the chromatograms could be recorded at appropriate λ_{max}, i.e., 285 nm for flavanones, 335 nm for flavones, and 520 nm for anthocyanins.

However, it is not possible to discriminate spectrophotometrically among the various flavanones, flavones, and anthocyanins simply on the basis of their absorption spectra, without the comparison with a reference standard or the aim of the mass spectra information. Mass spectra recorded in positive and negative ionization mode allow the determination of the molecular mass, generally providing the protonated $[M+H]^+$ and deprotonated $[M-H]^-$ quasi-molecular ions, respectively. The soft atmospheric electrospray ionization

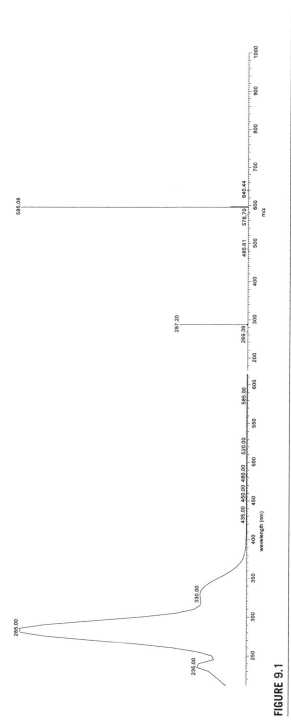

FIGURE 9.1

PDA spectrum and negative ESI/MS-MS fragmentation pattern of flavanone eriocitrin.

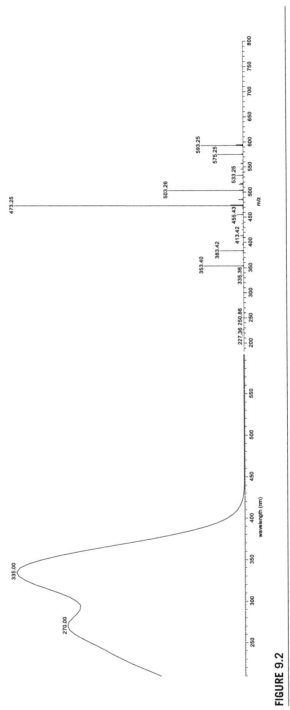

FIGURE 9.2

PDA spectrum and negative ESI/MS-MS fragmentation pattern of flavone vicenin-2.

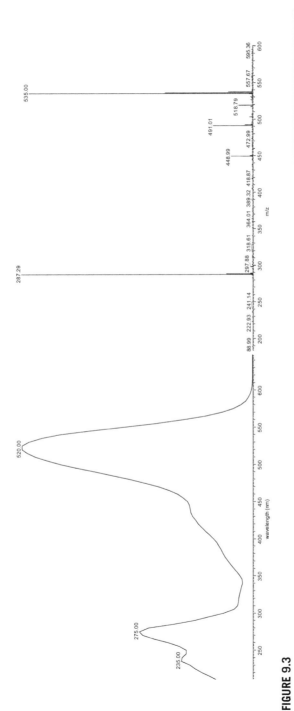

FIGURE 9.3

PDA spectrum and positive ESI/MS-MS fragmentation pattern of anthocyanin cyanidin 3-(6″-malonyl)-β-glucoside.

(ESI) could operate both in negative and positive ion modes using an ion spray LC/MS interface; depending on the nature of the studied flavonoids, the appropriate ionization mode must be chosen. Generally, flavanones and flavones could be detected both in positive and negative modes, providing the protonated $[M+H]^+$ and deprotonated $[M-H]^-$ quasi-molecular ions, respectively. The total ion chromatograms obtained in negative ionization generally have a lower background noise signal and are generally preferred for these molecules' investigation. The positive charge of anthocyanins at low pH values permits their easy detection only in positive ionization mode. Similarly, for the presence of numerous methoxyl and hydroxyl groups, the class of polymethoxylated flavones could be exclusively analyzed in positive ionization mode. For more information on the conformational structure of studied flavonoids, in recent times tandem mass spectrometry has been affirmed as the election technique. The MS/MS spectra allowed the establishment of an unambiguous structural assignment for all compounds because of the characteristic losses of aglyconic structure. The MS/MS spectra could be obtained generally in Selected Reaction Monitoring and in Product Ion Scan modes, depending on instrumental typology (triple quadrupole, ion trap, etc.), and system configuration. Preliminary positive and negative tunings carried out with continuous introduction of dilute solutions of reference molecules are recommended in order to optimize the instrumental conditions (tune lens, optics, collision energy, spray voltages, capillary heat, sheet flow, sheet gas, etc.) for better results. For these kinds of molecules, soft collision energy of about 15–35% of instrument maximum permitted the attainment of informative MS/MS spectra. Especially for diagnostic purposes, the Product Ion Scan mode, in which the fragmentation of a selected precursor ion to product ions is monitored, is chosen in mass spectrometry for qualitative reasons because a good degree of sensitivity and selectivity are achieved. In Product Ion Scan mode ions of all m/z ratios, except that of interest, are fragmented; only ions of selected m/z ratio are then excited to generate the production spectrum that is characteristic of the molecule of interest. Each analyte showed a typical mass spectrum profile. Under these conditions neutral loss of water, sugar moieties and the breakage of methyl and methoxy groups could be generate depending on the molecule structure. For each flavonoid a typical mass spectrum profile could be obtained and identification of the most meaningful fragments could be assigned.

The MS of flavanones showed ions at $[M+H]^+$ and $[M-H]^-$ in positive and negative modes, respectively, in agreement with the molecular weight. The common features of MS/MS data reported in negative ionization a common fragment of $[M-H-308]^-$ (Figure 9.1), and in positive ionization the characteristic two ions $[M+H-146]^+$ and $[M+H-308]^+$ (Gattuso *et al.*, 2007; Wang and Feng, 2009; Barreca *et al.*, 2011; Scordino *et al.*, 2011a, b). In particular neohesperidose favors the *O*-diglycoside breakage to yield the residue of 308 u, while rutinose favors the loss of the dehydrated glucose residue of 146 u. In the case of 3-hydroxy-3-methylglutaric acid conjugates of flavanones, diagnostic positive daughter ions ion, corresponding to the loss of 3-hydroxy-3-methylglutaryl moiety, could be observed together with a fragment of flavonoids diglycosides originating from the breakage of the *O*-C-1 and C-2/C-3 bonds of the hexose directly attached to the aglycon; moreover, the loss of the rhamnose unit could be observed in negative ionization (Di Donna *et al.*, 2009; Barreca *et al.*, 2011; Scordino *et al.*, 2011a, b).

Retro Dies-Alder reaction of the flavanone C ring can be also observed in case of deprotonated molecules (Wang and Feng, 2009).

Concerning *C*-glycosides and *O*-glycosides flavones, both positive and negative ionization could be used to obtain the quasi molecular ions $[M+H]^+$ and $[M-H]^-$, respectively. *C*-glycoside flavones break easily by the loss of sugar moiety. The MS/MS approach allows the elucidation of the type of the sugar moiety of *C*-glycosylation and the differentiation of position (C-6/C-8) of the sugar moieties. The type and number of fragments formed indicate the position of the link between the sugar and the aglycone. Ions produced from the internal cleavage of the sugars from *C*-glycosylation are generally $[M-H-18]^-$, $[M-H-90]^-$, $[M-H-120]^-$, $[M-H-210]^-$, and $[M-H-240]^-$ (Ferreres *et al.*, 2007; Gattuso *et al.*, 2007) (Figure 9.2).

As previously reported, polymethoxylated flavones could be analyzed only in ESI-positive ionization; mass spectra showed the quasi-molecular ions $[M+H]^+$ with a characteristic fragmentation pattern due to the breakage of methyl groups with principal ions $[M\text{-}n \times CH_3^{\bullet}]^+$ (mainly 15 u, 30 u, and 60 u). Moreover, other diagnostic fragments are due to the loss of 16 u (CH_4), 18 u (H_2O), 28 u (CO) and relative loss of methyl groups (Scordino *et al.*, 2011c; Zhang *et al.*, 2012).

Anthocyanins' positive MS spectrum produces an intense quasi-molecular ion $[M+H]^+$. MS/MS data obtained by the fragmentation of anthocyanins permit the detection of the corresponding anthocyanidin, with the loss of the sugar moiety and the production of the positively charged aglycon (Dugo *et al.*, 2003) (Figure 9.3).

As a result of numerous studies conducted on the flavonoids in *Citrus* fruits in recent years, a large number of quantitative data and exhaustive distribution patterns are available. Each type of *Citrus* fruit has a characteristic fingerprint that allows its use as a chemotaxonomic marker. The use of hyphenated techniques HPLC/PDA/ESI/MS-MS as election techniques allows us to avoid the time-consuming preparative purification step, facilitating the sample handling and decreasing the time of analysis and solvent consumption. Moreover, the reverse-phase separation coupled with MS/MS detection allows the secure identification of investigated molecules with a minimization of matrix effects. The sensitivity, precision, and accuracy achieved with the HPLC/PDA/ESI/MS-MS method also make it adequate for quantification purposes.

References

Abad-García, B., Berrueta, L.A., Garmón-Lobato, S., Gallo, B., Vicente, F., 2009. A general analytical strategy for the characterization of phenolic compounds in fruit juices by high-performance liquid chromatography with diode array detection coupled to electrospray ionization and triple quadrupole mass spectrometry. J. Chromatogr. A 1216, 5398–5415.

Barreca, D., Bellocco, E., Caristi, C., Leuzzi, U., Gattuso, G., 2010. Flavonoid composition and antioxidant activity of juices from Chinotto (*Citrus* x myrtifolia Raf.) fruits at different ripening stages. J. Agric. Food Chem. 58, 3031–3036.

Barreca, D., Bellocco, E., Caristi, C., Leuzzi, U., Gattuso, G., 2011. Distribution of *C*- and *O*-glycosyl flavonoids, (3-hydroxy-3-methylglutaryl)glycosyl flavanones and furocoumarins in *Citrus aurantium* L. juice. Food Chem. 124, 576–582.

Caristi, C., Bellocco, E., Gargiulli, E., Toscano, G., Lezzi, E., 2006. Flavone-di-*C*-glycosides in citrus juices from Southern Italy. Food Chem. 95, 431–437.

Di Donna, L., De Luca, G., Mazzotti, F., Napoli, A., Salerno, R., Taverna, D., et al., 2009. Statin-like principles of bergamot fruit (*Citrus bergamia*): isolation of 3-hydroxymethylglutaryl flavonoids glycosydes. J. Nat. Prod. 72, 1352–1354.

Dugo, G., Di Giacomo, A., 2002. Citrus: The Genus *Citrus*. CRC Press 2002.

Dugo, P., Mondello, L., Morabito, D., Dugo, G., 2003. Characterization of the anthocyanin fraction of sicilian blood orange juice by micro-HPLC-ESI/MS. J. Agric. Food Chem. 51, 1173–1176.

Fabre, N., Rustan, I., Hoffmann, E., Quetin-Leclercq, J., 2001. Determination of flavone, flavonol, and flavanone aglycones by negative ion liquid chromatography electrospray ion trap mass spectrometry. J. Am. Soc. Mass Spectrom. 12, 707–715.

Ferreres, F., Gil-Izquierdo, A., Andrade, P.B., Valentão, P., Tomás-Barberán, F.A., 2007. Characterization of *C*-glycosyl flavones *O*-glycosylated by liquid chromatography-tandem mass spectrometry. J. Chromatogr. A 1161, 214–223.

Food and Agricultural Organization of United Nations: Economic and Social Department: The Statistical Division. http://faostat.fao.org/site/567/DesktopDefault.aspx

Gattuso, G., Barreca, D., Gargiulli, C., Leuzzi, U., Caristi, C., 2007. Flavonoid composition of *Citrus* juice. Molecules 12, 1641–1673.

Gattuso, G., Caristi, C., Gargiulli, C., Bellocco, E., Toscano, G., Leuzzi, U., 2006. Flavonoid glycosides in bergamot juice (*Citrus bergamia* Risso). J. Agric. Food Chem. 54, 3929–3935.

González-Molina, E., Domínguez-Perles, R., Moreno, D.A., García-Viguera, C., 2010. Natural bioactive compounds of *Citrus limon* for food and health. J. Pharmaceut. Biomed. Anal. 51, 327–345.

Horowitz, R.M., 1961. The *Citrus* flavonoids. In: Sinclair, W.B. (Ed.), The orange. Its biochemistry and physiology. University of California, Division of Agricultural Sciences, pp. 334–372.

Hwang, S.L., Shih, P.H., Yen, G.C., 2012. Neuroprotective effects of *Citrus* flavonoids. J. Agric. Food Chem. 60, 877–885.

Maccarone, E., Rapisarda, P., Fanella, F., Arena, E., Mondello, L., 1998. Cyanidin 3-(6″-malonyl)-β-glucoside. One of the major anthocyanins in blood orange juice. Ital. J. Food Sci. 10, 367–372.

Mencherini, T., Campone, L., Piccinelli, A.L., García Mesa, M., Sánchez, D.M., Aquino, R.P., et al., 2012. HPLC-PDA-MS and NMR characterization of a hydroalcoholic extract of *Citrus aurantium* L. var. *amara* peel with antiedematogenic activity. J. Agric. Food Chem. 2013, 61(8), 1686–1693.

Mouly, P., Gaydou, E.M., Auffray, A., 1998. Simultaneous separation of flavanone glycosides and polymethoxylated flavones in citrus juices using liquid chromatography. J. Chromatogr. A 800, 171–179.

Peterson, J.J., Dwyer, J.T., Beecher, G.R., Bhagwat, S.A., Gebhardt, S.E., et al., 2006. Flavanones in oranges, tangerines (mandarins), tangors, and tangelos: a compilation and review of the data from the analytical literature. J. Food Comp. Anal. 19, S66–S73.

Scordino, M., Sabatino, L., Belligno, A., Gagliano, G., 2011a. Characterization of polyphenolic compounds in unripe chinotto (*Citrus myrtifolia*) fruit by HPLC/PDA/ESI/MS-MS. Nat. Prod. Commun. 6, 1857–1862.

Scordino, M., Sabatino, L., Belligno, A., Gagliano, G., 2011b. Flavonoids and furanocoumarins distribution of unripe Chinotto (*Citrus* × myrtifolia Rafinesque) fruit: beverage processing homogenate and juice characterization. Eur. Food Res. Technol. 233, 759–767.

Scordino, M., Sabatino, L., Traulo, P., Gargano, M., Pantó, V., Gagliano, G., 2011c. HPLC-PDA/ESI-MS/MS detection of polymethoxylated flavonoids highly degraded citrus juice: a quality control case study. Eur. Food Res. Technol. 232, 275–280.

Simirgiotis, M.J., Silva, M., Becerra, J., Schmeda-Hirschmann, G., 2012. Direct characterisation of phenolic antioxidants in infusions from four Mapuche medicinal plants by liquid chromatography with diode array detection (HPLC-DAD) and electrospray ionisation tandem mass spectrometry (HPLC-ESI–MS). Food Chem. 131, 318–327.

Tripoli, E., La Guardia, M., Giammanco, S., Di Majo, D., Giammanco, M., 2007. Citrus flavonoids: Molecular structure, biological activity and nutritional properties: A review. Food Chem. 104, 466–479.

Wang, H., Feng, F., 2009. Identification of components in Zhi-Zi-Da-Huang decoction by HPLC coupled with electrospray ionization tandem mass spectrometry, photodiode array and fluorescence detectors. J. Pharmaceut. Biomed. Anal. 49, 1157–1165.

Wu, H.W., Lei, H.M., Li, Q., Bi, W., 2007. HPLC determination of three polymethoxylated flavones in the fraction of orange peel. Chin. J. Pharm. Anal. 27, 1895–1897.

Zhang, J.Y., Li, N., Che, Y.Y., Zhang, Y., Liang, S.X., Zhao, M.B., et al., 2011. Characterization of seventy polymethoxylated flavonoids (PMFs) in the leaves of *Murraya paniculata* by on-line high performance liquid chromatography coupled to photodiode array detection and electrospray tandem mass spectrometry. J. Pharm. Biomed. Anal. 56, 950–961.

Zhang, J.Y., Zhang, Q., Zhang, H.X., Ma, Q., Lu, J.Q., Qiao, Y.J., 2012. Characterization of polymethoxylated flavonoids (PMFs) in the peels of 'Shatangju' mandarin (*Citrus reticulata Blanco*) by online high-performance liquid chromatography coupled to photodiode array detection and electrospray tandem mass spectrometry. J. Agric. Food Chem. 60, 9023–9034.

Isolation and Extraction Techniques

Non-Extractable Polyphenols in Plant Foods: Nature, Isolation, and Analysis

10

Jara Pérez-Jiménez, M Elena Díaz-Rubio, Fulgencio Saura-Calixto

Department of Metabolism and Nutrition, Institute of Food Science, Technology and Nutrition, Madrid, Spain

CHAPTER OUTLINE HEAD

10.1 Introduction

Separate from the role of polyphenols as plant secondary metabolites (defending against attack by herbivores, infection and UV radiation (Winkel-Shirley, 2002), among other funtions), interest in food polyphenols has shifted from traditional studies of their contribution to food sensory properties to an increasing interest in their potential health effects.

The polyphenol content of foods is usually considered to be that found in aqueous–organic extracts obtained from plant foods. Nevertheless, the polyphenols found in such extracts would correspond to only a fraction of these phytochemicals: the extractable polyphenols (EPP). The extractions leave a residue which is commonly disregarded but which may present a high content in the understudied fraction of plant food polyphenols, the so-called non-extractable polyphenols (NEPP) (Bravo *et al.*, 1994).

NEPP consist of polyphenols that belong to different classes, such as macromolecular polyphenols or single polyphenols associated with cell wall macromolecules

Polyphenols in Plants. http://dx.doi.org/10.1016/B978-0-12-397934-6.00010-3

(Bravo *et al.*, 1994; Arranz *et al.*, 2009; Saura-Calixto, 2011). Although only a small part of the existing literature on polyphenols in plant foods deals with NEPP, the existing data show that they are even more abundant than EPP in many plant foods (Arranz *et al.*, 2009; Pérez-Jiménez *et al.*, 2009). The NEPP may have potential applications in animal nutrition (Brenes *et al.*, 2008; Viveros *et al.*, 2011), the food industry (Sánchez-Alonso *et al.*, 2006, 2008; Sáyago-Ayerdi *et al.*, 2009a, b) and in human health (Pérez-Jiménez *et al.*, 2008b; Lizarraga *et al.*, 2011). This chapter aims to provide an overview of NEPP in plant foods. It includes a description of their nature, procedures for their isolation, some methods currently used in their analysis, data concerning their content in different foods and extracts, as well as a summary of their potential applications.

10.2 Nature of NEPP

Polyphenols are synthesized as part of the shikimic acid pathway. That pathway is the origin of several phenolic acids (a class of polyphenols) and *p*-coumaroyl Co-A, the latter of which then gives rise to lignans and, by combination with the malonic acid pathway, to stilbenes and flavonoids (Saltveit, 2010), the other three most common classes of polyphenols. In these processes, several kinds of structures are synthesized, some of them found mainly as EPP and others mainly as NEPP. NEPP mainly include: (a) a fraction of phenolic acids and some flavonoid subclasses that are associated with the cell wall; and (b) high-molecular-weight tannins, which are formed at the end of this biosynthesis route and consist of hydrolyzable tannins (derived from the polymerization of phenolic acids and simple sugars), and condensed tannins or proanthocyanidins (from the polymerization of flavanols via a process that could be spontaneous, enzyme catalyzed, or a mixture of the two (Tsao and McCallum, 2010). As examples of NEPP, Figure 10.1 shows the structures of a phenolic compound associated with the cell wall (A) and a high-molecular-weight proanthocyanidin (B).

So, NEPP are present in plant foods as high-molecular-weight molecules and compounds that are weakly associated with the food matrix (class I NEPP); or as high-molecular-weight molecules and compounds that are strongly associated with the different constituents of the cell wall (class II NEPP), i.e., neutral polysaccharides, acidic pectic substances and structural proteins (Pinelo *et al.*, 2006).

The interactions between NEPPs and the cell wall depend on certain characteristics of the cell wall (porosity, proportions of the different constituents, flexibility, etc.) as well as on certain aspects of the NEPP (class of polyphenol, stereochemistry, conformational flexibility, molecular weight, percentage of galloylation in the case of non-extractable proanthocyanidins, etc.) (Le Bourvellec *et al.*, 2004, 2005; Pinelo *et al.*, 2006; Bindon *et al.*, 2010). Specific combinations of NEPP and cell wall variables may give rise to the formation of hydrogen or covalent (ester and ether) bonds, or to the encapsulation of the NEPP within hydrophobic pockets, among other associations between the two (Matthews *et al.*, 1997; Faulds and Williamson, 1999; Sun *et al.*, 2002; Le Bourvellec *et al.*, 2004, 2005; Burr and Fry, 2009; Bunzel, 2010).

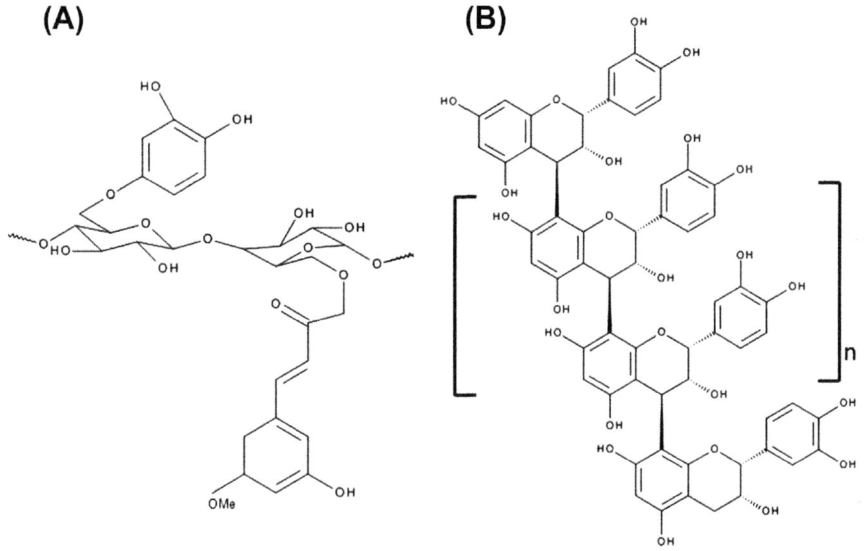

(A)　　　　　　　　　**(B)**

FIGURE 10.1

Examples of some non-extractable polyphenols. (A), Phenolic compound associated with cell wall. (B), High molecular weight proanthocyanidin

Although most interactions between NEPP and the plant food matrix take place within the cell wall, some associations between polyphenols and proteins have been described in vacuoles, which give rise to vacuolar inclusions (Markham *et al.*, 2001), as well as in the cell nucleus (Grandmaison and Ibrahim, 1996). However, little is currently known about these associations.

10.3 **Isolation of NEPP**

The isolation of NEPP from plant foods requires several steps, in order to remove low-molecular-weight phenolic compounds and other soluble substances, and to disrupt macromolecules—mainly protein and starch (Figure 10.2). Due to the complex nature of NEPP, it is quite difficult to obtain them as pure compounds and the product of isolation is a concentrate of these kinds of compounds. Although specific adaptations may be performed for particular plant foods, the general process for the isolation of NEPP from plant foods includes the following steps:

(1) Removal of EPP and other soluble substances.

　(A) *Aqueous–organic extraction.* Different solid–liquid extractions, combining water with various organic solvents, may be carried out to remove soluble substances, including EPP, soluble sugars, minerals, vitamins, organic acids, and others; Table 10.1 shows the recommended solvents for the extraction of different specific polyphenol subclasses. For instance, a

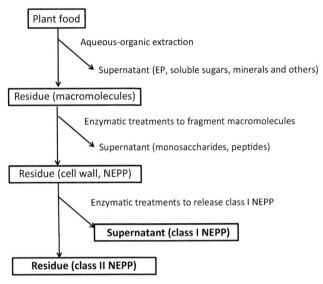

FIGURE 10.2

Scheme of NEPP isolation from plant foods.

Table 10.1 Recommended Solvents for the Extraction of Specific Polyphenol Classes

Class	Subclass	Solvent	References
Flavonoids	Flavanols	Acetone: water: acetic acid	Gu et al., 2002
	Other flavonoids	Methanol, methanol: water	Umphress et al., 2005; Touriño et al., 2008; Ferreres et al., 2011
Phenolic acids	Hydroxybenzoic acids	Ethyl acetate	de Simón et al., 1990
	Hydroxycinnamic acids	Diethyl ether	de Simón et al., 1990
Stilbenes		Ethanol: water 80:20	Chukwumah et al., 2007
Lignans		Methanol: water 70:30	Milder et al., 2004
Others	Tyrosols	Dimethylformamide	García et al., 2002
	Alkylresorcinols	Dichloromethane	Knödler et al., 2007

common procedure for the separation of EPP from plant foods includes a first extraction with acidified methanol/water, in order to extract the most polar polyphenols, such as phenolic acids. This is then followed by an extraction with acetone/water, to extract more apolar polyphenols, e.g., proanthocyanidins (Pérez-Jiménez et al., 2008a). Such a procedure has been applied to several plant foods and derived products, such as mango peel, nuts, cereal products, edible seaweeds, or tropical fruits (Larrauri et al., 1997; Arranz et al., 2008; Díaz-Rubio et al., 2009; Arranz and Saura Calixto, 2010; Rufino et al., 2010).

The same solvents will also extract most low-molecular-weight food components, although specific extraction methods may be required for some of them, depending on the particular composition of the food. For example, extraction with 80% ethanol, applied to carob pods, released 47% of soluble sugars, in addition to EPP (Saura Calixto, 1987).

- **(B)** *Enzymatic extraction.* The residue from the previous step is treated with several enzymes (proteases, amylases, lipases, etc.) that are able to hydrolyze macromolecular nutrients (starch, proteins, lipids, etc.), releasing polyphenols associated with the macromolecules or encapsulated in the food matrix. The resulting supernatant contains fragments from the macromolecules (peptides, monosaccharides, etc.) as well as soluble polyphenols; for instance, acid treatment followed by amylase treatment of barley released some phenolic acids (p-hydroxybenzoic acid, caffeic acid) that were not released by the acid treatment alone (Yu et al., 2001).

(2) Enzymatic treatments to release class I NEPP. Class I NEPP are those NEPP that can be extracted from the food matrix by the action of several cell wall enzymes. To release them, the residue from step 1B is treated with a combination of degrading cell wall enzymes, such as pectinases, cellulases, xylanases, esterases or others. For instance, treatment of barley with acid, amylase, and pectinase released twice as much chlorogenic acid as that released by treatment with just acid and amylase (Yu et al., 2001).

(3) Class II NEPP concentrate. Class II NEPP are polyphenols that are strongly associated with the cell wall or high-molecular-weight molecules. Therefore, the solid that remains after all the previous procedures is a class II NEPP concentrate that may be dried by different techniques.

Class II NEPP constitutes the major NEPP fraction. For instance, when the residue from an aqueous–organic extraction of nectarine was treated with a mixture of enzymes, the class I NEPP content was 4 mg/100 g dw, while when both class I and class II NEPP were determined (by performing different hydrolysis on the residue of the aqueous–organic extraction of the sample), the NEPP content was 450 mg/100 g dw (Arranz et al., 2009; Pérez-Jiménez et al., 2009).

Since work on NEPP and in particular NEPP isolation is still scarce, the optimization of the procedure described for obtaining them has not yet been studied. However, some steps of the process may be improved by combining enzymatic treatments with novel strategies, such as microwaves or ultrasound.

10.4 Analysis of NEPPS

After the isolation of NEPP from plant foods, either by the procedure described above or by some similar procedure, their content and structure can be analyzed, taking into account that the procedure will be different for class I and class II NEPP.

10.4.1 Analysis of class I NEPP

Since class I NEPP are obtained as a supernatant, they are analyzed in the same way as EPP, using HPLC-DAD methods or, as is becoming more common, HPLC-MS/MS methods (which allow unequivocal identification of compounds even in the absence of commercial standards). Nevertheless, due to the low concentration of some NEPP in the supernatant, a prior concentration step may be needed.

The class I NEPP content of apple, peach and nectarine has been evaluated by HPLC-MS after they were released by a combination of a pectinase, an esterase, and a protease (Pérez-Jiménez et al., 2009). One possible problem in the analysis of class I NEPP arises from the fact that they may include, besides low-molecular-weight compounds detectable by HPLC, some large polymers with molecular weights of several thousand Daltons, which may not be detected by HPLC-MS. MALDI-TOF MS, which has been shown to be useful in the analysis of polymeric EPP (Monagas et al., 2010) might be used for the analysis of polymeric class I NEPP.

It should also be noted that there are many reports of treatments with enzymes that degrade the cell wall directly in the plant food, without previous separation of EPP (Pinelo et al., 2008; Kapasakalidis et al., 2009; Chamorro et al., 2012). The results obtained in such cases would therefore correspond to both EPP and class I NEPP.

10.4.2 Analysis of class II NEPP

In contrast to class I NEPP, class II NEPP includes high-molecular-weight polyphenols and phenolic compounds strongly associated with the cell wall. They cannot be directly analyzed in the concentrates that result from the treatment of plant foods described above, making it necessary to carry out different hydrolysis procedures in order to analyze the corresponding products in the hydrolyzates.

The analysis of class II NEPP involves two main categories of compounds: (a) polymeric proanthocyanidins; and (b) hydrolyzable phenolic compounds, including polymeric hydrolyzable tannins, and both of them can be analyzed by different methodologies (Figure 10.3). It is quite remarkable that most work on the analysis of class II NEPP focuses on only one of these categories while both of them are present in significant amounts in many plan foods (Pérez-Jiménez and Torres, 2011). Consequently, studies that provide a complete approach for the analysis of all class II NEPP are very scarce (Arranz et al., 2009, 2010).

As regards the analysis of non-extractable proanthocyanidins, several methods have been proposed: treatment with butanol–HCl, followed by spectrophotometric

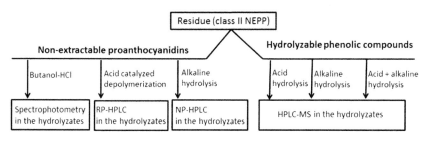

FIGURE 10.3

Scheme of class II NEPP analysis.

Table 10.2 Main Characteristics of the Existing Methods for the Analysis of Non-Extractable Proanthocyanidins in Plant Foods

Method	Analyzed Compounds	Analytical Technique	Drastic Conditions	Information on Degree of Polymerization
Butanol–HCl treatment	Released anthocyanins and flavylium compounds	Spectrophotometry	X	
Alkaline hydrolysis	Proanthocy-anidins	NP-HPLC		X
Acid-catalyzed depolimer-ization with nucleophlic reagents	Extension units and terminal units of proantho-cyanidins	DP-HPLC		X[1]

determination of the anthocyanins released (Porter *et al.*, 1985; Pérez-Jiménez *et al.*, 2009) or the anthocyanins and the xanthyllium compounds (Zurita *et al.*, 2012); alkaline hydrolysis, followed by NP-HPLC of the proanthocyanidins released, according to their degree of polymerization (White *et al.*, 2010); or acid-catalyzed depolymerization, using several nucleophilic reagents, followed by RP-HPLC of the terminal and extension units of the proanthocyanidins chain released (Downey *et al.*, 2003; Verries *et al.*, 2008).

The main advantages and disadvantages of these methods are shown in Table 10.2. Briefly, the butanol–HCl treatment (Porter method) is the most drastic, and therefore allows the most efficient release of proanthocyanidins, although it results in the breakdown of the original structures, which cannot then be characterized. In addition, since the butanol medium is not appropriate for HPLC analysis, the method has to be based on a less sensitive spectrophotometric determination. Alkaline hydrolysis and acid-catalyzed depolymerization, although followed by a more sensitive analytical technique (HPLC), use less drastic conditions than the Porter method and therefore give rise to a lower yield of non-extractable proanthocyanidins. These two methods provide structural information on the structures of the non-extractable proanthocyanidins, but with certain limitations [mean degree of polymerization in acid-catalyzed polymerization and not detailed information on polymers of more than 10 units in alkaline hydrolysis followed by NP-HPLC (Gu *et al.*, 2002)].

Conversely, as already mentioned, hydrolyzable phenolic compounds are another category of class II NEPP. They are released from the food matrix by chemical hydrolysis (either acid or alkaline) which can be followed by HPLC, HPLC-MS or GC-MS analysis of the individual compounds (Hatcher and Kruger, 1997; Bunzel *et al.*, 2003; Arranz *et al.*, 2009), most of which turn out to be phenolic acids, although there are also some phenolic compounds that belong to other classes (Arranz *et al.*, 2009). Alkaline hydrolysis has been the most common method used to date to release these compounds, particularly from cereal products (Chandrasekara and Shahidi, 2010). However, it has been reported that enhanced overall extraction of these compounds results from the use of acid hydrolysis (Arranz and Saura Calixto, 2010). Other authors have used the two kinds of chemical hydrolysis sequentially, considering that acid hydrolysis may be more appropriate for some hydrolyzable phenolic compounds and alkaline hydrolysis for others (Verma *et al.*, 2009).

The combination of chemical hydrolysis with HPLC has also been used to determine the content of a particular class of hydrolyzable phenolic compounds, the non-extractable hydrolyzable tannins, determining their content of gallic acid or ellagic acid (Li *et al.*, 2006; John and Shahidi, 2010).

Although a general approach to the analysis of class II NEPP from different samples has been described (Arranz *et al.*, 2010), it should be borne in mind that a combination of different procedures may be necessary to analyze them.

Finally, there is a need for new analytical approaches to the characterization of class II NEPP that are able to:

- Minimize the losses of class II NEPP that may occur due to the drastic conditions currently needed for their analysis (heat, acid, etc.)
- Analyze class II NEPP by physico-chemical methods directly as solids (i.e., spectrometric assays applied directly in the sample, such as MALDI-TOF or MALDI-FT-ICR mass spectrometry techniques), thereby avoiding chemical hydrolysis of the original molecules. This may help to elucidate some currently unknown aspects of the associations between NEPP and the food matrix, as well as to identify new structures specific to NEPP, as in the case of some derivatives of ferulic and sinapic acids (Bunzel *et al.*, 2001, 2003).

10.5 **NEPP content in plant foods**

The NEPP content of several plant foods and extracts has been determined using the methodologies described above. For instance, non-extractable proanthocyanidins, which are widespread among plant foods, have been determined (as NEPP-class II) in cocoa powder (602 mg/100 g dw) (Hellström *et al.*, 2009), cranberry pomace

(1685 mg/100 g dw) (White *et al.*, 2010), cider waste (300 mg/100 g dw) and carob pods (18,000 mg/100 g dw) (Saura Calixto, 1988). Similarly, the hydrolyzable phenolic compounds content (as class II NEPP) of several plant foods and by-products has been estimated, such as blackcurrant pomace (41 mg/100 g dw) (Kapasakalidis *et al.*, 2006) and red ginseng (2 mg/100 g dw) (Jung *et al.*, 2002). They have been found to be particularly abundant in cereals, such as cornflour (200 mg/100 g dw) (Krygier *et al.*, 1982) and wheat bran (2300 mg/100 g dw) (Verma *et al.*, 2009).

Despite the necessity to analyze many more samples in order to gain a clear overview of the NEPP content of plant foods, the available data show that they may be considered as the major fraction of plant food polyphenols, as shown by the examples in Table 10.3. In particular, non-extractable proanthocyanidin content

Table 10.3 Extractable and Non-Extractable Polyphenols Belonging to Several Classes in Different Plant Foods

Polyphenol Class	Sample	EPP (mg/100 g d.m.)	NEPP (mg/100 g d.m.)	Contribution of NEPP (%)	References
Proanthocy-anidins	Nectarine	74	401	85	Pérez-Jiménez *et al.*, 2009
	Peach	8.6*	59*	87	Arranz *et al.*, 2009
	Red grape	140	146	51	Hellström *et al.*, 2009
	Apple pomace	85	69	48	Tow *et al.*, 2011
	Mango peel	5.5	0.4	7	Larrauri *et al.*, 1997
Phenolic acids	Oat bran	2.8	85	97	Hosseinian and Mazza, 2009
	Wheat bran	53	2319	98	Verma *et al.*, 2009
	Wheat flour	112	262	70	Arranz and Saura Calixto, 2010
	Millet	220	2364	91	Chandrasekara and Shahidi, 2010
Hydrolyzable tannins	Walnut	162	121	43	Li *et al.*, 2006
	Brazil nut	1.1	1.4	56	John and Shahidi, 2010

d.m., dry matter.
*Content expressed as mg/100 g fresh matter.

has been estimated to be between 7% of total proanthocyanidin content in mango peel and 87% in peach (Larrauri *et al.*, 1997; Arranz *et al.*, 2009). Similarly, non-extractable phenolic acids account for more than 90% of the total phenolic acids in cereals (Hosseinian and Mazza, 2009; Arranz and Saura Calixto, 2010), and non-extractable hydrolyzable tannins are about half of the hydrolyzable tannins present in nuts (Li *et al.*, 2006).

10.6 **Potential applications of NEPP**

Our increasing knowledge of the content and composition of NEPP in plant foods is leading to the emergence of applications for them both in food products and specific products for nutrition and health, especially gastrointestinal health.

To date, the application of plant food polyphenols has focused on those derived from EPP, as with most polyphenol research. Extracts from plant foods containing EPP, including some industrial foodstuffs (Vinson *et al.*, 2001; Rohdewald, 2002) have been used in skin protection (Saliou *et al.*, 2001; Yamakoshi *et al.*, 2003), the improvement of antioxidant status (Devaraj *et al.*, 2002; Vigna *et al.*, 2003), and to modify some cardiovascular disease risk factors (Clifton, 2004; Bogdanski *et al.*, 2012).

In contrast, the potential applications of NEPP, as with other aspects related to NEPP, have been studied to a much lesser extent. However, increasing evidence suggests that NEPP concentrates may have several applications:

- As natural food antioxidants. Several NEPP concentrates from grape have been shown to prevent lipid oxidation in several foodstuffs, such as chicken breast hamburgers (Sáyago-Ayerdi *et al.*, 2009), chicken patties (Sáyago-Ayerdi *et al.*, 2009), minced fish (Sánchez-Alonso *et al.*, 2006), and restructured fish products (Sánchez-Alonso *et al.*, 2008).
- In animal nutrition. Supplementation of broiler chicks with NEPP concentrates from grape has shown several beneficial effects, such as an improvement in antioxidant status (Goñi *et al.*, 2007; Brenes *et al.*, 2008) and an increase in the degree of biodiversity of intestinal bacteria (Viveros *et al.*, 2011). In addition, this supplementation gave rise to meat with reduced lipid oxidation (Goñi *et al.*, 2007).
- For human health. Due to the metabolic fate of NEPP, which arrive nearly intact at the colon where they are extensively transformed by the microbiota, their potential health-related properties mainly involve gastrointestinal health. In particular, NEPP concentrates have been shown to potentially reduce colorectal cancer risk factors (Goñi *et al.*, 2005; Goñi and Serrano, 2005; López-Oliva *et al.*, 2006, 2010, in press; Tow *et al.*, 2011; Pozuelo *et al.*, 2012), as well as to modulate colonic microbiota (Hervert-Hernández *et al.*, 2009; Pozuelo *et al.*, 2012). Promising results on other systemic effects, such as modulation of cardiovascular disease risk factors or glucose metabolism from NEPP-derived metabolites after absorption, still require confirmation (Pérez-Jiménez *et al.*, 2008; Vitaglione and Fogliano, 2010; Lizarraga *et al.*, 2011).

10.7 **Concluding remarks**

- NEPP are the major fraction of food polyphenols and yet they are commonly ignored. They consist of high-molecular-weight molecules or lower-molecular-weight compounds associated with the food matrix.
- NEPP concentrates are obtained from plant foods by carrying out different extraction procedures and enzymatic treatments.
- NEPP concentrates can then be analyzed using different techniques, including in some cases previous chemical hydrolysis.
- Several applications of NEPP are currently emerging, in the food industry (as natural antioxidants), animal nutrition (improvement of antioxidant status) and human health (mainly in gastrointestinal health). This shows the potential of this understudied class of phytochemicals in nutrition and health.

References

Arranz, S., Pérez-Jiménez, J., Saura-Calixto, F., 2008. Antioxidant capacity of walnut (*Juglans regia* L.): Contribution of oil and defatted matter. Eur. Food Res. Technol. 227, 425–431.

Arranz, S., Saura-Calixto, F., Shaha, S., Kroon, P.A., 2009. High contents of nonextractable polyphenols in fruits suggest that polyphenol contents of plant foods have been underestimated. J. Agric. Food Chem. 57, 7298–7303.

Arranz, S., Saura Calixto, F., 2010. Analysis of polyphenols in cereals may be improved performing acidic hydrolysis: A study in wheat flour and wheat bran and cereals of the diet. J. Cereal Sci. 51, 313–318.

Arranz, S., Silván, J.M., Saura-Calixto, F., 2010. Nonextractable polyphenols, usually ignored, are the major part of dietary polyphenols: A study on the Spanish diet. Mol. Nutr. Food Res. 54, 1646–1658.

Bindon, K.A., Smith, P.A., Holt, H., Kennedy, J.A., 2010. Interaction between grape-derived proanthocyanidins and cell wall material. 2. Implications for vinification. J. Agric. Food Chem. 58, 10736–10746.

Bogdanski, P., Suliburska, J., Szulinska, M., Stepien, M., Pupek-Musialik, D., Jablecka, A., 2012. Green tea extract reduces blood pressure, inflammatory biomarkers, and oxidative stress and improves parameters associated with insulin resistance in obese, hypertensive patients. Nutr. Res. 32, 421–427.

Bravo, L., Abia, R., Saura-Calixto, F., 1994. Polyphenols as dietary fiber associated compounds. Comparative study on *in vivo* and *in vitro* properties. J. Agri. Food Chem. 42, 1481–1487.

Brenes, A., Viveros, A., Goñi, I., Centeno, C., Sáyago-Ayerdy, S.G., Arija, I., et al., 2008. Effect of grape pomace concentrate and vitamin E on digestibility of polyphenols and antioxidant activity in chickens. Poult. Sci. 87, 307–316.

Bunzel, M., 2010. Chemistry and occurrence of hydroxycinnamate oligomers. Phytochem. Rev. 9, 47–64.

Bunzel, M., Ralph, J., Marita, J.M., Hatfield, R.D., Steinhart, H., 2001. Diferulates as structural components in soluble and insoluble cereal dietary fibre. J. Sci. Food Agri. 81, 653–660.

Bunzel, M., Ralph, J., Kim, H., Lu, F.C., Ralph, S.A., Marita, J.M., et al., 2003. Sinapate dehydrodimers and sinapate-ferulate heterodimers in cereal dietary fiber. J. Agri. Food Chem. 51, 1427–1434.

Burr, S.J., Fry, S.C., 2009. Extracellular cross-linking of maize arabinoxylans by oxidation of feruloyl esters to form oligoferuloyl esters and ether-like bonds. Plant J. 58, 554–567.

Clifton, P.M., 2004. Effect of grape seed extract and quercetin on cardiovascular and endothelial parameters in high-risk subjects. J. Biomed. Biotechnol. 2004, 272–278.

Chamorro, S., Viveros, A., Alvarez, I., Vega, E., Brenes, A., 2012. Changes in polyphenol and polysaccharide content of grape seed extract and grape pomace after enzymatic treatment. Food Chem. 133, 308–314.

Chandrasekara, A., Shahidi, F., 2010. Content of Insoluble Bound Phenolics in Millets and Their Contribution to Antioxidant Capacity. J. Agri. Food Chem. 58, 6706–6714.

Chukwumah, Y.C., Walker, L.T., Verghese, M., Bokanga, M., Ogutu, S., Alphonse, K., 2007. Comparison of extraction methods for the quantification of selected phytochemicals in peanuts (*Arachis hypogaea*). J. Agri. Food Chem. 55, 285–290.

de Simón, B.F., Pérez-Ilzarbe, J., Hernández, T., Gómez-Cordovés, C., Estrella, I., 1990. HPLC study of the efficiency of extraction of phenolic compounds. Chromatographia 30, 35–37.

Devaraj, S., Vega-López, S., Kaul, N., Schönlau, F., Rohdewald, P., Jialal, I., 2002. Supplementation with a pine bark extract rich in polyphenols increases plasma antioxidant capacity and alters the plasma lipoprotein profile. Lipids 37, 931–934.

Díaz-Rubio, M.E., Pérez-Jiménez, J., Saura-Calixto, F., 2009. Dietary fiber and antioxidant capacity in Fucus vesiculosus products. Int. J. Food Sci. Nutr. 60, 23–34.

Downey, M.O., Harvey, J.S., Robinson, S.P., 2003. Analysis of tannins in seeds and skins of Shiraz grapes throughout berry development. J. Grape Wine Res. 9, 15–27.

Faulds, C.B., Williamson, G., 1999. The role of hydroxycinnamates in the plant cell wall. J. Sci. Food Agri. 79, 393–395.

Ferreres, F., Gil-Izquierdo, A., Vinholes, J., Grosso, C., Valentão, P., Andrade, P.B., 2011. Approach to the study of C-glycosyl flavones acylated with aliphatic and aromatic acids from *Spergularia rubra* by high-performance liquid chromatography–photodiode array detection/electrospray ionization multi-stage mass spectrometry. Rapid Commun. Mass Spectrom. 25, 700–712.

García, A., Brenes, M., Romero, C., García, P., Garrido, A., 2002. Study of phenolic compounds in virgin olive oils of the Picual variety. Eur. Food Res. Technol. 215, 407–412.

Goñi, I., Serrano, J., 2005. The intake of dietary fiber from grape seeds modifies the antioxidant status in rat cecum. J. Sci. Food Agri. 85, 1877–1881.

Goñi, I., Jiménez-Escrig, A., Gudiel, M., Saura-Calixto, F.D., 2005. Artichoke (*Cynara scolymus* L) modifies bacterial enzymatic activities and antioxidant status in rat cecum. Nutr. Res. 25, 607–615.

Goñi, I., Brenes, A., Centeno, C., Viveros, A., Saura-Calixto, F., Rebolé, A., et al., 2007. Effect of dietary grape pomace and vitamin E on growth performance, nutrient digestibility, and susceptibility to meat lipid oxidation in chickens. Poult. Sci. 86, 508–516.

Grandmaison, J., Ibrahim, R.K., 1996. Evidence for nuclear protein binding of flavonol sulfate esters in *Flaveria chloraefolia*. J. Plant Physiol. 147, 653–660.

Gu, L., Kelm, M., Hammerstone, J.F., Beecher, G., Cunningham, D., Vannozzi, S., et al., 2002. Fractionation of polymeric procyanidins from lowbush blueberry and quantification of procyanidins in selected foods with an optimized normal-phase HPLC-MS fluorescent detection method. J. Agri. Food Chem. 50, 4852–4860.

Hatcher, D.W., Kruger, J.E., 1997. Simple phenolic acids in flours prepared from Canadian wheat: Relationship to ash content, color, and polyphenol oxidase activity. Cereal Chem. 74, 337–343.

Hellström, J.K., Törrönen, A.R., Mattila, P.H., 2009. Proanthocyanidins in common food products of plant origin. J. Agri. Food Chem. 57, 7899–7906.

Hervert-Hernández, D., Pintado, C., Rotger, R., Goñi, I., 2009. Stimulatory role of grape pomace polyphenols on *Lactobacillus acidophilus* growth. Int. J. Food Microbiol. 136, 119–122.

Hosseinian, F.S., Mazza, G., 2009. Triticale bran and straw: Potential new sources of phenolic acids, proanthocyanidins, and lignans. J. Funct. Foods 1, 57–64.

John, J.A., Shahidi, F., 2010. Phenolic compounds and antioxidant activity of Brazil nut (*Bertholletia excelsa*). J. Funct. Foods 2, 196–209.

Jung, M.Y., Jeon, B.S., Bock, J.Y., 2002. Free, esterified, and insoluble-bound phenolic acids in white and red Korean ginsengs (*Panax ginseng* C.A. Meyer). Food Chem. 79, 105–111.

Kapasakalidis, P.G., Rastall, R.A., Gordon, M.H., 2006. Extraction of polyphenols from processed blackcurrant (*Ribes nigrum* L.) residues. J. Agri. Food Chem. 54, 4016–4021.

Kapasakalidis, P.G., Rastall, R.A., Gordon, M.H., 2009. Effect of a Cellulase Treatment on Extraction of Antioxidant Phenols from Blackcurrant (*Ribes nigrum* L.) Pomace. J. Agri. Food Chem. 57, 4342–4351.

Knödler, M., Berardini, N., Kammerer, D.R., Carle, R., Schieber, A., 2007. Characterization of major and minor alk(en)ylresorcinols from mango (*Mangifera indica* L.) peels by high-performance liquid chromatography/atmospheric pressure chemical ionization mass spectrometry. Rapid Commun. Mass Spectrom. 21, 945–951.

Krygier, K., Sosulski, F., Hogge, L., 1982. Free, esterified and insoluble-bound phenolic acids2. Composition of phenolic acids in rapeseed flour and hulls. J. Agri. Food Chem. 30, 334–336.

Larrauri, J.A., Rupérez, P., Saura-Calixto, F., 1997. Mango peel fibres with antioxidant activity. Zeitschrift fur Lebensmittel – Untersuchung und – Forschung 205, 39–42.

Le Bourvellec, C., Guyot, S., Renard, C.M.G.C., 2004. Non-covalent interaction between procyanidins and apple cell wall material: Part I. Effect of some environmental parameters. Biochimica et Biophysica Acta – General Subjects 1672, 192–202.

Le Bourvellec, C., Bouchet, B., Renard, C.M.G.C., 2005. Non-covalent interaction between procyanidins and apple cell wall material. Part III: Study on model polysaccharides. Biochimica et Biophysica Acta – General Subjects 1725, 10–18.

Li, L., Tsao, R., Yang, R., Liu, C., Zhu, H., Young, J.C., 2006. Polyphenolic profiles and antioxidant activities of heartnut (*Juglans ailanthifolia* var. cordiformis) and Persian walnut (*Juglans regia* L.). J. Agri. Food Chem. 54, 8033–8040.

Lizarraga, D., Vinardell, M.P., Noé, V., van Delft, J.H., Alcarraz–Vizán, G., van Breda, S.G., et al., 2011. A Lyophilized red grape pomace containing proanthocyanidin–rich dietary fiber induces genetic and metabolic alterations in colon mucosa of female C57Bl/6J mice. J. Nutr. 141, 1597–1604.

López-Oliva, M.E., Agis-Torres, A., García-Palencia, P., Goñi, I., Muñoz-Martínez, E., 2006. Induction of epithelial hypoplasia in rat cecal and distal colonic mucosa by grape antioxidant dietary fiber. Nutr. Res. 26, 651–658.

López-Oliva, M.E., Agis-Torres, A., Goñi, I., Muñoz-Martínez, E., 2010. Grape antioxidant dietary fibre reduced apoptosis and induced a pro-reducing shift in the glutathione redox state of the rat proximal colonic mucosa. Br. J. Nutr. 103, 1110–1117.

López-Oliva, M.E., Pozuelo, M.J. Rotger, R. Muñoz-Martínez, E. and Goñ, I., 2013. Grape antioxidant dietary fibre prevents mitochondrial apoptotic pathways by enhancing Bcl-2 and Bcl-xL expression and minimising oxidative stress in rat distal colonic mucosa. Br. J. Nutr. 109, 4–16.

Markham, K.R., Gould, K.S., Ryan, K.G., 2001. Cytoplasmic accumulation of flavonoids in flower petals and its relevance to yellow flower colouration. Phytochemistry 58, 403–413.

Matthews, S., Mila, I., Scalbert, A., Donnelly, D.M.X., 1997. Extractable and non-extractable proanthocyanidins in barks. Phytochemistry 45, 405–410.

Milder, I.E.J., Arts, I.C.W., Venema, D.P., Lasaroms, J.J.P., Wähälä, K., Hollman, P.C.H., 2004. Optimization of a liquid chromatography-tandem mass spectrometry method for quantification of the plant lignans secoisolariciresinol, matairesinol, lariciresinol, and pinoresinol in foods. J. Agri. Food Chem. 52, 4643–4651.

Monagas, M., Quintanilla-López, J.E., Gómez-Cordovés, C., Bartolomé, B., Lebrón-Aguilar, R., 2010. MALDI-TOF MS analysis of plant proanthocyanidins. J. Pharma. Biomed. Anal. 51, 358–372.

Pérez-Jiménez, J., Arranz, S., Tabernero, M., Díaz- Rubio, M.E., Serrano, J., Goñi, I., et al., 2008a. Updated methodology to determine antioxidant capacity in plant foods, oils and beverages: Extraction, measurement and expression of results. Food Res. Int. 41, 274–285.

Pérez-Jiménez, J., Serrano, J., Tabernero, M., Arranz, S., Díaz-Rubio, M.E., García-Diz, L., et al., 2008b. Effects of grape antioxidant dietary fiber in cardiovascular disease risk factors. Nutrition 24, 646–653.

Pérez–Jiménez, J., Arranz, S., Saura-Calixto, F., 2009. Proanthocyanidin content in foods is largely underestimated in the literature data: An approach to quantification of the missing proanthocyanidins. Food Res. Int. 42, 1381–1388.

Pérez-Jiménez, J., Torres, J.L., 2011. Analysis of nonextractable phenolic compounds in foods: The current state of the art. J. Agri. Food Chem. 59, 12713–12724.

Pinelo, M., Arnous, A., Meyer, A.S., 2006. Upgrading of grape skins: Significance of plant cell-wall structural components and extraction techniques for phenol release. Trends Food Sci. Technol. 17, 579–590.

Pinelo, M., Zornoza, B., Meyer, A.S., 2008. Selective release of phenols from apple skin: Mass transfer kinetics during solvent and enzyme-assisted extraction. Sep. Purif. Technol. 63, 620–627.

Porter, L.J., Hrstich, L.N., Chan, B.G., 1985. The conversion of procyanidins and prodelphinidins to cyanidin and delphinidin. Phytochemistry 25, 223–230.

Pozuelo, M.J., Agis-Torres, A., Hervert-Hernández, D., López-Oliva, M.E., Muñoz-Martínez, E., Rotger, R., et al., 2012. Grape Antioxidant Dietary Fiber Stimulates Lactobacillus Growth in Rat Cecum. J. Food Sci. 77, H59–H62.

Rohdewald, P., 2002. A review of the French maritime pine bark extract (Pycnogenol®), a herbal medication with a diverse clinical pharmacology. Int. J. Clin. Pharmacol. Ther. 40, 158–168.

Rufino, M.D.S.M., Pérez-Jiménez, J., Tabernero, M., Alves, R.E., De Brito, E.S., Saura-Calixto, F., 2010. Acerola and cashew apple as sources of antioxidants and dietary fibre. Int. J. Food Sci. Technol. 45, 2227–2233.

Saliou, C., Rimbach, G., Moini, H., McLaughlin, L., Hosseini, S., Lee, J., et al., 2001. Solar ultraviolet-induced erythema in human skin and nuclear factor-kappa-B-dependent gene expression in keratinocytes are modulated by a French maritime pine bark extract. Free Radic. Biol. Med. 30, 154–160.

Saltveit, M.E., 2010. Synthesis and metabolism of phenolic compounds. In: De la Rosa, L.A., Álvarez-Parrilla, E., González-Aguilar, G.A. (Eds.), Fruit and vegetable phytochemicals. Blackwell Publishing, Iowa, US, pp. 89–100.

Sánchez-Alonso, I., Jiménez-Escrig, A., Saura-Calixto, F., Borderías, A.J., 2006. Effect of grape antioxidant dietary fibre on the prevention of lipid oxidation in minced fish: Evaluation by different methodologies. Food Chem. 101, 372–378.

Sánchez-Alonso, I., Jiménez-Escrig, A., Saura-Calixto, F., Borderías, A.J., 2008. Antioxidant protection of white grape pomace on restructured fish products during frozen storage. LWT – Food Sci. Technol. 41, 42–50.

Saura Calixto, F., 1987. Determination of chemical composition of carob pods (Ceratoia siliqua). Sugars, tannins, pectins and aminoacids. Anales de Bromatología XXXXIX, 81–93.

Saura Calixto, F., 1988. Effect of condensed tannins in the analysis of dietary fiber in carob pods. J. Food Sci. 53, 1769–1771.

Saura-Calixto, F., 2011. Dietary Fiber as a Carrier of Dietary Antioxidants: An Essential Physiological Function. J. Agri. Food Chem. 59, 43–49.

Sáyago-Ayerdi, S.G., Brenes, A., Goñi, I., 2009a. Effect of grape antioxidant dietary fiber on the lipid oxidation of raw and cooked chicken hamburgers. LWT – Food Sci. Technol. 42, 971–976.

Sáyago-Ayerdi, S.G., Brenes, A., Viveros, A., Goñi, I., 2009b. Antioxidative effect of dietary grape pomace concentrate on lipid oxidation of chilled and long-term frozen stored chicken patties. Meat Sci. 83, 528–533.

Sun, R.C., Sun, X.F., Wang, S.Q., Zhu, W., Wang, X.Y., 2002. Ester and ether linkages between hydroxycinnamic acids and lignins from wheat, rice, rye, and barley straws, maize stems, and fast-growing poplar wood. Ind. Crops Prod. 15, 179–188.

Touriño, S., Fuguet, E., Jauregui, O., Saura-Calixto, F., Cascante, M., Torres, J.L., 2008. High-resolution liquid chromatography/electrospray ionization time-of-flight mass spectrometry combined with liquid chromatography/electrospray ionization tandem mass spectrometry to identify polyphenols from grape antioxidant dietary fiber. Rapid Commun. Mass Spectrom. 22, 3489–3500.

Tow, W.W., Premier, R., Jing, H., Ajlouni, S., 2011. Antioxidant and Antiproliferation Effects of Extractable and Nonextractable Polyphenols Isolated from Apple Waste Using Different Extraction Methods. J. Food Sci. 76, T163–T172.

Tsao, R., McCallum, J., 2010. Chemistry of flavonoids. In: De la Rosa, L.A., Álvarez-Parrilla, E., González-Aguilar, G.A. (Eds.), Fruit and vegetable phytochemicals. Blackwell Publishing, Iowa, US, pp. 131–154.

Umphress, S.T., Murphy, S.P., Franke, A.A., Custer, L.J., Blitz, C.L., 2005. Isoflavone content of foods with soy additives. J. Food Composition Anal. 18, 533–550.

Verma, B., Hucl, P., Chibbar, R.N., 2009. Phenolic acid composition and antioxidant capacity of acid and alkali hydrolysed wheat bran fractions. Food Chem. 116, 947–954.

Verries, C., Guiraud, J.L., Souquet, J.M., Vialet, S., Terrier, N., Olle, D., 2008. Validation of an extraction method on whole pericarp of grape berry (*Vitis vinifera* L. cv. Shiraz) to study biochemical and molecular aspects of flavan-3-ol synthesis during berry development. J. Agri. Food Chem. 56, 5896–5904.

Vigna, G.B., Costantini, F., Aldini, G., Carini, M., Catapano, A., Schena, F., et al., 2003. Effect of a standardized grape seed extract on low-density lipoprotein susceptibility to oxidation in heavy smokers. Metabolism 52, 1250–1257.

Vinson, J.A., Proch, J., Bose, P., 2001. MegaNatural® Gold grapeseed extract: In vitro antioxidant and *in vivo* human supplementation studies. J. Med. Food 4, 17–26.

Vitaglione, P., Fogliano, V., 2010. Cereal fibres, antioxidant activity and health. In: Van Der Kamp, J.W., Jones, J., McCleary, B., Topping, D. (Eds.), Dietary fibre: new frontiers for food and health. Wageningen Academic Publishers, Wageningen, pp. 379–390.

Viveros, A., Chamorro, S., Pizarro, M., Arija, I., Centeno, C., Brenes, A., 2011. Effects of dietary polyphenol-rich grape products on intestinal microflora and gut morphology in broiler chicks. Poult. Sci. 90, 566–578.

White, B.L., Howard, L.R., Prior, R.L., 2010. Release of Bound Procyanidins from Cranberry Pomace by Alkaline Hydrolysis. J. Agri. Food Chem. 58, 7572–7579.

Winkel-Shirley, B., 2002. Biosynthesis of flavonoids and effects of stress. Curr. Opin. Plant Biol. 5, 218–223.

Yamakoshi, J., Otsuka, F., Sano, A., Tokutake, S., Saito, M., Kikuchi, M., et al., 2003. Lightening Effect on Ultraviolet-Induced Pigmentation of Guinea Pig Skin by Oral Administration of a Proanthocyanidin-Rich Extract from Grape Seeds. Pigment Cell Res. 16, 629–638.

Yu, J., Vasanthan, T., Temelli, F., 2001. Analysis of phenolic acids in barley by high-performance liquid chromatography. J. Agri. Food Chem. 49, 4352–4358.

Zurita, J., Díaz-Rubio, M.E., Saura-Calixto, F., 2012. Improved procedure to determine non-extractable polymeric proanthocyanidins in plant foods. Int. J. Food Sci. Nutri. 63, 936–939.

Resin Adsorption and Ion Exchange to Recover and Fractionate Polyphenols

11

Dietmar Rolf Kammerer, Judith Kammerer, Reinhold Carle

Hohenheim University, Institute of Food Science and Biotechnology, Stuttgart, Germany

CHAPTER OUTLINE HEAD

11.1 Adsorption and ion exchange technology—from past to present and future applications

Adsorption phenomena have a long history, reaching back to the Egyptians and Sumerians who by 3750 BC already applied charcoal to reduce the contents of certain metals in bronze manufacturing. The first medicinal applications of charcoal for the sorption and inactivation of toxic components *in vivo* were described by the Egyptians and later by Hippocrates, before potable water treatment with such sorbent materials was described by the Phoenicians. More systematic and quantitative studies of adsorption phenomena were not performed until the 18th century. Nowadays, industrial applications based on sorption processes are highly diverse, and range from gas purification in industrial exhausts to the purification of liquid materials both in the food and non-food sectors. For these purposes, charcoal, clay, and zeolites have been commonly used. More recently, high-performance activated carbons, synthetic zeolites and synthetic resins with polystyrene, polyacrylic esters or phenolic backbones have been designed making sorption processes much more efficient (Inglezakis and Poulopoulos, 2006; LeVav and Carta, 2007; Zaganiaris, 2011; Zagorodni, 2007). Similarly, ion exchange technology also has a long history, and first reports can probably be found in the Old Testament. However, thorough investigations were performed in the 19th century, probably without knowledge

Polyphenols in Plants. http://dx.doi.org/10.1016/B978-0-12-397934-6.00011-5

of the underlying physical phenomena. As in the case of adsorption, ion exchange technology proceeded very rapidly with the development of modern synthetic resins with well-designed characteristics, which replaced natural mineral ion exchangers, such as clay, glauconite, humic acid, and zeolite, followed by synthetic inorganic exchanger materials, which were produced at the beginning of the 20th century (Dorfner, 1970; Helfferich, 1995; Inglezakis and Poulopoulos, 2006; Kammerer *et al.*, 2011a; Zaganiaris, 2011; Zagorodni, 2007).

11.2 Adsorbent and ion exchange materials

Activated carbon is among the best-known and most frequently applied adsorbent materials, and can be produced from animal and plant carbonaceous materials, such as bones, coals, petroleum coke, nutshells, peat, wood, and lignite. Activated carbon is produced in a two-step manufacturing process with a first phase being characterized by carbonization, i.e., the removal of undesirable by-products from the raw materials. The second phase brings about activation of the material. The selection of the raw material and control of the carbonization and activation conditions determine the pore size distribution and, consequently, the properties of the adsorbent material. Optimized manufacturing processes allow the production of materials with surface areas ranging up to $3000\,m^2g^{-1}$ and pore volumes of up to $1.8\,cm^3g^{-1}$, bringing about an immense diversity of applications. Activated carbons are mainly used for the removal of unwanted compounds from gases, vapors, and liquids in the chemical industry, in medicine, in the food industry, and for water and wastewater treatment (Crittenden and Thomas, 1998; Inglezakis and Poulopoulos, 2006; Kümmel and Worch, 1990; Yang, 2003).

Further commonly applied adsorbents and ion exchangers are derived from zeolites. Among these, around 40 natural and more than 150 synthetic crystalline aluminosilicates of alkali and earth alkali elements are known today. Zeolites are based on tetrahedron units of SiO_4 and AlO_4, which form secondary polyhedral units of cubes, hexagonal prisms, octahedral or truncated octahedral systems being linked via oxygen atoms. The three-dimensional network of these solid materials is built up of secondary units, where these secondary structures form cages, which are connected through channels crossing this three-dimensional structure. The size of these channels is determined by the number of silicon and aluminum atoms linked with each other and also by the ratio of Si and Al atoms. Furthermore, the counter ion of the negatively charged aluminum has a significant impact on the channel size. This provides the opportunity to produce zeolites with tailor-made properties as regards surface chemistry and structure. Accordingly, zeolites may be applied for a wide range of separation and purification processes and as molecular sieves, which are used for the drying and dehydrating of gases and organic solvents (Breck, 1974; Crittenden and Thomas, 1998; Yang, 2003).

Major progress in the field of adsorption and ion exchange has been made with the development of synthetic resins. These are characterized by their polymeric structure and large internal surface areas and a much more homogeneous structure compared to the aforementioned materials. Such resins are formed in polycondensation and polyaddition reactions as well as through radical polymerization. Among these, styrene, acrylic acid, or methacrylic acid may polymerize with divinylbenzene or other divinyl monomers as crosslinking agents. In addition, crosslinking of polymeric styrene chains may also be realized after their formation, e.g., through chloromethylation of the aromatic bodies, which allows methylene bridge formation. Such polymers may be obtained either in gel type form, in macroreticular form, or as hypercrosslinked resins, all of them significantly differing in their properties and application areas (Belfer and Golzman, 1979; Davankov and Tsyurupa, 1990; Kammerer *et al.*, 2011a; Odian, 2004; Zaganiaris, 2011).

Synthetic ion exchangers are composed of a three-dimensional high molecular network with charged functional groups attached to this network via chemical bonds. The variability of synthetic adsorbent resin goes along with an even more pronounced structural diversity of ion exchangers, because a number of different functional groups may be attached to the apolar resin networks. Resins produced by crosslinking acrylic and methacrylic acids with divinylbenzene carry carboxylic groups, which act as weak cation exchangers without further modification of the resins (Dorfner, 1970; Helfferich, 1995). Crosslinked apolar resins may be functionalized, and thus transformed into ion exchange resins by treatments such as sulfonation and chloromethylation as a result of the Friedel–Crafts/Blanc reaction followed by amination of the intermediate reaction products (Akelah and Sherrington, 1981; Zaganiaris, 2011; Zagorodni, 2007).

11.3 **Principles of adsorption and ion exchange**

Adsorption may be generally described as an enrichment of compounds, mainly from fluids and gases, on the surface of solid-state bodies. This accumulation goes along with interactions between the atoms and molecules of the fluid phase, the adsorptive, and the solid, the adsorbent. Depending on the nature of the interactions between the adsorbent matrix and the adsorptive, three different sorption types are generally distinguished, physisorption, chemisorption, and ionosorption (Akelah and Sherrington, 1981; Kümmel and Worch, 1990).

Ion exchange phenomena exhibit a range of similarities with adsorption processes; however, there are also some significant differences. The most pronounced difference is the fact that ionic species are considered in this type of process. Furthermore, these ions are not removed from the solutions but replaced by ions bound by the solid phase via electrostatic interactions in order to achieve electroneutrality. Consequently, two ionic fluxes need to be considered, one into the ion exchange particles, and the other in the opposite direction out of the resin particles. Nevertheless, adsorption and ion exchange phenomena are often not differentiated in practical applications, and most theories and models established to describe the kinetics and equilibrium conditions of sorption and ion exchange processes have been deduced by using adsorbent resins. This approach is due to less complexity of sorption processes due to the absence of functional groups. Thus, the findings of studies performed with adsorbent resins are frequently translated into ion exchange processes (Inglezakis and Poulopoulos, 2006).

The types of realization of sorption and ion exchange processes on an industrial scale are highly variable. They may comprise bed processes with fixed-bed reactors, batch processes in agitated reactors, and moving-bed processes performed in reactors with moving solid phase. These process types have been reviewed elsewhere (Kammerer *et al.*, 2011a; Soto *et al.*, 2011).

11.4 Application of adsorption and ion exchange technology

Water purification and softening is by far the most important application of sorbent and ion exchange resins, thus providing high-quality water, not only for the food industry, but also for a wide range of other industrial branches. This also covers wastewater treatment, which is becoming of increasing interest to reduce the organic loads of industrial effluents accruing in huge amounts. In addition, numerous biotechnological processes have been described making use of resin-based technologies for the enrichment and purification of valuable components, such as polyphenols, which may then be used in enriched foods, beverages, cosmetics, and pharmaceuticals (Kammerer *et al.*, 2011a; Soto *et al.*, 2011).

In the past, the application of adsorbents and ion exchangers in the food industry was dominated by processes established mainly for the removal of phenolic compounds for various reasons. In this context, fruit juice technology is one of the most important application areas of such resins in the food industry, where they are mainly applied for the stabilization and de-coloration of juices, and for reducing the bitterness of citrus juices, i.e., to overcome problems often caused by phenolic compounds. In this regard, adsorber technology is an appropriate alternative to conventional fining methods using, e.g., bentonite, gelatin, and silica sol, respectively. Thus, juices may be standardized, and light-colored almost water-clear juices may be obtained by binding both Maillard reaction products formed upon thermal treatment of the juices as well as phenolic compounds, which also contribute to cloud formation and brownish hues of the juices (Lyndon, 1996; Weinand, 1996). Apart from discoloration, off-tastes and off-flavors are frequent problems associated with juice production, which may arise immediately after processing or during storage. In this context, the most prominent example is the bitterness of grapefruits and navel oranges, which is observed pre-storage and caused by both limonin and either naringin or hesperidin, respectively (Chandler *et al.*, 1968; Johnson and Chandler, 1988; Shaw and Buslig, 1986). Such unwanted bitterness can be effectively mitigated by selective binding of the bitter compounds onto the surface of adsorbent resins as a final step of juice production. Consequently, off-tastes and off-flavors developed during storage remain unaffected by sorptive treatment. Such phenomena may be observed when precursors, which themselves do not negatively affect the sensory properties of the products, are degraded or form undesired reaction products with other juice components. The aforementioned precursor compounds may result from heating steps performed for thermal preservation of the juices or from degradation processes such as the Strecker reaction. This has been demonstrated for ferulic acid, a hydroxycinnamic acid frequently occurring in plant materials, which may give rise to an enhanced release of *para*-vinylguajacol, the latter contributing to an "old fruit" or "rotten" flavor. Thus, strategies to cope with such problems have been developed, which are based on selective binding of the precursor compounds from juices using macromolecular resins. In such cases, resin-based technologies may significantly enhance the sensory quality of the juices without affecting their nutritional values, and at the same time extend the shelf life of the products (Ma and Lada, 2001a, b).

11.5 Application of adsorbent resins and ion exchangers for the recovery of bio- and techno-functional phenolic compounds from by-products of food processing

Along with the steadily increasing body of literature reporting potential health-beneficial properties of numerous phenolic compounds, adsorption and ion exchange processes have now also been described to be valuable tools not only for the removal of unwanted phenolics, but also for their selective recovery and enrichment. Therefore, resin-based technologies provide an ideal tool to recover preparations rich in valuable phenolic compounds. Generally, the by-products of food processing are still very rich in secondary plant metabolites, and consequently also in phenolic compounds. Given this fact and the large amounts of by-products arising from food production, adsorption and ion exchange technology allow the recovery of polyphenols, thus creating added value from still underutilized by-products and contributing to sustainable agricultural production (Kammerer *et al.*, 2007a; Schieber *et al.*, 2001). Preparations rich in phenolic compounds are of particular interest due to their techno-functional properties, such as their antioxidant and antimicrobial potential or their protective and stabilizing effects on food color and aroma. Furthermore, the valorization of phenolic by-products appears to be particularly promising for the food, pharmaceutical, and cosmetics industry due to the bio-functional characteristics of polyphenols, including their anticarcinogenic, antithrombotic, anti-inflammatory, antimicrobial, and antioxidant properties.

As an example, citrus fruits and products derived therefrom are known to be particularly rich in flavanones and flavanone glycosides. Thus, Di Mauro *et al.* (1999) developed and optimized a process for the recovery of hesperidin from orange peel by concentration on a styrene-divinylbenzene copolymerisate. This process comprises crude extract preparation using an aqueous calcium hydroxide solution to precipitate colloidal pectins as calcium pectate. The neutralized crude extract is subsequently applied onto an adsorbent resin, and the target compound eluted with an alkaline hydroalcoholic solution. Hesperidin may be obtained in high purity from the column eluate by acidification, resulting in flavanone crystallization. Furthermore, the application of adsorbent resins for the recovery of anthocyanins from food processing by-products, such as the pulp wash of pigmented oranges, has been described. In this process, the pigments may be purified and concentrated by sorptive binding onto a resin surface and subsequent desorption with ethanol. The eluates recovered under such conditions further contain hesperidin and hydroxycinnamic acids, which may act as co-pigments and stabilize anthocyanins (Di Mauro *et al.*, 2002). The same group further studied the potential of liquid residues, i.e., wastewater originating from citrus processing, for the recovery of valuable phenolic compounds. Such wastewater arises from essential oil recovery being comparatively rich in hesperidin. Consequently, hesperidin may be recovered by solubilization under alkaline pH conditions and filtration to remove insolubles from the wastewater. After neutralization (pH 6), the solution is applied to an adsorbent resin, and hesperidin is recovered by elution with 10% aqueous ethanol containing 0.46 M sodium hydroxide. Subsequent precipitation of the target compound at pH 5 provides the flavanone in high purity (Di Mauro *et al.*, 2000).

Apple juice production is also of particular economic importance. This process is associated with large amounts of wet pomace, causing high disposal costs for the juice-producing companies and posing significant environmental problems. So far, apple pomace is mainly used as forage, for biogas or fertilizer production as well as for land filling, even though these by-products are known to be especially rich in pectins and phenolic compounds. Thus, apple pomace is nowadays commercially exploited for pectin recovery (Carle and Schieber, 2006). However, apple pectins are inferior to citrus pectins with regard to their visual appearance due to co-extraction of oxidized polymeric brown polyphenols imparting a brownish shade to apple pectins, which is a major drawback when producing light-colored or colorless gelled products. Thus, a resin-based process for the simultaneous recovery of apple pectins and polyphenols applicable on an industrial scale has been developed. For this purpose, dried apple pomace is extracted with diluted mineral acid, releasing both pectins and phenolic compounds into the extract. This extract is subsequently applied to an adsorption column filled with a hydrophobic styrene divinylbenzene copolymerizate. The phenolics are selectively bound by the resin, whereas pectins can be quantitatively removed by washing the column with de-ionized water. Finally, the phenolics are desorbed with methanol. This process allows the recovery of almost colorless pectins through precipitation with alcohols from the eluates collected during column loading with the crude apple pomace extract. At the same time purified phenolic preparations are obtained which may be applied as functional ingredients (Carle *et al.*, 2001; Schieber *et al.*, 2003).

Similarly, the by-products of pomegranate processing have been identified as rich sources of phenolic compounds, which are still underutilized so far. Thus, a large-scale polyphenol isolation process was established by producing aqueous polyphenol extracts from pomegranate husks, which were purified using a column filled with Amberlite XAD-16. This procedure allowed the recovery of pomegranate tannins, such as punicalagin, ellagic acid, punicalin, and ellagic acid glycosides in substantial amounts (Seeram *et al.*, 2005).

The recovery of natural food colorants has especially been boosted in recent years, which is due to increasing consumer rejection of synthetic food additives. This trend was further intensified as a result of the so-called Southampton study, revealing a correlation between the consumption of beverages colored with synthetic dyes and the occurrence of attention-deficit hyperactivity disorder in children (McCann *et al.*, 2007). As a consequence, foods colored with some azo dyes need to be labeled with a warning according to latest European Food Safety Authority regulations.

Thus, there is an enhanced search for natural pigment sources to cope with the needs of the food industry. Grape pomace is one potential source for the recovery of natural colorants, that is available in large amounts, being particularly rich in anthocyanins that are only poorly extracted upon vinification. Consequently, grape pomace is an ideal matrix for pigment recovery in sufficient amounts, and methods for efficient anthocyanin purification and concentration with adsorbent resins revealing high recovery rates have been successfully applied (Kammerer *et al.*, 2004). Moreover, a wide range of further plant matrices has been described, which may be suitable for anthocyanin recovery for their use as natural food colorants. Among these, purple-fleshed potatoes (Liu *et al.*, 2007) and mulberries (Liu *et al.*, 2004) have been suggested, and resin-based technologies have been developed for pigment recovery from these sources.

Furthermore, vegetables may also be attractive sources of phenolic compounds, the recovery of which by resin-based processes has been reported. Among these, pigeon pea (*Cajanus cajan* (L.) Millsp.) has been described with its major phenolic compound luteolin, which may be recovered by applying various macroporous resins. This latter compound has been associated with a range of putative beneficial properties, such as anti-allergenic, antioxidant, anti-proliferative, and anti-inflammatory properties, as well as cytotoxicity towards certain cancer cells, which renders the recovery of this component highly interesting (Fu *et al.*, 2006).

In a similar way, adsorbent resins were applied for the preparation of natural antioxidant preparations rich in phenolic compounds from aqueous spinach extracts as a result of the increasing commercial value of spinach wastes (Aehle *et al.*, 2004). This study demonstrated careful adsorber selection to be an indispensable tool for the systematic optimization of selective polyphenol enrichment.

The cell wall matrix of fruit and vegetable by-products, such as sugar-beet pulp or wheat and grain bran, is another attractive source for the recovery of phenolic antioxidants, since hydroxycinnamates, such as ferulic acid, are often covalently bound to cell wall polysaccharides. These may be cleaved by alkaline hydrolysis of the cell wall polymers prior to further purification of the hydrolyzate using activated charcoal and a strongly basic anion exchanger (Ou *et al.*, 2007).

Soy isoflavones are further examples of phenolic compounds of high economic importance. These phenolics are known to play a significant role in the prevention of certain types of cancer and for their estrogenic activity. Consequently, efforts have been made to selectively enrich isoflavones from soy protein extracts, soy molasses or soy whey feed stream. For this purpose, the corresponding solutions were concentrated by ultrafiltration or reverse osmosis and subsequently treated with an adsorbent resin to selectively bind isoflavones. The resin was subsequently washed and the target compounds eluted with hydroalcoholic solutions which were finally dried (Gugger and Grabiel, 2001).

This broad range of applications of adsorbent and ion exchange resins demonstrates the high potential of these technologies for the recovery of techno- and bio-functional phenolic components. Consequently, ion exchange and adsorber technologies have been markedly improved in recent years, mainly by developing tailor-made resins which allow novel applications. Nevertheless, most adsorber and ion exchange processes which are established even on an industrial scale are based on empirical approaches and have only rarely been systematically optimized. Thus, there is still a lack of knowledge with regard to systematic adaption of process parameters for cost-efficiently and selectively enriching and purifying individual phenolic compounds from crude plant or by-product extracts.

For this reason, a range of studies was performed, bringing solutions of isolated phenolic compounds in contact with food-grade macroporous resins. Quercetin 3-O-rutinoside, caffeic and chlorogenic acids as well as catechin and phloridzin were studied in these model systems to assess the impact of compound structure on the adsorption behavior. D-optimal experimental designs proved to be very helpful tools allowing mathematical modeling and the prediction of sorption phenomena depending on the temperature, pH value, polyphenol concentration and resin amount. This provides the opportunity to systematically adapt sorption conditions and optimize resin-based processes. Common sorption isotherms, such as the Langmuir and Freundlich isotherm, were applied to describe the sorption systems under equilibrium conditions. Further, the calculation of free energy, enthalpy, and entropy changes upon adsorption allowed the thermodynamic interpretation of polyphenol sorption. The comparison of the aforementioned phenolic compounds revealed different affinities towards the adsorbent resin (Bretag *et al.*, 2009a, b). To study these different affinities in more detail, mixtures of these five phenolics were brought into contact with the adsorbent resin to mimic more complex plant extracts, which are commonly highly diverse with regard to their phenolic profile. The results demonstrated that the hydrophobic properties of the compounds are not solely determining adsorption affinity and that careful adjustment of sorption conditions may allow fractionating crude phenolic extracts by adsorption. However, the findings also showed that sorption phenomena in multi-compound systems are highly complex (Kammerer *et al.*, 2010a). To further elucidate the complex situation in plant crude extracts, model phenolic solutions were supplemented with saccharides and amino acids to assess the impact of these compound classes on polyphenol binding, demonstrating that these components may not be neglected in the systematic adaption and optimization of sorption and ion exchange processes (Kammerer *et al.*, 2010b). Further studies applying phenolic model solutions to ion exchange resins with differing structures and under differing pH conditions revealed the opportunity to exploit different affinities to selectively enrich certain target compounds. Thus, this allows the fractionation of crude phenolic mixtures to obtain phenolic preparations with well-defined composition and functional properties (Kammerer *et al.*, 2011b).

These findings were subsequently transferred to the recovery of phenolics from plant extracts. For this purpose, extracts from apple pomace and red grape peels were prepared and applied to columns filled with adsorbent and ion exchange resins, respectively. The results were compared to those obtained with corresponding model solutions mimicking the composition of the complex plant extracts. The findings demonstrated binding rates of the model systems to be largely transferable to the crude plant extracts, except that in the latter case resin capacities were exhausted more rapidly due to the presence of oligomeric and polymeric compounds, which were not covered by the model experiments (Kammerer *et al.*, 2011c).

To study the polyphenol sorption and desorption behavior from crude plant extracts in more detail, an apple juice concentrate was applied to a polymethylmethacrylate resin. The findings clearly demonstrated the compound structure to markedly affect the binding onto the resin surface, which, thus, allows systematic fractionation of crude phenolic extracts by careful adjustment of sorption and desorption conditions (Kammerer *et al.*, 2007b). Such differences in the binding behavior of individual compounds were even observed upon upscaling of the process. Accordingly, this technology is also applicable for industrial production of phenolic preparations and fractionation of complex phenolic mixtures (Kammerer *et al.*, 2010c; Scordino *et al.*, 2005).

The potential of adsorption and ion exchange technology to valorize by-products of food processing was further demonstrated by exploiting the press residues originating from sunflower oil production. These by-products are rich in high-value proteins and phenolics. The latter are readily oxidized upon conventional alkaline protein extraction, followed by their covalent binding to reactive protein side chains. This goes along with a decreased nutritional value and brown discoloration of the proteins obtained under conventional extraction conditions. For this reason, sunflower proteins are still not exploited for human consumption. Consequently, a novel process was developed extracting sunflower proteins under mild acidic conditions, thus precluding polyphenol oxidation (Pickardt *et al.*, 2011). The extracts, being rich both in proteins and polyphenols, were subsequently applied to an anion exchange and adsorber column connected in series. These enabled selective binding of the phenolic acids and polymerized brown polyphenols, whereas proteins passed the columns. The overall process may thus be applied to recover sunflower proteins with improved sensory and nutritional characteristics. By careful adjustment of the sorption and ion exchange step, this process also allows the recovery of purified phenolic preparations as ingredients of functional or enriched foods (Weisz *et al.*, 2010, 2013).

In summary, ion exchange and adsorber processes may be successfully applied for the recovery, purification, and fractionation of phenolic compounds. Systematic studies allow the deduction of optimal sorption and desorption conditions and, thus, the optimization of processes, which so far have hardly been assessed in detail. Consequently, these recent investigations together with the development of tailor-made resin materials will contribute to broader application of this technology in the field of by-product valorization and recovery of functional secondary plant metabolites.

References

Aehle, E., Grandic, S.R.-L., Ralainirina, R., Balatora-Rosset, S., Mesnard, F., Prouillet, C., Mazière, J.-C., Fliniaux, M.-A., 2004. Development and evaluation of an enriched natural antioxidant preparation obtained from aqueous spinach (*Spinacia oleracea*) extracts by an adsorption procedure. Food Chem. 86, 579–585.

Akelah, A., Sherrington, D.C., 1981. Application of functionalized polymers in organic synthesis. Chem. Rev. (Washington, DC, U.S.) 81, 557–587.

Belfer, S., Golzman, R., 1979. Anion exchange resins prepared from polystyrene crosslinked via a Friedel–Crafts reaction. J. Appl. Polym. Sci. 24, 2147–2157.

Breck, D.W., 1974. Zeolite molecular sieves. John Wiley & Sons, New York, London, Sydney, Toronto.

Bretag, J., Kammerer, D.R., Jensen, U., Carle, R., 2009a. Adsorption of rutin onto a food-grade styrene-divinylbenzene copolymer in a model system. Food Chem. 114, 151–160.

Bretag, J., Kammerer, D.R., Jensen, U., Carle, R., 2009b. Evaluation of the adsorption behavior of flavonoids and phenolic acids onto a food-grade resin using a D-optimal design. Eur. Food Res. Technol. 228, 985–999.

Carle, R., Schieber, A., 2006. Functional food components obtained from waste of carrot and apple juice production. Ernaehr. Umsch. 53, 348–352.

Carle, R., Keller, P., Schieber, A., Rentschler, C., Katzschner, T., Rauch, D., Fox, G., Endress, H.-U., 2001. Method for obtaining useful materials from the by-products of fruit and vegetable processing. Patent WO 01/78859 A1.

Chandler, B.V., Kefford, J.K., Ziemelis, G., 1968. Removal of limonin from bitter orange juice. J. Sci. Food Agric. 19, 83–86.

Crittenden, B., Thomas, W.J., 1998. Adsorption technology and design. Elsevier Science & Technology Books, Oxford, Boston, Johannesburg, Melbourne, New Deli, Singapore.

Davankov, V.A., Tsyurupa, M.P., 1990. Structure and properties of hypercrosslinked polystyrene – the first representative of a new class of polymer networks. React. Polym. 13, 27–42.

Di Mauro, A., Fallico, B., Passerini, A., Rapisarda, P., Maccarone, E., 1999. Recovery of hesperidin from orange peel by concentration of extracts on styrene-divinylbenzene resin. J. Agric. Food Chem. 47, 4391–4397.

Di Mauro, A., Fallico, B., Passerini, A., Maccarone, E., 2000. Waste water from citrus processing as a source of hesperidin by concentration on styrene-divinylbenzene resin. J. Agric. Food Chem. 48, 2291–2295.

Di Mauro, A., Arena, E., Fallico, B., Passerini, A., Maccarone, E., 2002. Recovery of anthocyanins from pulp wash of pigmented oranges by concentration on resins. J. Agric. Food Chem. 50, 5968–5974.

Dorfner, K., 1970. Ionenaustauscher. Walter de Gruyter & Co, Berlin.

Fu, Y., Zu, Y., Liu, W., Efferth, T., Zhang, N., Liu, X., Kong, Y., 2006. Optimization of luteolin separation from pigeonpea [*Cajanus cajan* (L.) Millsp.] leaves by macroporous resins. J. Chromatogr. A 1137, 145–152.

Gugger, E., Grabiel, R., 2001. Production of isoflavone enriched fractions from soy protein extracts. U.S. Patent 6,171,638.

Helfferich, F., 1995. Ion exchange. Dover Publications, INC, New York.

Inglezakis, V., Poulopoulos, S., 2006. Adsorption, ion exchange and catalysis: design of operations and environmental applications. Elsevier, Amsterdam.

Johnson, R.L., Chandler, B.C., 1988. Adsorptive removal of bitter principles and titratable acid from citrus juices. Food Technol. 42, 130–137.

Kammerer, D., Gajdos Kljusuric, J., Carle, R., Schieber, A., 2004. Recovery of anthocyanins from grape pomace extracts (*Vitis vinifera* L. cv. Cabernet Mitos) using a polymeric adsorber resin. Eur. Food Res. Technol. 220, 431–437.

Kammerer, D.R., Carle, R., 2007a. Current and emerging methods for the recovery, purification and fractionation of polyphenols from fruit and vegetable by-products. Fruit Process. 17, 89–94.

Kammerer, D.R., Saleh, Z.S., Carle, R., Stanley, R.A., 2007b. Adsorptive recovery of phenolic compounds from apple juice. Eur. Food Res. Technol. 224, 605–613.

Kammerer, J., Kammerer, D.R., Jensen, U., Carle, R., 2010a. Interaction of apple polyphenols in a multi-compound system upon adsorption onto a food-grade resin. J. Food Eng. 96, 544–554.

Kammerer, J., Kammerer, D.R., Carle, R., 2010b. Impact of saccharides and amino acids on the interaction of apple polyphenols with ion exchange and adsorbent resins. J. Food Eng. 98, 230–239.

Kammerer, D.R., Carle, R., Stanley, R.A., Saleh, Z.S., 2010c. Pilot-scale resin adsorption as a means to recover and fractionate apple polyphenols. J. Agric. Food Chem. 58, 6787–6796.

Kammerer, J., Carle, R., Kammerer, D.R., 2011a. Adsorption and ion exchange: basic principles and their application in food processing. J. Agric. Food Chem. 59, 22–42.

Kammerer, J., Boschet, J., Kammerer, D.R., Carle, R., 2011b. Enrichment and fractionation of major apple flavonoids, phenolic acids and dihydrochalcones using anion exchange resins. LWT – Food Sci. Technol. 44, 1079–1087.

Kammerer, J., Schweizer, C., Carle, R., Kammerer, D.R., 2011c. Recovery and fractionation of major apple and grape polyphenols from model solutions and crude plant extracts using ion exchange and adsorbent resins. Int. J. Food Sci. Technol. 46, 1755–1767.

Kümmel, R., Worch, E., 1990. Adsorption aus wässrigen Lösungen. VBE Deutscher Verlag für Grundstoffindustrie, Leipzig.

LeVav, M.D., Carta, G., 2007. Perry´s chemical engineers´ handbook: adsorption and ion exchange. McGraw–Hill Professional, New York.

Liu, X., Xiao, G., Chen, W., Xu, Y., Wu, J., 2004. Quantification and purification of mulberry anthocyanins with macroporous resins. J. Biomed. Biotechnol. 5, 326–331.

Liu, X., Xu, Z., Gao, Y., Yang, B., Zhao, J., Wang, L., 2007. Adsorption characteristics of anthocyanins from purple-fleshed potato (*Solanum tuberosum* Jasim) extract on macroporous resins. Int. J. Food Eng. 3, Art 4, pp. 18.

Lyndon, R., 1996. Kommerzialisierung der Adsorbertechnologie in der Fruchtsaftindustrie. Flüssiges Obst 63, 499–503.

Ma, S.X., Lada, M.W., 2001a. Quality fruit juice beverages having extended quality shelf-life and methods of making the same. Patent WO 01/87097 A2.

Ma, S.X., Lada, M.W., 2001b. Fruit juices having enhanced flavor and extended quality shelf-life. Patent WO 01/87092 A2.

McCann, D., Barrett, A., Cooper, A., Crumpler, D., Dalen, L., Grimshaw, K., Kitchin, E., Lok, K., Porteous, L., Prince, E., Sonuga-Barke, E., Warner, J.O., Stevenson, J., 2007. Food additives and hyperactive behaviour in 3-year-old and 8/9-year-old children in the community: a randomised, double-blinded, placebo-controlled trial. Lancet 370, 1560–1567.

Odian, G., 2004. Principles of polymerization. John Wiley & Sons, Hoboken.

Ou, S., Luo, Y., Xue, F., Huang, C., Zhang, N., Liu, Z., 2007. Separation and purification of ferulic acid in alkaline-hydrolysate from sugarcane bagasse by activated charcoal desorption/anion macroporous resin exchange chromatography. J. Food Eng. 78, 1298–1304.

Pickardt, C., Hager, T., Eisner, P., Carle, R., Kammerer, D.R., 2011. Isoelectric protein precipitation from mild-acidic extracts of de-oiled sunflower (*Helianthus annuus* L.) press cake. Eur. Food Res. Technol. 233, 31–44.

Schieber, A., Stintzing, F.C., Carle, R., 2001. By-products of plant food processing as a source of functional compounds – recent developments. Trends Food Sci. Technol. 12, 401–413.

Schieber, A., Hilt, P., Streker, P., Endreß, H.U., Rentschler, C., Carle, R., 2003. A new process for the combined recovery of pectin and phenolic compounds from apple pomace. Innovative Food Sci. Emerg. Technol. 4, 99–107.

Scordino, M., Di Mauro, A., Passerini, A., Maccarone, E., 2005. Selective recovery of anthocyanins and hydroxycinnamates from a byproduct of citrus processing. J. Agric. Food Chem. 53, 651–658.

Seeram, N., Lee, R., Hardy, M., Heber, D., 2005. Rapid large scale purification of ellagitannins from pomegranate husk, a by-product of the commercial juice industry. Sep. Purif. Technol. 41, 49–55.

Shaw, P.E., Buslig, B.S., 1986. Selective removal of bitter compounds from grapefruit juice and from aqueous solution with cyclodextrin polymers and with Amberlite XAD-4. J. Agric. Food Chem. 34, 837–840.

Soto, M.L., Moure, A., Domínguez, H., Parajó, J.C., 2011. Recovery, concentration and purification of phenolic compounds by adsorption: A review. J. Food Eng. 105, 1–27.

Weinand, R., 1996. Apple juice stabilization and decoloration by modern adsorption techniques. Flüssiges Obst 96, 495–496, 498.

Weisz, G.M., Carle, R., Kammerer, D.R., 2012. Sustainable sunflower processing – II. Recovery of phenolic compounds as a by-product of sunflower protein extraction. Innov. Food Sci. Emerg. Technol. 17, 169–179.

Weisz, G.M., Schneider, L., Schweiggert, U., Kammerer, D.R., Carle, R., 2010. Sustainable sunflower processing – I. Development of a process for the adsorptive decolorization of sunflower (*Helianthus annuus* L.) protein extracts. Innov. Food Sci. Emerg. Technol. 11, 733–741.

Yang, R.T., 2003. Adsorbents: fundamentals and applications; John Wiley & Sons. Hoboken, New Jersey.

Zaganiaris, E.J., 2011. Ion exchange resins and synthetic adsorbents in food processing. Books on Demand GmbH, Paris.

Zagorodni, A.A., 2007. Ion exchange materials – properties and applications. Elsevier, Amsterdam.

Polyphenolic Compounds from Flowers of *Hibiscus*: Characterization and Bioactivity

12

Josline Y. Salib

Department of Chemistry of Tanning Materials, National Research Center, Dokki, Egypt

Nature has provided us with a variety of pleasant things like flowers, fruits, and more, which bring us happiness. Many flowers have been eaten since ancient times, and some have medicinal properties as well as nutritional value (Wongwattanasathien *et al.*, 2010). Regular consumption of different colors and types of flower, fruits, and vegetables ensures a consistent supply of a variety of health-promoting nutrients that have been shown to protect cells. Phytonutrients represent a wide array of diverse compounds, including flavonoids and polyphenols. Polyphenols and other food phenolics are the subject of increasing scientific interest because of their possible beneficial effects on human health. One of the most interesting flowers is the *Hibiscus* flower; recently scientific studies have begun to provide evidence for some of the claimed health benefits of *Hibiscus* extract. Therefore, the objective of the present study was to provide information about the phenolic compounds, biological properties, and nutritional value of different *Hibiscus* flower species and to shed light on their potential health benefits, which could be useful for consumers and public health workers.

Hibiscus is a genus of more than 200 flowering plants. Located primarily in warm-temperate, tropical, and subtropical regions, the plants produce showy flowers that are a staple in decorations for many cultures. Known for its ability to attract bees and butterflies, *Hibiscus* has been used to help other flowers to lure pollinators.

The flowers are of a magnificent variety of colors indicating their country of origin (Figure 12.1); for example, *Hibiscus syriacus* is the national flower of South Korea, *Hibiscus rosa-sinensis* is the national flower of Malaysia, and the red *Hibiscus* is the flower of the Hindu goddess Kali and appears frequently in depictions of her in the art of Bengal, India. In the Philippines, the *gumamela* (local name for *Hibiscus*) is used by children as part of a bubble-making pastime. The red *Hibiscus* flower is traditionally worn by Tahitian women to indicate the wearer's availability for marriage. A Nigerian author named her first novel "Purple *Hibiscus*" after the delicate flower.

Polyphenols in Plants. http://dx.doi.org/10.1016/B978-0-12-397934-6.00012-7

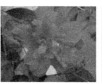

| *Hibiscus syriacus* (South Korea) | *Hibiscus rosa-sinensis* (Malaysia) | Red hibiscus (India) | Yellow hibiscus (Hawaii) |

FIGURE 12.1

Hibiscus flowers from different origins.

With a range of uses spanning from food coloring to medicinal applications, there are many folk remedies attributed to *Hibiscus* flowers, including help with stomach or digestive problems, and to help soothe the nerves. They are also the main ingredient in wonderfully refreshing teas made around the world, especially in Mexico, Latin America, and North Africa. Known as "Agua de Jamaica", or simply "Jamaica" in Mexico, this tea is usually served chilled with copious amounts of sugar to sweeten the natural tartness of the *Hibiscus*. Recently it has been added to many ready-made teas due to its high levels of antioxidants, and has even become the main ingredient in certain sodas. The *Journal of Human Hypertension* published an article that showed that drinking *Hibiscus* tea can reduce the blood pressure in people with type 2 diabetes. Members of the genus *Hibiscus* produce a variety of bioactive compounds, such as lignanamides, naphthalenes, polyphenols, carotenoids, tocopherols, flavonoids, anthocyanins, phytosterols, and long-chain fatty esters (Holser *et al.*, 2004). Consequently, studies are continually conducted to discover some of the plant's additional potential benefits and medicinal properties.

Roselle, *Hibiscus sabdariffa L.* is a tropical plant which belongs to the family Malvaceae and is known in Egypt as Karkadah. It is probably a native of West Africa and is now widely cultivated throughout the tropics and subtropics, e.g., Sudan, China, Thailand, Egypt, Mexico, and the West Indies (El-Saidy *et al.*, 1992). In addition, Roselle juice, which is conventionally made from water extraction of fresh or dried Roselle calyxes, has been reported as being a popular soft drink with daily consumption in many countries including Egypt, Sudan, Mexico, Nigeria, and Thailand (Aurelio *et al.*, 2007). The chemical components contained in the flowers of *Hibiscus sabdariffa* include anthocyanins, flavonoids, and polyphenols (Lin *et al.*, 2007). The petals are potentially a good source of antioxidant agents (mainly anthocyanins like delphinidin-3-glucoside, sambubioside, and cyanidin-3-sambubioside) (Figure 12.2) contributing to their antioxidant properties (Aurelio *et al.*, 2007; Prenesti *et al.*, 2007). The anthocyanins are responsible for the red color, while the acid taste is due to the presence of some organic acids such as ascorbic acid. Above all, it has been reported that *Hibiscus sabdariffa* Linne has anticancerous activity. *Hibiscus* polyphenol rich extract (HPE) induces cell death of eight kinds of carcinoma cell lines and it is most effective against human gastric carcinoma (Lin *et al.*, 2005). These polyphenols include flavonoids like gossypetin, hibiscetin, and their respective glycosides; protocatechuic acid, eugenol, and

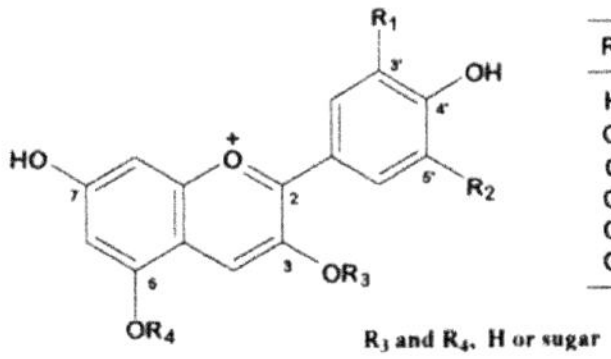

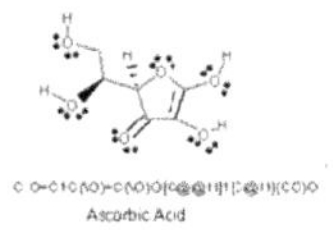

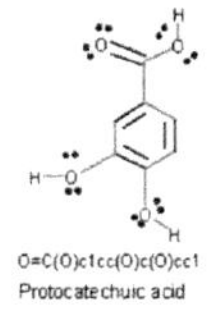

FIGURE 12.2

Structures of some anthocyanins and phenolic acids of *H. sabdariffa, H.mutabilis,* and *H. syriacus* L.

sterols like β-sitoesterol and ergoesterol (Ali-Bradeldin *et al.*, 2005). *Hibiscus* anthocyanin, delphinidin-3-sambubioside (Dp3-Sam) isolated from dried calices of *Hibiscus sabdariffa* may induce a dose-dependent apoptosis in human leukemia cells (HL-60) (Hou *et al.*, 2005). Another phenolic compound, *Hibiscus* protocatechuic acid (PCA), isolated from the dried flower of *Hibiscus sabdariffa*, has an antioxidant, antitumor suppressor activity and is also effective against human promyelocytic leukemia HL-60 cells (Tseng *et al.*, 2000).

The herb *Hibiscus rosa-sinensis* L. is native to China. It is a shrub widely cultivated in the tropics as an ornamental plant and has several forms with varying colors of flowers. The red-flowered variety is preferred in medicine. The flowers have wound-healing properties (Shivananda *et al.*, 2007), have been found to be effective in the treatment of arterial hypertension (Dwivedi *et al.*, 1977), and to have significant anti-fertility effects (Singh *et al.*, 1982; Sethi *et al.*, 1986). The leaves and flowers are observed to be promoters of hair growth (Ali and Ansari, 1997; Adhirajan *et al.*, 2003). *Hibiscus rosa-sinensis* Linne is also effective against diabetes (Sachdewa and Khemani, 2003). *Hibiscus rosa-sinensis* contains numerous compounds including anthocyanins and flavonoids (cyanidin-3,5-diglucoside, cyanidin-3-sophoroside-5-glucoside, quercetin-3,7-diglucoside, quercetin-3-diglucoside) as major compounds (Figures 12.2 and 12.3) (Gupta *et al.*, 2005), together with a cyclopeptide alkaloid (Khokhar and Ahmed, 1992), cyanidin chloride, quercetin, hentriacontane, and vitamins (Shrivastava, 1974). Also, the investigation into the water-soluble fraction of the methanolic extract of flowers of *Hibiscus rosa-sinensis* led to the identification of 10 polyphenolic compounds, namely vitexin, quercetin-7-*O*-galactoside, gallic acid, phydroxybenzoic acid, neochlorogenic acid, and the aglycones apigenin, quercetin, and kaempferol, together with two new compounds, kaempferol-7-*O*-[6‴-*O*-p-hydroxybenzoyl-β-D-glucosyl-(1→6)β-D-glucopyranoside] and scutellarein-6-*O*-α-L-rhamnopyranoside-8-C-β-D-glucopyranoside (Salib *et al.*, 2011), whereby both the methanolic extract and its water-soluble fraction showed significant inhibitory effects on the enzyme activity *in vitro* (among which the quercetin-7-*O*-galactoside showed a high potent inhibition of the enzyme activity reaching 100% at 100mg/ml).

It has been reported that *Hibiscus tiliaceus* flowers possess properties that are useful in the treatment of bronchitis, as well as in the treatment of fevers and coughs, ear infections and abscesses, *postpartum* disorders, and skin diseases

(Brondegaard, 1973; Holdsworth and Wamoi 1982; Singh *et al.*, 1984; Whistler, 1985; Holdsworth, 1991). This species, whose tree is native to the shores of the Pacific and Indian Oceans, is consumed as a beverage (tea) in southern Brazil; however, its pharmacological effects and chemical composition are still poorly studied. Recently, it was found that vitamin E and several phytosterols, such as stigmasterol (Figure 12.4), stigmastadienol and stigmastadienone, are present in the methanol extract of *H. tiliaceus* (Rosa *et al.*, 2007). This extract showed antigenotoxic and antimutagenic effects against oxidative DNA damage induced by hydrogen peroxide (H_2O_2) and tert-butyl hydroperoxide in V79 cells; these cells are frequently used for studies on DNA damage and DNA repair. Additionally, the methanol extract of *Hibiscus tiliaceus* prevented the increase in lipid peroxidation and decrease in GSH content (Holdsworth, 1991; Kumar and Robson 1984). Methanol extract of *H. tiliaceus* flowers also showed an antidepressant-like profile of action without sedative side effects. Another study isolated triterpenoid compounds: friedelin, pachysandiol, glutinol, lupeol, germanicol, and sterols as stigmast-4-en-3-one, stigmast-4, 22-dien-3-one, β-sitosterol, and stigmasterol.

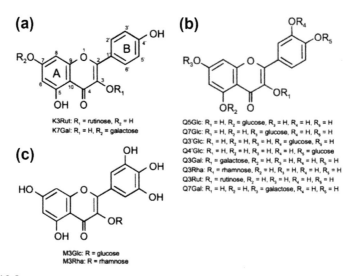

FIGURE 12.3

Representative structures of flavonoid glycosides from *H. rosa-sinensis* and *H. esculents*.

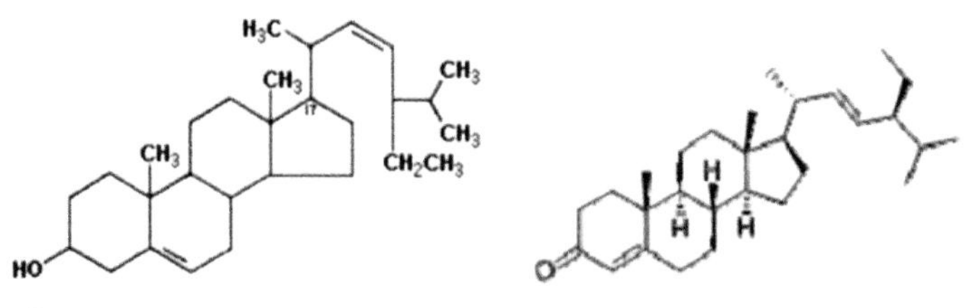

FIGURE 12.4

Stigmasterol and derivatives from *H. tiliaceus* and *H. cannabinus*.

Kenaf, *Hibiscus cannabinus*, is probably native to southern Asia, though its exact natural origin is unknown. Kenaf was grown in Egypt over 3000 years ago. The kenaf leaves were consumed in human and animal diets, the bast fiber was used for bags, cordage, and the sails for Egyptian boats. This crop was not introduced into southern Europe until the early 1900s. Today, principal farming areas are China, India, and it is also grown in many other countries such as the US, Mexico, and Senegal (Gove and Webster, 1993). Previous studies show that *Hibiscus cannabinus* possesses anticomplimentary, antidiarrhetic, and antipholo-gistic activities (Reddy, 1997). The leaves and flowers have been found to be effective in the treatment of heart disorders. Studies of the methanol extract of *H. cannabinus* in streptozotocin-induced diabetic rats showed significant lower-ing of blood glucose. Phytochemical investigations of its extract yielded phytos-terols, flavonoids, and glycosides (Figures 12.3, 12.4).

Hibiscus syriacus L. is widely distributed all around the world as ornamental and green plants (Sung *et al.*, 1998). Common names include Rose of Sharon (espe-cially in North America), Shrub Althea, Rose Althea and St Joseph's rod (Italy). Furthermore, it is also a medicinal plant used as an antipyretic, antihelminthic, and antifungal agent in the orient (Chen and Chen, 1993; Li, 2000; Huang, 1993; Xu *et al.*, 2000). There have been some reports on the active constituents of its buds and flowers (Yun *et al.*, 2001). The methanolic formic acid extract of petals yielded 3-*O*-malonylglucosides of delphinidin, cyanidin, pentunidin, pelargonidin, peoni-din, and malvidin (Figure 12.2) (Kim and Nonaka, 1989). However, there are few investigations into the antioxidative activity of the pigment from its petals where research indicates that extracting pigment from *H. syriacus* L. shows efficient radi-cals scavenging activity on hydroxyl, DPPH, and LPO. Anthocyanins, Saponarin, mucilage, and isovitexin (Figure 12.5) may be a few of the active chemicals in Rose of Sharon.

Hibiscus trionum L. is a common weed in summer crops at the Nile banks, Oases, Mediterranean, and Sinai phytogeographical regions. It is widely spread especially in the tropical regions of Asia and Africa (Täckholm, 1974). It was reported that it

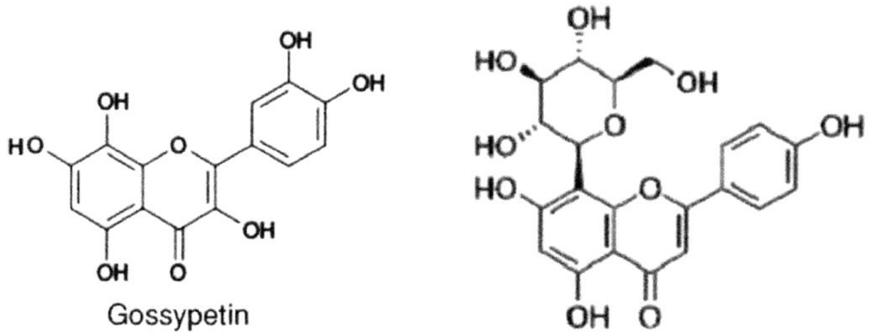

FIGURE 12.5

8-C-substiuted apgenin, isovitexin and 8-*O*-hydroxyquercetin, and gossypetin from *H. syriacus* and *H. cannabinus*.

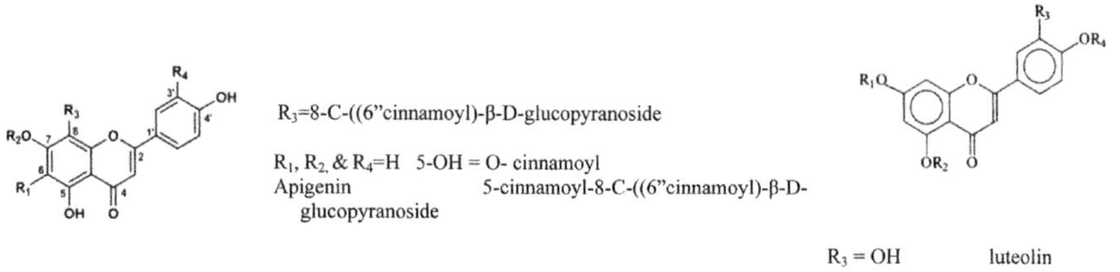

R$_3$=8-C-((6"cinnamoyl)-β-D-glucopyranoside

R$_1$, R$_2$ & R$_4$=H 5-OH = O- cinnamoyl
Apigenin   5-cinnamoyl-8-C-((6"cinnamoyl)-β-D-
glucopyranoside

R$_3$ = OH   luteolin

FIGURE 12.6

Representative structures of flavone glycosides from *H. trionum* L.

possesses a pronounced cytotoxicity, antioxidant, hypoglycemic, hepatoprotective, and hypotensive effects. Apigenin 5-cinnamoyl-8-C-((6″cinnamoyl)-β-D-glucopyranoside, luteolin 3-benzoyl-5-*O*-β-D-glucopyranosyl-β-D-glucopyranoside (Figure 12.6), 2-hydroxy-5-(3-*O*-glucosyl-but-1-enyl)-4,6, 6-trimethyl-cyclohexa-2,4-dienone and 3, 4-dihydroxybenzoic acid were isolated from methanol extract. From the petroleum ether extract, taraxasterol acetate, β-friedoolean-5-en-3β-ol-29acetate, β-amyrin, β-sitosterol were isolated (Ateya *et al.*, 2012).

The hepatoprotective effect of Gossypitrin, a flavonoid extracted from *Hibiscus elatus* S.W, was investigated against the CCl4-induced *in vivo* hepatotoxicity. *Hibiscus elatus* is native to Cuba and Jamaica; it is widely grown as an ornamental or a reforestation tree, and is grown in gardens (Adams, 1971). Beside gossypitrin both (E)-cinnamaldehyde and (Z)-cinnamaldehyde isomers are present in the *Hibiscus elatus* flower.

The Confederate rose is also known as *Hibiscus* Mutabilis or the Cotton Rose-mallow. Contrary to the name of the flower itself, it originated in Southern China thousands of years ago. It was said to have grown quite commonly in the areas that were once known as the Confederate States of America, hence the name. The astounding aspect of these Confederate flowers is how quickly they can change color. These plants, for instance, will emerge a pure white or light pink in the morning and transform drastically to an intense or hot pink, or possibly even red, by the afternoon (Ishikura, 1982). *Hibiscus mutabilis* flower is one of the components of Chinese medicine for treating nasopharyngeal carcinoma (Duke and Ayensu, 1985). Modern pharmacological studies have shown that a decoction of *Hibiscus mutabilis* flower inhibited streptococcus *in vitro*. Lowry (1971) isolated free cyanidin in flowers of *Hibiscus mutabilis* whereby the pink basal blotch in petals of *H. mutabilis* is due to the presence of cyanidin. This may be the first unequivocal case of free anthocyanidin occurring in flowers.

Okra, *Hibiscus esculentus*, also known as lady's finger, is a flowering plant in the mallow family. Even though the plant is cultivated in tropical and warm temperate regions around the world, the species is still poorly studied (Maganha *et al.*, 2010). There is not much informatiom available on the bioactive properties of the plant extract despite its wide usage as a medicinal plant, especially for its antiseptic benefits for cleaning and healing wounds. *Hibiscus esculentus* extract is used in cosmetics to improve skin wrinkles by inhibiting muscle contraction and removing oxygen free radicals (Kang *et al.*, 2010). Later research revealed the plant extract's ability to act as a remedy in the management of diabetes mellitus. *H. esculents* tea is also a popular natural remedy for weight loss. Quercetin 3-*O*-xylosyl (1‴→2″) glucoside, quercetin 3-*O*-glucosyl-(1‴→6″)-glucoside, quercetin 3-*O*-glucoside and quercetin 3-*O*-(6″-*O*-malonyl)-glucoside were first identified and characterized as major antioxidants in lady's finger (Figure 12.3).

Hibiscus is one of the most popular tropical flowers, and certainly among the most pleasantly fragrant. However, did you know that consumption of these amazing flowers can actually be beneficial to your health and that these fragrant powerhouses are loaded with free-radical-fighting polyphenolic antioxidants to help keep your cardiovascular system purring along?

References

Adams, C.D., 1971. The Blue Mahoe & Other Bushes: An Introduction to Plant Life in Jamaica. McGraw–Hill, Far Eastern publishers, Singapore.

Adhirajan, N., Kumar, T.R., Shanmugasundaram, N., Babu, M., 2003. *In vivo* and *in vitro* evaluation of hair growth potential of *Hibiscus rosa-sinensis* Linn. J. Ethnopharmacol. 88, 235–239.

Ali-Bradeldin, H., Al-Wabel, N., Gerald, B., 2005. Phytochemical, pharmacological and toxicological aspects of *Hibiscus sabdariffa*. L: A review. Phytother. Res. 19, 369–375.

Ali, M., Ansari, S.H., 1997. Hair care and herbal drugs. Ind. J. Nat. Prod. 13, 3–5.

Ateya, A.-M. El, Sayed, Z.I., Fekry, M., 2012. Chemical Constituents, Cytotoxicity, Antioxidant, Hypoglycemic and Antihypertensive Activities of Egyptian *Hibiscus trionum*. Aust. J. Basic & Appl. Sci. 6 (3), 756–766.

Aurelio, D., Edgardo, R.G., Navarro, G.S., 2007. Thermal kinetic degradation of anthocyanins in a roselle (*Hibiscus sabdariffa* L. cv. ´Criollo´) infusion. Int. J. food Sci. and Tech. 43, 322–325.

Brondegaard, V.J., 1973. Contraceptive plant drugs. Plant Med. 23, 167–172.

Chen, R.T., Chen, L., 1993. On the chemical constituents of cotton rose Hibiscus. Chinese Traditional and Herbal Drugs 5, 227–229.

Duke, J.A., Ayensu, E.S., 1985. Medicinal Plants of China2 Vols. Reference Publ., Inc, Algonac. Michigan.

Dwivedi, R.N., Pandey, S.P., Tripathi, V.J., 1977. Role of japapushpa (*Hibiscus rosa-sinensis*) in the treatment of arterial hypertension. A trial study. J. Res. Indian Med., Yoga & Homeopathy 12, 13–36.

El-Saidy, S.M., Ismail, I.A., EL-Zoghbi, M., 1992. A study on Roselle extraction as a beverage or as a source for anthocyanins. Zagazig J. Agric. Res. 19, 831–839.

Gove, P.B., Webster, M., 1993. Webster's Third New International Dictionary. MA. Merriam–Webster Inc, Springfield.

Gupta, A.K., Sharma, M., Tandon, N., 2005. Quality standards of Indian medicinal plants. Vol 2. New Delhi: Indian Council of Medical Research, pp. 25–33.

Holdsworth, D., 1991. Traditional medicinal plants of Rarotonga. Cook Island. Part II. Pharm Biol. 29, 71–79.

Holdsworth, D., Wamoi, B., 1982. Medicinal plants of the Admiralty Island, Papua New Guinea. Part I. Int. J. Crude Drug Res. 20, 169–181.

Holser, R.A., Bost, G., Boven, M., 2004. Phytosterol composition of hybrid *Hibiscus* seed oils. J. Agric. Food Chem. 52, 2546–2548.

Hou, D.X., Tong, X., Terahara, N., Luo, D., Fujii, M., 2005. Delphinidin 3-sambubioside, a *Hibiscus* anthocyanin, induces apoptosis in human leukemia cells through reactive oxygen species-mediated mitochondrial pathway. Arch. Biochem. Biophys. 440 (1), 101–109.

Huang, K.C., 1993. The pharmacology of Chinese herbs. FL. CRC press, Tokyo. Boca Raton pp. 193–194.

Ishikura, N., 1982. Flavonol glycosides in the flowers of *Hibiscus mutabilis* F. versicolor. Agric. Biol. Chem pp. 46, 1705–1706.

Kang, C.K., Lee, Y.J., Han, S.H., Kim, S.H., 2010. Anti-aging cosmetic composition. Amorepacific Corporation. US20100119628.

Khokhar, I., Ahmed, A., 1992. Studies in medicinal plants of Pakistan: new cyclopeptide alkaloids from the flowers of *Hibiscus rosa-sinensis*. Sci. Int. (Lahore) 4, 147–150 .

Kim, J.H., Nonaka, G.-I., 1989. Anthocyanidin malonylglucosides in flowers of *Hibiscus syriacus*. Phytochemistry 28, 1503–1506.

Kumar, R., Robson, K.A., 1984. Prospective study of emotional disorders in childbearing women. Br J. Psychiatry 144, 35–47.

Li, S.Z., 2000. Compendium of material medica, 3rd ed. People's Health Publisher, Beijing pp. 2128–2129.

Lin, H.H., Huang, H.P., Huang, C.C., Chen, J.H., Wang, C.J., 2005. *Hibiscus* polyphenol-rich extract induces apoptosis in human gastric carcinoma cells via p53 phosphorylation and p38 MAPK/FasL cascade pathway. Mol. Carcinog 43 (2), 86–99.

Lin, T.-L., Lin, H.H., Chen, C.C., Lin, Chou, M.C., Wang, C.J., 2007. *Hibiscus sabdariffa* extract reduces serum cholesterol in men and women. Nutr. Res. 27, 140–145.

Lowry, J.B., 1971. Free cyanidin in flowers of *Hibiscus* mutabilis. Phytochemistry 10 (3), 673–674.

Maganha, E.G., Halmenschlager, R.C., Rosa, R.M., Henriques, J.A.P., Ramos, A.L.P., Saffi, J., 2010. Pharmacological evidences for the extracts and secondary metabolites from plants of the genus *Hibiscus*. Food Chem. 118, 1–10.

Prenesti, E., Berto, S., Daniele, P.G., Toso, S., 2007. Antioxidant power quantification of decoction and cold infusions of *Hibiscus sabdariffa* flowers. Food Chem. 100, 433–438.

Reddy, C.M., 1997. Antispermatogenic and androgenic activities of various extracts of *Hibiscus rosa sinensis* in albino mice. Indian J. Exp. Biol. 35, 1170–1174.

Rosa, R.M., Moura, D.J., Melecchi, M.I., dos Santos, R.S., Richter, M.F., Camarão, E.B., et al., 2007. Protective effects of *Hibiscus tiliaceus* L methanolic extract to V79 cells against cytotoxicity and genotoxicity induced by hydrogen peroxide and tert-butyl-hydroperoxide. Toxicol. In Vitro. 21, 1442–1452.

Sachdewa, A., Khemani, L.D., 2003. Effect of *Hibiscus rosa sinensis* Linn. ethanol flower extract on blood glucose and lipid profile in streptozotocin induced diabetes in rats. J. Ethnopharmacol. 89 (1), 61–66.

Salib, J.Y., Daniel, E.N., Hifnawy, M.S., Azzam, S.M., shaheed, I.B., Abdel-Latif, S.M., 2011. Polyphenolic Compounds from Flowers of *Hibiscus rosa sinensis* Linn. and their Inhibitory Effect on Alkaline Phosphatase Enzyme Activity *in vitro*. Z. *Naturforsch*, 453–459 66 c.

Sethi, N., Nath, D., Singh, R.K., 1986. Teratological study of an indigenous antifertility medicine, *Hibiscus rosa sinensis* in rats. Arogya J. Health Sci. 12, 86–88.

Shivananda, N.B., Sivachandra, R.S., Orette, F.A., Chalapathi, R.A.V., 2007. Effects of *Hibiscus rosa sinensis* L (Malvaceae) on wound healing activity: a preclinical study in a Sprague Dawley rat. International Journal of Lower Extremity Wounds 6 (2), 76–81.

Shrivastava, D.N., 1974. Phytochemical analysis of Japakusum. J. Res. Ind. Med. 9, 103–104.

Singh, M.P., Singh, R.H., Udupa, K.N., 1982. Antifertility activity of a benzene extract of *Hibiscus rosa sinensis* flowers on female albino rats. Planta Medica 44, 171–174.

Singh, Y.H., Ikahihifo, T., Panuve, M., Slatter, C., 1984. Folk medicine in Tonga: a study on the use of herbal medicines for obstetric and gynecological conditions and disorders. J. Ethnopharmacol. 12, 305–329.

Sung, C.H., Wang, Y.N., Sun, E.J., Ho, L.H., 1998. Evaluation of tree species for absorption and tolerance to ozone and nitrogen dioxide (III). Quarterly Journal of the Experimental Forest Nat. Taiwan University 12 (2), 269–288.

Täckholm, V., 1974. Student's Flora of Egypt. Cairo University, Cairo. pp. 888.

Tseng, T.H., Kao, T.W., Chu, C.Y., Chou, F.P., Lin, W.L., Wang, C.J., 2000. Induction of apoptosis by *Hibiscus* protocatechuic acid in human leukemia cells via reduction of retinoblastoma (RB) phosphorylation and Bcl-2 expression. Biochem. Pharmacol. 60 (3), 307–315.

Whistler, W.A., 1985. Traditional and herbal medicine in the Cook Islands. J. Ethnopharmacol. 13, 239–280.

Wongwattanasathien, O., Kangsadalampai, K., Tongyonk, L., 2010. Antimutagenicity of some flowers grown in Thailand. Food Chem. Toxicol. 48, 1045–1051.

Xu, G.J., Wang, Q., Yu, B.Y., Pan, M.J., 2000. Coloured illustrations of antitumour Chinese traditional and herbal drugs. Science & Technology Publishing House, Fuzhou: Fujian pp. 141–143.

Yun, B.S., Lee, I.K., Ryoo, I.J., Yoo, I.D., 2001. Coumarins with monoamine oxidase inhibitory activity and antioxidatie coumarins-lignans from *Hibiscus syriacus*. J. Nat. Prod. 64 (9), 1238–1240.

Hydrothermal Processing on Phenols and Polyphenols in Vegetables

13

Elżbieta Sikora, Barbara Borczak

Department of Human Nutrition, Faculty of Food Technology, Agricultural University in Kraków, Kraków, Poland

Among food products that are rich sources of phenolic compounds, fruit and vegetables are usually the first to be mentioned (Cieślik *et al.*, 2006; Sikora *et al.*, 2008b). In particular, the importance of vegetables is very high. These generally low-energy products provide people with valuable and often scarce nutrients such as vitamin C, folic acid, and β-carotene, minerals with an alkaline action, and dietary fiber. Many previous studies have also shown that vegetables are compsed of many non-nutritive substances, which are beneficial to humans (Lampe, 1999; Liu, 2003; Prior, 2003; Moreno *et al.*, 2006). These substances include polyphenol compounds whose multiple health benefits on humans are mainly attributed to their antioxidant properties (Soengas *et al.*, 2011).

Vegetables may be eaten raw or cooked. They are often frozen and stored in that state for an extended period of time. The technological procedures to which vegetables are subjected prior to their consumption affect qualitative and quantitative changes in their chemical composition. Similarly, the polyphenolic compounds and their related antioxidant activity are both dependent on the technological treatments. These compounds are water soluble; it is therefore expected that hydrothermal processes in particular may cause polyphenolic losses. Wachtel-Galor *et al.* (2008) confirmed a decrease in the level of polyphenols in cauliflower after boiling, and a simultaneous increase of their content in the water in which the vegetables were cooked.

The type and the content of phenolic compounds in vegetables, and changes in them generated by various factors, are the subject of numerous scientific studies. Special attention is paid to cruciferous vegetables, which are considered as very valuable sources of bioactive compounds. Literature data indicated that the main polyphenolic compounds in cruciferous vegetables are: hydroxycinnamic acids and flavonoids, especially flavonols (Vallejo *et al.*, 2003a, b; Sakakibara *et al.,* 2003; Heimler et al., 2006; Podsędek, 2007; Sikora *et al.*, 2008b; Soengas *et al.*, 2011; Cartea *et al.*, 2011). For example, kale is abundant in polyphenols. In the kale's leaves, a considerable amount

Polyphenols in Plants. http://dx.doi.org/10.1016/B978-0-12-397934-6.00013-9

of hydroxycinnamic acid derivatives were found (chlorogenic acid, caffeic, ferulic, p-coumaric. and acylated derivatives of quercetin and kaempferol-glycosides) (Nilsson *et al.*, 2006; Ayaz *et al.*, 2008; Olsen *et al.*, 2009; Zietz *et al.*, 2010; Schmidt *et al.*, 2010; Korus, 2011; Olsen *et al.*, 2012). Another vegetable in this group is broccoli, which has been thoroughly studied in terms of its polyphenol content (Price *et al.*, 1998; Zhang and Hamauzu, 2004; Moreno *et al.*, 2006). Vallejo *et al.* (2003) showed the presence of caffeoyl-quinic's derivatives (chlorogenic acid and neochlorogenic), sinapic acid, and feruloyl derivative acid in broccoli; and in the progress of further research, the presence of flavonoids (glycosides of kaempferol, quercetin, and the traces of isorhamnetin) has been shown (Vallejo *et al.*, 2004). According to Matilla and Hellstron (2007), the most commonly occurring phenolic acids in fresh broccoli are synapic, chlorogenic, and coumaric acids, while quercetin is its main flavonoid. These data have been confirmed in studies by Pellegrini *et al.* (2010) and Sikora *et al.* (2012). The same compounds, although in smaller quantities, were found in cauliflower, white cabbage, Brussels sprouts, and other cruciferous vegetables (Llorach *et al.*, 2003; Nilsson et al., 2006; Podsędek, 2007; Filipiak-Florkiewicz, 2011; Cartea *et al.*, 2011; Soengas *et al.*, 2011; Sikora *et al.*, 2012).

Cruciferous vegetables are generally cooked traditionally in a large amount of boiling water or steam. Broccoli, cauliflower, Brussels sprouts, and kale are often stored in the frozen state, but the freezing process is preceded by blanching, used to inactivate phenyloxidase responsible for the darkening of the vegetables, and ascorbinase, which causes the loss of vitamin C. The effect of these processes on the content of polyphenolic compounds in the plant's material was the subject of many studies. Most of the research indicated that the traditional cooking of cruciferous vegetables, mainly carried out under home conditions, contributed to the losses of polyphenols. Zhang and Hamauzu (2004) determined the retention of total phenolics in cooked broccoli to be 28–68%, depending on the cooking time, compared with a level of 94% determined by Turkmen *et al.* (2005). In the studies by Sikora *et al.* (2008a, b), the total losses of polyphenols in broccoli, cauliflower, and Brussels sprouts reached 30–40% as a result of traditional cooking (15 min). Similar results were obtained by Puupponen-Pimia *et al.* (2003), and by Miglio *et al.* (2008), where only 27% of the polyphenols determined in the raw vegetable were retained in the cooked broccoli. Faller and Fiahlo (2009) presented some changes as a result of cooking broccoli and cabbage in relation to soluble and hydrolysable polyphenols, and found that soluble polyphenols were maintained to a greater extent (73.9% in broccoli, 104.5% in white cabbage) than hydrolysable polyphenols (65% in broccoli, 34.5% in white cabbage). Interesting results were obtained by Pellegrini *et al.* (2010) in cruciferous vegetables, cooked for 8–10 min. In the case of broccoli, polyphenol content was not changed significantly compared to raw vegetables. Conversely, in Brussels sprouts, a significant increase in the level of these components was observed, while a significant decrease was seen in the case of white cauliflower. Borowski *et al.* (2005) found a 12-fold decrease in the concentration of polyphenols in broccoli boiled for 15 min in water, while Filipiak-Florkiewicz (2011) noted losses of these compounds by 31 and 43% in the two traditionally cooked varieties of cauliflower, white and green, respectively. Smaller, but also statistically significant losses of polyphenols in the same vegetables through their cooking were reported by Gębczyński and Kmiecik (2007) and Mazzeo *et al.* (2011). Sikora *et al.* (2008a) reported 70% losses of these components in cooked kale, which was explained by a greater degree of the vegetables' fragmentation. Subsequent studies conducted by Sikora and Bodziarczyk (2012) confirmed a decrease in the level of these compounds by 56%, in comparison to the raw vegetable.

By examining the impact of hydrothermal processes, Vallejo *et al.* (2003a, b) showed that during conventional cooking of broccoli, the amount of caffeic acid's derivatives decreased by about 62% in relation to the content in the raw vegetable, while ferulic and sinapic acids' derivatives decreased by about 51%. Similarly, Pellegrini *et al.* (2010) reported that cooking of broccoli in boiling water caused 50% losses of flavonoids. The content of phenolic acids was also reduced, with the exception of chlorogenic acid, whose level increased four-fold, and eventually the content of total polyphenols remained virtually unchanged. Price *et al.* (1998) tracked the changes of quercetin's and kaempferol's glycosides and reported that the retention of these compounds stayed at the level of 14–28%. Miglio *et al.* (2008) observed that, as a result of cooking broccoli, the content of chlorogenic acid decreased by almost two-fold, sinapic acid decreased by two-fold, and ferulic acid decreased by almost nine-fold. But the content of caffeic acid did not change, which was explained by the authors as probably being due to hydrolysis of chlorogenic acid yielding the formation of this new compound. Decrease of phenolic acids was generally attributed to their autooxidation. The same study demonstrated that cooking contributed to losses of flavonols by 85–90%. According to Olivera *et al.* (2008), flavonol glycosides are located in the epidermal layers of plants in the hydrophilic areas. This means that they are readily soluble in water, explaining their high losses during cooking in a large amount of water, depending on the cooking duration.

Many of the cited studies supported a close relationship between the duration for which the vegetables are subjected to hydrothermal processes and the losses of phenolic compounds. The blanching process performed before freezing is a process of short duration; therefore many authors showed that in this process, losses of the tested polyphenols were less than in the case of conventional cooking (Price *et al.*, 1998; Ewald *et al.*, 1999; Vallejo *et al.*, 2003a, b; Gębczyński and Lisiewska, 2006; Gębczyński and Kmiecik, 2007; Viña *et al.*, 2007; Olivera *et al.*, 2008). Viña *et al.* (2007) subjected Brussels sprouts to blanching for 1, 3 and 4 min and found no significant changes in flavonoid content compared to the raw material. Gębczyński and Kmiecik (2007) observed a reduction by 10% in the level of polyphenols in the course of blanching of white and green cauliflower. Three-times more losses of polyphenols upon blanching of vegetables were shown by Filipiak-Florkiewicz (2011). Sikora *et al.* (2008a) reported 18–30% losses of polyphenols through blanching of broccoli, Brussels sprouts, cauliflower, white and green kale for 3 min, while the greatest losses were reported for the kale. Amin *et al.* (2006) showed that, in the progress of milder hydrothermal treatment, which was blanching, leafy vegetables lose up to 50% of their antioxidant activity, which is also caused by the losses of polyphenols.

Jaiswal *et al.* (2012) studied the effects of not only time, but also blanching temperatures on the content of total polyphenols and flavonoids in Irish York cabbage. They showed that losses were greatest and amounted to 43% (total polyphenols) and 45% (flavonoids) after blanching at 80, 85 and 90°C for 2 min, while at 95–100°C, the losses reached 47–50%, respectively. At the same time, it was shown that blanching time in the range of 6–14 min did not affect the content of the tested compounds.

Steam cooking is a highly recommended heat treatment because it is not associated with such losses of nutrients as observed in traditional cooking. Literature data showed that upon steam cooking, polyphenols were generally well recovered in cruciferous vegetables; this was confirmed in the studies by Vallejo *et al.* (2003) in the case of broccoli. As a result of steaming, this vegetable preserved approximately 89% of total flavonoids and 91% of caffeoyl-quinic derivatives. A similar effect was observed by Wachtel-Galor *et al.* (2008), Borowski *et al.* (2005), and Roy *et al.* (2009). They found 6–18% increases in the total phenolics and up to three- to four-fold increases in the content of flavonoids depending on the duration of broccoli steaming. Mazzeo *et al.* (2011) did not observe significant changes in the content of polyphenols in the steamed cauliflower, while Turkmen *et al.* (2005) reported a significantly higher content of total phenolics in the steamed broccoli, in relation to the dry weight of the vegetables. In the studies by Miglio *et al.* (2008), about 60% of total phenolic compounds were retained in broccoli. At the same time, the authors claimed that the greatest losses were related to the chlorogenic, sinapic, and ferulic acids, but the contents of caffeic acid and kaempferol did not change. Pellegrini *et al.* (2010) applied two types of steaming, an air/steam impingement oven and a domestic cooker equipped with a mesh basket, to the following cruciferous vegetables: broccoli, Brussels sprouts, cauliflower. They found that in the case of broccoli, steaming in a basket in a conventional oven contributed to significant losses of both flavonoids and phenolic acids. Oven steaming contributed to the growth of the above-mentioned components, confirming the greater stability of polyphenols, probably as a result of oxidative enzyme inactivation and/or due to the lack of direct contact with water, which may protect polyphenols from dissolving in aqueous environments. A similar trend was observed for cauliflower. However, surprisingly oven steaming resulted in a large decline in several cinnamic acids (i.e., synapic and caffeic) in the case of Brussels sprouts, as well as of flavonoids, mainly naringenin. The increase of polyphenols (45%) in the steamed cauliflower was reported by Wachtel-Galor *et al.* (2010), while in the case of steamed cabbage, they observed slight losses of these compounds at the same time.

A convenient and relatively quick method of heat treatment is microwaving. In the course of the studies investigating the effect of this process on the retention of phenolic compounds in broccoli, Zhang and Hamauzu (2004) placed 10 g of product in 200 ml of boiling water and subjected it to microwaving for 30, 60, 90, 120, and 300 s. They showed that the level of polyphenols decreased proportionally to the duration of microwaving and the losses reached 50–70%. Considerable losses of flavonoids and caffeoyl-quinic derivatives were also reported by Vallejo *et al.* (2003a, b) in the microwaved broccoli. They applied the highest power of 1000 W and placed 150 g of product in 150 ml of water. Lopez-Berenguer *et al.* (2007) used different durations of broccoli microwaving in varying volumes of water at different microwave powers, observing similarly the decline of polyphenols with a simultaneous increase in the cooking water. Losses were relatively small and depended significantly on the duration of microwaving and interdependence variables: power × water, time × water, power × time × water. Turkmen *et al.* (2005) observed a significant increase of polyphenols in the dry weight of broccoli after microwaving for 1.5 min, at a power of 1000 W, in small amount of water (100 g/6 ml of water). In turn, Faller and Fiahlo (2009) observed a significant decrease of soluble and hydrolysable polyphenols in microwaved white cabbage. In the case of broccoli, a recovery of more than 100% of soluble polyphenols was recorded. Pellegrini *et al.* (2010) subjected broccoli, Brussels sprouts, and cauliflower to microwaving lasting for over a dozen minutes, using a domestic microwave oven with a capacity of 300 W, and observed a significant decrease of polyphenols in broccoli and cauliflower, and a significant increase in Brussels sprouts. In the case of broccoli, the content of all polyphenolic compounds decreased, while in the cauliflower increased the level of sinapic acid and kaempferol, while microwaving of Brussels sprouts led to increased content of each individual polyphenol, except of kaempferol and naringenin, whose levels were significantly decreased.

The maintenance of phenolic compounds in frozen and cooked vegetables is an important issue. Before freezing, cruciferous vegetables such as broccoli, Brussels sprouts, cauliflower, are blanched. The effect of this process, which causes a decrease in the content of phenolic compounds, has been described previously. For this reason, different authors declared lower polyphenol content in frozen vegetables compared to fresh vegetables (Sikora *et al.*, 2008; Gębczyński and Kmiecik, 2007; Pellegrini *et al.*, 2010; Korus and Lisiewska, 2011; Filipiak-Florkiewicz, 2011). At the same time, both Gębczyński and Kmiecik (2007) as well as, Korus and Lisiewska (2011) observed a gradual decrease in the content of polyphenols in white and green cauliflower, as well as in kale, along the length of frozen storage. Olivera *et al.* (2008) showed no differences in the content of polyphenols in Brussels sprouts, just after freezing and after 8 months of frozen storage. Before consumption, frozen vegetables are usually subjected to heat treatment by boiling, steaming, or microwaving. As the research shows, the content of phenolic compounds may undergo various changes. Sikora *et al.* (2008a) reported 60% losses of polyphenols in the fresh weight of frozen cruciferous vegetable after traditional cooking compared with raw, frozen vegetables. Similar losses were observed by Filipiak-Florkiewicz (2011) upon cooking of frozen white and green cauliflower. Conversely, the results of studies conducted by Pellegrini *et al.* (2010) were different. The authors subjected frozen vegetables (broccoli, Brussels sprouts, and cauliflower) to different heat treatment and found that in the case of broccoli, cooking in a large amount of water (1 : 5, food/water) resulted in 55% losses of total phenols. As a result of both microwaving without water and oven steaming, losses were at the level of 16 and 32%, respectively. Conversely, hand cooking of the broccoli by using basket steaming led to significant increases in the content of the tested compounds. In the case of frozen Brussels sprouts, cooking did not result in any changes, while after submission of the vegetables to the remaining process, there was a significant rise in the content of polyphenols as a result of microwaving (up to two-fold). Similarly, a two-and-a-half-fold growth in the content of phenolic compounds in cooked and frozen broccoli was recorded by Gawlik-Dziki (2008). In the frozen, boiled and oven-steamed cauliflower, Pellegrini *et al.* (2010) found a significant decline of the polyphenols compared with frozen, raw product. Conversely, there were no significant changes in the microwaved vegetables and steamed ones by using a basket. Therefore, it is evident that in the case of frozen and cooked cruciferous vegetables, changes in the polyphenol content may vary depending on the conditions used, the thermal processes, as well as the type of vegetable. The freezing process itself may contribute to the damage of plant cell walls, which allows the release of cellular components. If they are readily soluble in water, they are easily able to go into solution during the hydrothermal process. The more water and the longer the cooking time (traditional cooking), the greater their loss. Conversely, according to Gawlik-Dziki (2008), high temperatures may contribute to the release of phenolics from insoluble complexes, and when water losses are avoided (microwaving without water, steaming), then a larger amount will be measured in the final product than in the starting product.

It is generally known that the greatest damage to the components of leafy vegetables occurrs during hydrothermal treatment, due to the large contact surface with an aqueous medium. Spinach (*Spinacia oleracea*) is an example of such a vegetable, which is considered to be a valuable source of minerals, carotenoids, and polyphenols. Various authors reported that the major phenolic compounds found in spinach are: glucuronic acid derivatives of flavonoids and p-coumaric acid derivatives (Bergman *et al.*, 2001; Aehle *et al.*, 2004; Hait-Darshan *et al.*, 2009; Bunea *et al.*, 2008, Khanam *et al.*, 2012). Interestingly, research on phenolic compounds in spinach subjected to hydrothermal processes did not indicate significant losses. Mazzeo *et al.* (2011) reported a reduction by 5% of the total phenols in cooked spinach compared with uncooked vegetables, while the content of these compounds increased significantly in the steamed product. Turkmen *et al.* (2005) noted no significant changes in polyphenols content in spinach subjected to cooking, steaming, and microwaving, while Bunea *et al.* (2008) observed 10% and 50% losses of these compounds by cooking of raw and frozen spinach leaves, respectively. Simultaneously, the authors showed no significant effects of the freezing process preceded by blanching on the level of the tested compounds.

The most commonly consumed root vegetable is carrot (*Daucus carota*). Primarily, this vegetable is regarded as a rich source of carotenoids, especially β-carotene, but many authors are also interested in the polyphenol content. Gonçalves *et al.* (2010) reported the content of polyphenols at the level of 84 mg GAE/100 g of fresh tissue in raw carrots. Similar amounts were observed by Miglio *et al.* (2008) and Mazzeo *et al.* (2011). Carrot's polyphenols include flavonoids—mainly quercetin, luteolin, kaempferol, and naringerin, as well as phenolic acids (Mazzeo *et al.*, 2011). According to Søltoft *et al.* (2010), the major phenolic compound that constituted up to 80% of the total phenolic acids was 5-*O*-caffeoylquinic acid. In turn,

Miglio *et al.* (2008) reported that chlorogenic acid was dominant. Hydrothermal treatment of carrot affected polyphenol content in varied degrees. In the study by Mazzeo *et al.* (2011), 50% losses of these compounds were observed during conventional cooking, while Miglio *et al.* (2008) did not find any polyphenols present in the carrots after cooking, explained by their diffusion into the boiling water. These authors also showed about 50% losses of these compounds due to the steaming process. Conversely, Mazzeo *et al.* (2011) indicated a significant increase in the level of these componets. Carrot's blanching caused losses in polyphenols and it was time and temperature dependent (Gonçalves *et al.*, 2010).

Red beet is another root vegetable which has received much attention in the literature. Its high nutritional value is characterized by betalains' pigments and phenolic compounds, among which are the following: conjugates of ferulic acid, cinnamic, vanillic, chlorogenic acid, and flavonoids—betagarin, betavulgarin, cochliophilin, and dihydroisoramnetin (Kujala *et al.*, 2002; Ravichandran *et al.*, 2012). The latter study reported some losses of phenolic acids as a result of cooking of beets cut into slices, while microwaving and baking resulted in an increased phenolic acid content or remained without an effect. A striking result of this study was that almost all processes contributed to an elevated antioxidant activity of the product. In the other studies, the impact of different time and temperature preservation of shredded beets was tested, and a slight change in the content of phenolic compounds was shown in the context of these parameters (Jiratanan and Liu, 2004).

The health benefits of garlic and onions have been known for a long time, and they have been used widely in folk and herbal medicines. Garlic and onions contain different active ingredients, among which are polyphenols (Price *et al.*, 1997; Bozin *et al.*, 2008; Kim *et al.*, 2013). According to Gorinstein *et al.* (2008), there are phenolic acids such as p-hydroxybenzoic acid, caffeic acid, and vanillic acid in onion and garlic, but it is mainly quercetin that prevails in onion. The authors found that red onion may contain more than two-fold more of this flavonoid than white onion, and also red onion contained 10-times more anthocyanins. After subjecting these vegetables to blanching for 90 s at 100°C and cooking for 10 min, they showed losses of the tested compounds. In the case of garlic, those losses were about 10% by blanching, but flavanols were reduced by 23%; while upon cooking a reduction of almost 50% was observed. Slightly smaller losses of the total polyphenols, flavonoids, and flavanols occurred as a result of blanching and cooking of white and red onions; however, in the case of red onions 23% and 40% losses of anthocyanins were shown for blanching and cooking, respectively.

Similarly, Gorinstein *et al.*, (2005) examined the content of bioactive compounds in garlic coming from Poland, Ukraine, and Israel, subjecting it to high temperature (100°C) in the oven for 20, 40, or 60 min. In the case of polyphenols, losses were observed depending on the length of the process, approximately 10% (insignificantly), 30%, and 35%, respectively. Conversely, a different effect was observed in the studies by Kim *et al.* (2013), in which garlic was subjected to high temperature, without the water share but in presence of air with different degrees of humidity. As a result of these treatments, a very large four- to 10-fold growth in total phenolics was observed, and a smaller, gradual increase in the content of flavonoids, inversely proportional to the temperature and humidity.

In the case of onion, quercetin is the main flavonoid. Generally, it occurs in greater quantities in red onion than in other varieties. Price *et al.* (1997) examined the effect of hydrothermal processes on the content of this compound in the onion with red–brown husk and found some losses of approximately 20% and, simultaneously, the presence of this component in the water in which the vegetables were previously cooked. Ewald *et al.* (1999) subjected the onion to blanching by using steaming, cooking, and microwaving for 3 min in small amounts of water. All these processes contributed to the losses of both quercetin and kaempferol, the greatest loss (50% of quercetin and 66% of kaempferol) occurred as a result of cooking. Lombard *et al.* (2005) simulated the processes which onion is mostly subjected to in the household [browning in fat (sautéing), baking, and cooking], and showed 18% losses of quercetin in onion as a result of cooking, but a 7% increase in its concentration due to baking and a 25% increase as a result of sautéing. In turn, Siddiq *et al.* (2013) studied changes in the total phenolics of sliced onions, immersed for 1 min in water at 50, 60, and 70°C. They showed a significant increase (16%) of phenolic content in onions treated with water at a temperature of 60°C and a significant decrease by 18% at a temperature of 70°C.

Tomatoes and peppers are very popular vegetables all over the world that are often consumed in both raw and processed forms. They belong to the same family and are characterized by a high content of carotenoids, ascorbic acid, and polyphenols. In tomatoes, there are phenolic acids such as caffeic acid, p-coumaric, ferulic, and chlorogenic, as well as flavonoids, especially flavonols (quercetin, kaempferol, rutin) and flavanones (naringenin) (Luthria *et al.*, 2006; Vallverdú-Queralt *et al.*, 2011, 2012b). The study conducted by Stewart *et al.* (2000) showed that 98% of the flavonols were contained in the tomatoes' skin, 96% of which was quercetin. In the pulp and seeds, quercetin compounds accounted for approximately 70%, while kaempferol constituted 30% of the total flavonols. High content of polyphenolic compounds in the peel of tomatoes was confirmed by Četković *et al.* (2012). This implied that the removal of the tomatoes' skin greatly reduced the level of polyphenols. In turn, Zhuang *et al.* (2012) marked a large amount of gallic acid, 3,4-dihydroxybenzoic, benzoic acid, salicylic acid, and catechins in peppers.

Tomatoes are usually subjected to thermal treatment in order to obtain a sauce or soup. Another popular product is tomato juice, which is subjected to pasteurization or sterilization. Stewart *et al.* (2000) showed that tomato products such as juice, soup, sauté paste, ketchup, and puree may be also rich source of flavonols, suggesting that the flavonols contained in tomatoes withstand the industrial processes very well, while juice and puree may be particularly valuable sources of these compounds in the diet. Also, the study by Gahler *et al.* (2003) confirmed that baking tomatoes in order to obtain a soup or sauce was associated with an increased concentration of polyphenols in these products. Similar results were obtained by Vallverdú-Queralt *et al.* (2012a). Sahlin *et al.* (2004) found a decrease in the content of total phenolics as a result of cooking in a large volume of water (500 g of products/500 ml water), and as a result of frying (sautéing), while there were no changes as a result of baking. In general, it can be safely said that tomatoes belong to a group of very few vegetables that contain the highest amounts of bioactive polyphenols after processing compared with fresh vegetables.

Peppers also contain considerable amounts of phenolic compounds, among which the following were identified: hydroxycinnamic acids (trans-ferulic acid, trans-sinapic acid) and flavonoids (quercetin, luteolin, and apigenin, present in the form of glycosides) (Marin *et al.*, 2004; Materska and Perucka, 2005). The content of phenolic compounds varies depending on the variety and maturity of the pepper. Chuah *et al.* (2008) showed that the highest content of polyphenols is found in red and orange peppers, and the lowest in green.

Sweet pepper may be an integral component of a variety dishes, as well as spice, while cayenne pepper, because of its taste, is consumed in smaller quantities, usually as a seasoning. Changes in the content of phenolic compounds in pepper have been the subject of many studies. Turkmen *et al.* (2005) subjected pepper to short-duration cooking (5 min) in a small amount of water (100 g in 150 ml water), microwaving, and steaming, and found that steaming did not cause significant changes in total phenolic content, whereas in the cooked and microwaved products this content was substantially higher, by 14 and 26%, respectively. Ornelas-Paz *et al.* (2010) examined changes in the content of polyphenols in different types of Mexican pepper, and showed either an increase of these compounds as a result of cooking, or insignificant changes, while in the grilled pepper they found significant increases in polyphenol content in most of the samples. In turn, Chuah *et al.* (2008) did not observe significant changes in polyphenol content by microwaving and stir-frying of the pepper, while a significant decrease of these compounds was observed in the traditionally cooked product; the longer the cooking time, the greater this decrease was, and a simultaneous increase of these compounds in boiled water was observed. Relatively small losses of free phenolics (up to 20%), depending on the duration time, were observed by Dorantes-Alvarez *et al.* (2011) in pepper prepared and processed in a microwave oven for 10–30 s, together with an increase in its antioxidant activity.

Pulses are important and valuable components of the human diet, providing precious vegetable protein, minerals, fiber (e.g., oligosaccharides), and polyphenols (Lin and Lai, 2006). Generally, these products contain phenolic acids (ferulic, sinapic), quercetin, tannins, catechins, and anthocyanins (seeds with colored husks), and isoflavones (genistein, daidzein, glycitein), of which soy is a particularly rich source (Drużyńska and Klepacka, 2004; Amarowicz and Troszyńska, 2005).

Before consumption, pulses must be subjected to heat treatment, which facilitates in particular the bioavailability of starch, and increases the bioavailability of the protein. Dry legumes' seeds are usually subjected to soaking for several hours before the thermal treatment. This process contributes to a decline of compounds with an increasing effect. Bishnoi *et al.* (1994) reported that soaking of peas resulted in polyphenol losses of up to 50%, depending on the soaking time, while cooking of the soaked seeds did not result in further losses of these compounds. Conversely, cooking of the unsoaked seeds led to 9% losses of polyphenols. Particularly large losses of polyphenols (79%) were observed when soaked peas' seeds were dehulled. Investigations of the cooking effect and steaming of yellow and black soybean on the various types of polyphenols were conducted by Xu and Chang (2008b). They found that yellow beans were characterized by significantly decreased amounts of total polyphenols, increased content of flavonoids and condensed tannins, as a result of cooking in a conventional manner and under conventional pressure, while steaming under pressure resulted in a significant rise in all of these components. In the case of black beans, all processes contributed to losses of phenolic compounds, especially anthocyanins. In further studies, Xu and Chang (2009b) investigated the influence of cooking and steaming on the content of total phenolics, phenolic acids, isoflavones, and anthocyanins in pinto and black beans. They stated that all thermal processes differentially affected the content of individual phenolic acids, flavanols, flavonols, and anthocyanins (black bean), but the total content of phenolic acids and favonols was significantly reduced. The largest losses (approximately 10-fold) occurred in the case of anthocyanins contained in black beans. Xu and Chang (2008a, 2009a) observed similar effect, by subjecting of green peas, yellow peas, chickpeas, and lentils to the same heat treatment. On the basis of their research, they found that thermal processes may cause degradation of polyphenols and release of bound phenolic composition. Ranilla *et al.* (2009) studied the effects of different cooking conditions on phenolic compounds in two varieties of Brazilian beans. They applied two temperatures, 100 and 120°C, cooked soaked and unsoaked seeds, and marked the tested compounds in the seeds with and without water straining. They found that soaking prior to cooking and draining following the cooking had the greatest impact on the observed changes. Soaking of beans before cooking contributed to significantly less polyphenols than not soaking beans. Moreover, their content was nearly three-times higher in samples that were not drained. It was also shown that the variation in cooking temperature did not have a specific effect on quercetin and kaempferol derivatives, or on phenolic acids; but in the case of anthocyanins, the reduction of these compounds was significantly higher at 120°C than at 100°C. Nithiyanantham *et al.* (2012) observed a good retention of total phenolics in autoclaved seeds of chickpea and peas. In the case of soaked pea seeds before autoclaving, the polyphenol content was not significantly higher than in unsoaked seeds.

Potatoes are an important source of dietary polyphenols in many countries; their primary component is starch and they must be treated using heat before consumption. The main phenolic compounds in potato tubers are flavonoids, mainly catechins, and phenolic acids, of which approximately 80% constitute chlorogenic acid. The pigmented cultivars also contain anthocyanins (Mendez *et al.*, 2004; Mattila and Hellström, 2007; Leo *et al.*, 2008; Mulinacci *et al.*, 2008; Im *et al.*, 2008; Deußer *et al.*, 2012). Changes in the content of phenolic compounds occurring during the thermal treatment of potato were the subject of numerous studies. Tudela *et al.* (2002) found more than 50% losses of total flavonoids and 70% of caffeic acid derivatives during traditional cooking, steaming, microwaving or frying of potatoes. Similarly, total polyphenols losses were shown by Mäder *et al.* (2009) and by Perl *et al.* (2012) as a result of cooking, blanching, microwaving, and baking of tubers. Mattila and Hellström (2007) focused their attention on phenolic acids and showed insignificant changes of their content as a result of cooking unpeeled potatoes; however when cooking was followed by peeling, this resulted in a significant reduction of these compounds. Analyses of cooked, baked, and microwaved potatoes without peeling showed relatively small or insignificant changes in the total phenolics (Xu *et al.*, 2009) and phenolic acids (Mulinacci *et al.*, 2008). Mäder *et al.* (2009) and Im *et al.* (2008) confirmed that most polyphenols were focused in the skin of potato tubers. The latter authors showed that household methods of cooking affected chlorogenic acid content in different ways. The greatest losses (30–70%) were reported by boiling and steaming, losses were lower as a result of microwaving or frying, while most of the remaining compound (80–90%) was observed in the potatoes treated by oven heating.

A review of the data presented in the available articles implicated that the effect of hydrothermal processes on the content of phenolic compounds in vegetables depends on many factors. In general, in all cases where the vegetables were brought into contact with water, loss of phenolic compounds followed, due to their good solubility in water. These losses can be somewhat limited by cooking unpeeled vegetables (roots, tubers, tomatoes), but the process of skin disposal is generally associated with losses of polyphenols, because that is where they are most concentrated. In the hydrothermal process, an important issue is the degree of vegetables' grinding: the higher the degree of grinding, the greater the loss of water-soluble components. Thermal processes carried out without the participation of water often result in an increase in polyphenols, which is associated with the loss of water from the product. Frequently observed simultaneous increase in antioxidant activity of the product was explained by many authors as a result of changes in the properties of presented polyphenols under high temperatures.

Knowledge of the factors and their impact on the content of phenolic compounds in vegetables may allow selection of such technological processes and culinary processing, which will be associated with optimal retention of the components.

References

Aehle, E., Raynaud-Le Grandie, S., Ralainirina, R., Baltora-Rosset, S., Mesnard, F., Prouillet, C., et al., 2004. Development and evaluation of an enriched natural antioxidant preparation obtained from aqueous spinach (*Spinacia oleracea*) extracts by an adsorption procedure. Food Chem. 86, 579–585.

Amarowicz, R., Troszyńska, A., 2005. Antioxidant activity and reduction power of extract of red bean and its fractions. Bromat. Chem. Toksykol 38, 119–124. (in polish).

Amin, I., Norazaidah, Y., Hainida, K.I., 2006. Antioxidant activity and phenolic content of raw and blanched *Amaranthus* species. Food Chem. 94 (1), 47–52.

Ayaz, F.A., Hayırlıoglu-Ayaz, S., Alpay-Karaoglu, S., Grúz, J., Vatentová, K., Ulrichová, J., et al., 2008. Phenolic acid contents of kale (*Brassica oleraceae* L. var. *acephala* DC.) extracts and their antioxidant and antibacterial activities. Food Chem. 107, 19–25.

Bergman, M., Varshavsky, L., Gottlieb, H.E., Grossman, S., 2001. The antioxidant activity of aqueos spinach extract: chemical identification of active fractions. Phytochemistry 58, 143–152.

Bishnoi, S., Khetarpaul, N., Yadav, R.K., 1994. Effect of domestic processing and cooking methods on phytic acid and polyphenol contents of pea cultivars (*Pisum sativum*). Plant Foods Hum. Nutr. 45, 381–388.

Borowski, J., Borowska, E.J., Szajdek, A., 2005. The influence of heat treatment of broccoli (*Brassica oleracea* var. *italica*) on the scavenging of polyphenols and DPPH. Bromat. Chem. Toksykol 37, 125–131.

Bozin, B., Mimica-Dukic, N., Samojlik, I., Goran, A., Igic, R., 2008. Phenolic as antioxidants in garlic (*Allium sativum* L., *Alliaceae*). Food Chem. 111, 925–929.

Bunea, A., Andjelkovic, M., Socaciu, C., Bobis, O., Neascu, M., Verhe, R., et al., 2008. Total and individual carotenoids and phenolic acids content in fresh, refrigerated and processed spinach (*Spinacia oleracea* L.). Food Chem. 108, 649–656.

Cartea, M.E., Francisco, M., Soengas, P., Velasco, P., 2011. Phenolic Compounds in *Brassica* Vegetables. Molecules 16, 251–280.

Četković, G., Savatović, S., Čanadanović-Brunet, J., Djilas, S., Vulić, J., Mandić, A., et al., 2012. Valorisation of phenolic composition, antioxidant and cell growth acivities of tomato waste. Food Chem. 133, 938–945.

Chuah, A.M., Lee, Y.-C., Yamaguchi, T., Takamura, H., Yin, L.-J., Matoba, T., 2008. Effect of cooking on the antioxidant properties of coloured peppers. Food Chem. 111, 20–28.

Cieślik, E., Gręda, A., Adamus, W., 2006. Content of polyphenols in fruit and vegetables. Food Chem. 94, 135–142.

Deußer, H., Guignrd, C., Hoffmann, L., Evers, D., 2012. Polyphenol and Glycoalkaloid contents in potato cultivars grown in Luxembourg. Food Chem. 135, 2814–2824.

Dorantes-Alvarez, L., Jaramillo-Flores, E., González, K., Martinez, R., Parada, L., 2011. Blanching peppers using microwaves. Procedia Food Sci. 1, 178–183.

Drużyńska, B., Klepacka, M., 2004. Functional properties of the bean (*Phaseolus vulgaris*) seed preparations obtained using the crystallization and classical isolation methods. Żywność 4 (41), 69–78. (in polish).

Ewald, C., Fjelkner-Modig, S., Johansson, K., Sjöholm, I., Åkesson, B., 1999. Effect of processing on major flavonoids in processed onions, green beans, and peas. Food Chem. 64, 231–235.

Faller, A.L.K., Fialho, E., 2009. The antioxidant capacity and polyphenol content of organic and conventional retail vegetables after domestic cooking. Food Res. Int. 42, 210–215.

Filipiak_Florkiewicz A, 2011. The effect of hydrothermal processing on selected health properties of cauliflower (*Brassica oleracea var. botrytis*). Zesz. Nauk UR w Krakowie, 470, Rozprawy z. 347. (in polish).

Gahler, S., Otto, K., Böhm, V., 2003. Alterations of Vitamin C, Total Phenolics, and Antioxidant Capacity as Affected by Processing Tomatoes to Different Products. J. Agric. Food Chem. 51, 7962–7968.

Gawlik-Dziki, U., 2008. Effect of hydrothermal treatment on the antioxidant properties of broccoli (*Brassica oleracea* var. *botrytis italica*) florets. Food Chem. 109, 393–401.

Gębczyński, P., Kmiecik, W., 2007. Effects of traditional and modified technology, in the production of frozen cauliflower, on the contents of selected antioxidative compounds. Food Chem. 101, 229–235.

Gębczyński, P., Lisiewska, Z., 2006. Comparison of the level of selected antioxidative compounds in frozen broccoli produced using traditional and modified methods. Innovative Food Sci. Emerg. Technol. 7, 239–245.

Gonçalves, E.M., Pinheiro, J., Abreu, M., Brandão, T.R.S., Silva, C.L.M., 2010. Carrot (*Daucus carota* L.) peroxidase inactivation, phenolic content and physical changes kinetics due to blanching. J. Food Eng. 97, 574–581.

Gorinstein, S., Drzewiecki, J., Leontowicz, H., Leontowicz, M., Najman, K., Jastrzebski, Z., et al., 2005. Comparison of the Bioactive and Antioxidant Potentials of Fresh and Cooked Polish, Ukrainian, and Israeli Garlic. J. Agric. Food Chem. 53, 2726–2732.

Gorinstein, S., Leontowicz, H., Leontowicz, M., Namiesnik, J., Najman, K., Drzewiecki, J., et al., 2008. Comparison of the Main Bioactive Compounds and Antioxidant Activities in Garlic and White and Red Onions after Treatment Protocols. J. Agric. Food Chem. 56, 4418–4426.

Hait-Darshan, R., Grossman, S., Bergman, M., Deutsch, M., Zurgil, N., 2009. Synergistic activity between a spinach-derived natural antioxidant (NAO) and commercial antioxidants in a variety of oxidation systems. Food Res. Int. 42, 246–253.

Heimler, D., Vignolini, P., Dini, M.G., Vincieri, F.F., Romani, A., 2006. Antiradical activity and polyphenol composition of local *Brassicaceae* edible varieties. Food Chem. 99 (3), 464–469.

Im, H.W., Suh, B.-S., Lee, S.-U., Kozukue, N., Ohnisi-Kameyama, M., Levin, C.E., et al., 2008. Analysis of Phenolic Compounds by High-Performance Liquid Chamatography and Liquid Chromatography/Mass Spectrometry in Potato Plant Flowers, Leaves, Stems, and Tubers and in Home-Processed Potatoes. J. Agric. Food Chem. 56, 3341–3349.

Jaiswal, A.K., Gupta, S., Abu-Ghannm, N., 2012. Kinetic ecaluation of colour, texture, polyphenols and antioxidant capacity of Irish York cabbage after blanching treatment. Food Chem. 131, 63–72.

Jiratanan, T., Liu, R.A., 2004. Antioxidant Activity of Processed Table Beeta (*Beta vulgaris* var. *conditiva*) and Geer Beans (*Phaseolus vulgaris* L.). J. Agric. Food Chem. 52, 2659–2670.

Khanam, U.K.S., Oba, S., Yanase, E., Murakami, Y., 2012. Phenolic AIDS, flavonoids and total antioxidant capa city of selected leafy vegetables. J. Funct. Foods 4, 979–987.

Kim, J.-S., Kang, O.-J., Gweon O-.C., 2013. Comparison of phenolic AIDS and flavonoids in black garlic at different thermal processing steps. J. Funct. Foods 5, 80–86.

Korus, A., 2011. Effect of preliminary processing, method of drying and storage temperature on the level of antioxidants in kale (*Brassica oleracea* L. var. *acephala*). LWT 44, 1711–1716.

Korus, A., Lisiewska, Z., 2011. Effect of preliminary processing and method of preservation on the content of selected antioxidative compounds in kale (*Brassica oleracea* L. var. *acephala*) leaves. Food Chem. 129, 149–154.

Kujala, T.S., Vienola, M.S., Klika, K.D., Loponen, J.M., Pihlaja, K., 2002. Betalain and phenolic compositions of four beetroot (*Beta vulgaris*) cultivars. Eur. Food Res. Technol. 214, 505–510.

Lampe, J.W., 1999. Health effects of vegetables and fruit: assessing mechanisms of action in human experimental studies. Am. J. Clin. Nutr. 70 (Suppl), 475S–490S.

Leo, L., Leone, A., Longo, C., ombardi, D.A., Raimo, F., Zacheo, G., 2008. Antioxidant Compounds and Antioxidant Activity in "Early Potatoes". J. Agric. Food Chem. 56, 4154–4163.

Lin, P.-Y., Lai H-.M., 2006. Bioactive Compounds in Legumes and Their Germinated Products. J. Agric. Food Chem. 54, 3807–3814.

Liu, R.H., 2003. Health benefits of fruit and vegetables are from additive and synergistic combinations of phytochemicals. Am. J. Clin. Nutr. 78 (Suppl), 517S–520S.

Llorach, R., Espin, J.C., Tomás-Barberán, F.A., Ferreres, F., 2003. Valorization of Cauliflower (*Brassica oleracea* L. var. *botrytis*) By-Products as a Source of Antioxidant Phenolics. J. Agric. Food Chem. 51, 2181–2187.

Lombard, K., effley, E., Geoffriau, E., Thompson, L., Herring, A., 2005. Quercetin in onion (*Allium cepa* L.) after heat-treatment simulating home preparation. J. Food Compos. Anal. 18, 571–581.

López-Berenguer, C., Carvajal, M., Moreno, D.A., García-Viguera, C., 2007. Effects of Microwave Cooking Conditions on Bioactive Compounds Present in Broccoli Inflorescences. J. Agric. Food Chem. 55, 10001–10007.

Luthria, D.L., Mukhopadhyay, S., Krizek, D.T., 2006. Content of total phenolics and phenolic acids in tomato (*Lycopersicon esculentum* Mill.) fruits as influenced by cultivar and solar UV radiation. J. Food Compos. Anal. 19, 771–777.

Mäder, Rawel H., Kroh, L.W., 2009. Composition of Phenolic Compounds and Glycoalkaloids α-Solanine and α-Chaconine during Commercial Potato Processing. J. Agric. Food Chem. 57, 6292–6297.

Marín, A., Ferreres, F., Tomás-Barberán, F.A., Gil, M.I., 2004. Characterization andquantitation of antioxidant constituents of sweet pepper (*Capsicum annuum* L.). J. Agric. Food. Chem. 52, 3861–3869.

Materska, M., Perucka, I., 2005. Antioxidant Activity of the Main Phenolic Compounds Isolated from Hot Pepper Fruit (*Capsicum annuum* L.). J. Agric. Food Chem. 53, 1750–1756.

Mattila, P., Hellström, J., 2007. Phenolic acids in potatoes, vegetables, and some of their products. J. Food Compos. Anal. 20, 152–160.

Mazzeo, T., N'Dri, D., Chiavaro, E., Visconti, A., Fogliano, V., 2011. Effect of two cooking procedures on phytochemical compounds, total antioxidant capacity and colour of selected frozen vegetables. Food Chem. 128, 627–633.

Mendez, C.M.V., Delgado, M.A.R., Rodriguez, E.M.R., Romero, C.D., 2004. Content of Free Phenolic Compounds in Cultivar of Potatoes Harvested in Tenerife (Canary Islands). J. Agric. Food Chem. 52 1323–1227.

Miglio, C., Chiavaro, E., Visconti, A., Fogliano, V., Pellegrini, N., 2008. Effects of Different Cooking Methods on Nutritional and Physicochemical Characteristic of Selected Vegetables. J. Agric. Food Chem. 56, 139–147.

Moreno, D.A., Carvajal, M., López-Berenguer, C., Garcia-Viguera, C., 2006. Chemical and biological characterization of nutraceutical compounds of broccoli. J. Pharmaceut. Biomed. 41, 1508–1522.

Mulinacci, N., Ieri, F., Giaccherini, C., Innocenti, M., Andrenelli, L., Canova, G., et al., 2008. Effect of Cooking on the Anthocyanins, Phenolic Acids, Glycoalkaloids, and Resistant Starch Content in Two Pigmented Cultivars of *Solanum tuberosum* L. J. Agric. Food Chem. 56, 11830–11837.

Nilsson, J., Olsson, K., Engqvis, G., Ekvall, J., Olsson, M., Nyman, M., et al., 2006. Variation in the content of glucosinolates, hydroxycinnamic acids, carotenoids, total antioxidant capacity and low-molecular-weight carbohydrates in *Brassica* vegetables. J. Sci. Food Agric. 86, 528–538.

Nithiyanantham, S., Selvakumar, S., Siddhuraju, P., 2012. Total phenolic content and antioxidant activity of two different solvent extracts from raw and processed legumes, *Cicer arietinum* L. and *Pisum sativum* L. J. Food Compos. Anal. 27, 52–60.

Olivera, D.F., Vina, S.Z., Marani, C.M., Ferreyra, R.M., Mugridge, A., Chaves, A.R., et al., 2008. Effect of blanching on the quality of Brussels sprouts (*Brassica oleracea* L. *gemmifera* DC) after frozen storage. J. Food Eng. 84, 148–155.

Olsen, H., Aaby, K., Borge, G.I.A., 2009. Characterization and Quantification of Flavonoids and HYdroxycinnamic Acids in Curly Kale (*Brassica oleracea* L. Convar. *acephala* Var. *sabellica*) by HPLC-DAD- ESI-MS. J. Agric. Food Chem. 57, 2816–2825.

Olsen, H., Grimmer, S., Aaby, K., Saha, S., Borge, G.I.A., 2012. Antiproliferative Effects of Fresh and Thermal Processed Green and Red Cultivars of Curly Kale (*Brassica oleracea* L. Convar. *acephala* Var. *sabellica*). J. Agric. Food Chem. 60, 7375–7383.

Ornelas-Paz, J.J., Martínez-Burrola, J.M., Ruiz-Cruz, S., Santana-Rodríguez, V., Ibarra-Junquera, V., Olovas, G.I., et al., 2010. Effect of cooking on the capsaicinoids and phenolics contents of Mexican peppers. Food Chem. 119, 1619–1625.

Pellegrini, N., Chiavaro, E., Gardana, C., Fogliano, V., Porrini, M., 2010. Effect of Different Cooking Methods on Colour, Phytochemical Concentration, and Antioxidant Capacity of Raw and Frozen *Brassica* Vegetables. J. Agric. Food Chem. 58, 4310–4321.

Perla, V., Holm, D.G., Jayanty, S., 2012. Effects of cooking methods on polyphenols, pigments and antioxidant activity in potato tubers. LWT 45, 161–171.

Podsędek, A., 2007. Natural antioxidants and antioxidant capacity of *Brassica* vegetables: A review. LWT 40, 1–11.

Price, K.R., Bacon, J.R., Rhodes, M.J.C., 1997. Effect of Storage and Domestic Processing on the ontent and Composition of Flavonol Glucosides in Onion (*Allium cepa*). J. Agric. Food Chem. 45, 938–942.

Price, K.R., Casuscelli, F., Colquhoun, I.J., Rhodes, M.J.C., 1998. Composition and Content of Flavonol Glycosides in Broccoli florets (*Brassica oleracea*) and their Fate during Cooking. J. Sci. Food Agric. 77, 468–472.

Prior, R.L., 2003. Fruits and vegetables in the prevention of cellular oxidative damage. Am. J. Clin. Nutr. 78 (Suppl), 570S–578S.

Puupponen-Pimia, R., Häkkinen, S.T., Aarni, M., Suortti, T., Lampi, A.-M., Eurola, M., et al., 2003. Blanching and long-term freezing affect various bioactive compounds of vegetables in different ways. J. Sci. Food Agric. 83 (14), 1389–1402.

Ranilla., G., Genovese, M.I., Lajolo, F.M., 2009. Effect of Different Conditions on Phenolic Compounds and Antioxidant Capacity of Some Selected Brazilian Bean (*Phaseolus vulgaris* L.) Cultivars. J. Agric. Food Chem. 57, 5734–5742.

Ravichandran, K., Ahmed, A.R., Knorr, D., Smetanska, I., 2012. The effect of defferent processing methods on phenolic acid content and antioxidant activity of red beet. Food Res. Int. 48, 16–20.

Roy, M.K., Juneja, L.R., Isobe, S., Tsushida, 2009. Steam processed broccoli (*Brassica oleracea*) has higher antioxidant activity in chemical and cellular assay system. Food Chem. 114, 263–269.

Sahlin, E., Savage, G.P., Lister, C.E., 2004. Investigation of the antioxidant properties of tomatoes after processing. J. Food. Compos. Anal. 17, 635–647.

Sakakibara, H., Honda, Y., Nakagawa, S., Ashida, H., Kanazawa, K., 2003. Simultaneous Determination of All Polyphenols in Vegetables, Fruits, and Teas. J. Agric. Food Chem. 51, 571–581.

Schmidt, S., Zietz, M., Schreiner, M., Rohn, S., Kroh, L.W., 2010. Genotypic and climatic influences on the concentration and composition of flavonoids in kale (*Brassica oleracea* var. *sabellica*). Food Chem. 119, 1293–1299.

Siddiq, M., Roidoung, S., Sogi, D.S., Dolan, K.D., 2013. Total phenolics, antioxidant properties and quality of fresh-cut onions (*Allium cepa* L.) treated with mild-heat. Food Chem. 136, 803–806.

Sikora, E., Bodziarczyk, I., 2012. Composition and antioxidant activity of kale (*Brassica oleracea* L. *Var. acephala*) raw and cooked. Acta Sci. Pol., Technol. Aliment. 11 (3), 239–248.

Sikora, E., Cieślik, E., Leszczyńska, T., Filipiak-Florkiewicz, A., Pisulewski, P.M., 2008a. The antioxidant activity of selected cruciferous vegetables subjected to aquathermal processing. Food Chem. 107, 55–59.

Sikora, E., Cieślik, E., Topolska, K., 2008b. The sources of natural antioxidants. Acta Sci. Pol. Technol. Aliment. 7 (1), 5–17.

Sikora, E., Cieślik, E., Filipiak-Florkiewicz, A., Leszczyńska, T., 2012. Effect of hydrothermal processing on phenolic acids and flavonols contents in selected brassica vegetables. Acta Sci. Pol., Technol. Aliment. 11 (1), 45–51.

Soengas, P., Soleto, T., Velasco, P., Cartea, M.E., 2011. Antioxidant Properties of *Brassica* Vegetables. Funct. Plant Sci. Biotechnol. 5, (Special Issue 2), 43–55.

Søltoft, M., Nielsen, J., Laursen, K.H., Husted, S., Halekoh, U., Knuthsen, P., 2010. Effects of Organic and Conventional Growth System on the Content of Flavonoids in Onions and Phenolic Acids in Carrots and Potatoes. J. Agric. Food Chem. 58, 10323–10329.

Stewart, A.J., Bozonnet, S., Mullen, W., Jenkins, G.I., Lean, M.E.J., Crozer, A., 2000. Occurrence of Flavonols in Tomatoes and Tomato-Based Products. J. Agric. Food Chem. 48, 2663–2669.

Tudela, J.A., Cantos, E., Espin, J.C., Tomás-Barberán, F.A., Gil, M.I., 2002. Induction of Antioxinat Flavonol Biosynthesis in Fresh-Cut Potatoes. Effect of Domestic Cooking. J. Agric. Food Chem. 50, 5925–5931.

Turkmen, N., Sari, F., Velioglu, Y.S., 2005. The effect of cooking methods on total phenolics and antioxidant activity of selected green vegetables. Food Chem. 93, 713–718.

Vallejo, F., Tomás-Barberán, F.A., Garcia-Viguera, C., 2003. Phenolic compound contents in edible parts of broccoli inflorescences after domestic cooking. J. Sci. Food Agric. 83, 1511–1516.

Vallejo, F., Tomás-Berberán, F.A., Ferreres, F., 2004. Characterisation of flavonols in broccoli (*Brassica oleracea* L. var. *italica*) by liquid chromatography-UV diode-array detection-electrospray ionisation mass spectrometry. J. Chromatogr. 1054, 181–193.

Vallverdú-Queralt, A., Medina-Remón, A., Martinez-Huélamo, M., Jauregui, O., Andres-Lacueva, C., Lamuela-Raventos, R.M., 2011. Phenolic Profile and Hydrophilic Antioxidant Capacity as Chemotaxonomic Markers of Tomato Varieties. J. Agric. Food Chem. 59, 3994–4001.

Vallverdú-Queralt, A., Medina-Remón, A., Casals-Ribes, I., Andres-Lacueva, C., Waterhouse, A.L., Lamuela-Raventos, R.M., 2012a. Effect of tomato industrial processing on phenolic profile and hydrophilic antioxidant capacity. LWT 47, 154–160.

Vallverdú-Queralt, A., Medina-Remón, A., Casals-Ribes, I., Lamuela-Raventos, R.M., 2012b. Is there any difference between the phenolic content of organic and conventional tomato juices? Food Chem. 130, 222–227.

Viňa, S.Z., Olivera, D.F., Marani, C.M., Ferreyra, R.M., Mugridge, A., Chaves, A.R., et al., 2007. Quality of Brussels sprouts (*Brassica oleracea* L. *gemmifera* DC) as affected by blanching method. J. Food Eng. 80, 218–225.

Wachtel-Galor, S., Wong, K.W., Benzie, I.F.F., 2008. The effect of cooking on *Brassica* vegetables. Food Chem. 110, 706–710.

Xu, B., Chang, S.K.C., 2008a. Effect of soaking, boiling, and steaming on total phenolic content and antioxidant activities of cool season food legumes. Food Chem. 110, 1–13.

Xu, B., Chang, S.K.C., 2008b. Total Phenolics, Phenolic Acids, Isoflavones, and Anthocyanins and Antioxidant Properties of Yellow and Black Soybeans As Affected by Thermal Processing. J. Agric. Food Chem. 56, 7165–7175.

Xu, B., Chang, S.K.C., 2009a. Phytochemical Profiles and Health-Promoting Effects of Cool-Season Food Legumes As Influenced by Thermal Processing. J. Agric. Food Chem. 57, 10718–10731.

Xu, B., Chang, S.K.C., 2009b. Total Phenolic, Phenolis Acids, Anthocyanin, Flavan-3-ol, and Flavonol Profiles and Antioxidant Properties of Pinto and Black Beans (*Phaseolus vulgaris* L.) as Affected by Thermal Processing. J. Agric. Food Chem. 57, 4754–4764.

Xu, X., Li, W., Lu, Z., Beta, T., Hydamaka, A.W., 2009c. Phenolic Content, Composition, Antioxidant Activity, and Their Changes Turing Domestic ooking of Potatoes. J. Agric. Food Chem. 57, 10231–10238.

Zhang, D., Hamauzu, Y., 2004. Phenolics, ascorbic acid, carotenoids and antioxidant activity of broccoli and their changes during conventional and microwave cooking. Food Chem. 88, 503–509.

Zhuang, Y., Chen, L., Sun, L., Cao, J., 2012. Bioactive characteristics and antioxidant activities of nine peppers. J. Funct. Foods 4, 331–338.

Zietz, M., Weckmüller, A., Schmidt, S., Rohn, S., Schreiner, M., Krumbein, A., et al., 2010. Genottpic and Climatic Influence on the Antioxidant Activity of Flavonoids in Kale (*Brassica oleracea* var. *sabellica*). J. Agric. Food Chem. 58, 2123–2130.

Polyphenols Identification and Occurrence

Improved Characterization of Polyphenols Using Liquid Chromatography

14

Rosa María Lamuela-Raventós[*†], **Anna Vallverdú-Queralt**[*†], **Olga Jáuregui**[‡],
Miriam Martínez-Huélamo[*†], **Paola Quifer-Rada**[*†]

*CIBEROBN Fisiopatología de la Obesidad y la Nutrición and RETIC, Instituto de Salud Carlos III, Spain, †Nutrition and Food Science Department, Pharmacy School, University of Barcelona, Barcelona, Spain, ‡Scientific and Technical Services, University of Barcelona, Barcelona, Spain

CHAPTER OUTLINE HEAD

14.1 Introduction

Polyphenols are plant secondary metabolites and the most abundant dietary bioactive compounds. Nowadays, it is estimated that 100,000 to 200,000 secondary metabolites exist (Metcalf, 1987). Despite their extreme variety, polyphenols possess a common carbon skeleton building block: the C6–C3 phenylpropanoid unit. Biosynthesis by this pathway leads to a wide range of plant phenols: cinnamic acids (C6–C3),

Polyphenols in Plants. http://dx.doi.org/10.1016/B978-0-12-397934-6.00014-0

261

benzoic acids (C6–C1), flavonoids (C6–C3–C6), proanthocyanidins [(C6–C3–C6)n], coumarins (C6–C3), stilbenes (C6–C2–C6), lignans (C6–C3–C3–C6) and lignins [(C6–C3)n] (Seabra *et al.*, 2006).

An exhaustive identification of polyphenols in food and biological samples is of great interest due to their health-promoting effects. Notably, they have an important protective role against a number of pathological disturbances, such as atherosclerosis, brain dysfunction, and cancer (Ignat *et al.*, 2011). It is well known that the protective effects of polyphenols *in vivo* depend on their accessibility and extractability from food, intestinal absorption, metabolism, final biological action in the human body, and potential interaction with target tissues (Tulipani *et al.*, 2012). Phenolics may also act as antifeedants, contributors to plant pigmentation and protective agents against UV light, amongst other activities (Ignat *et al.*, 2011). Nevertheless, the lack of commercially available standards and the wide range of phenolic structures found in nature make identification of phenolic compounds a complex task.

Polyphenol extraction is a crucial step in the development of an analytical method sensitive enough to determine these substances at low concentrations. Several extraction methods are described in the literature (Ignat *et al.*, 2011), but the most common are liquid–liquid extraction (Baydar *et al.*, 2004; Vallverdú-Queralt *et al.*, 2010), solid–liquid extraction (Martinez-Huelamo *et al.*, 2012; Medina-Remon *et al.*, 2009), and extraction with supercritical fluid (Palenzuela *et al.*, 2004; Palma and Taylor, 1999).

Diverse methods have been reported for the identification and quantification of phenolic compounds (Ignat *et al.*, 2011), including spectrophotometry (Huang *et al.*, 2009; Medina-Remon *et al.*, 2009), capillary electrophoresis (CE) (Herrero-Martinez *et al.*, 2005), nuclear magnetic resonance spectroscopy (NMR) (Slimestad *et al.*, 2008), near-infrared spectroscopy (NIR) (Chen *et al.*, 2009), and chromatographic techniques like high-performance liquid chromatography (HPLC) (Martinez-Huelamo *et al.*, 2012; Vallverdú-Queralt *et al.*, 2010), ultra-high-performance liquid chromatography (UHPLC) (Epriliati *et al.*, 2010; Gruz *et al.*, 2008), high-speed counter-current chromatography (HSCCC) (Cao *et al.*, 2009; Yanagida *et al.*, 2006), supercritical fluid chromatography (SFC) (Kamangerpour *et al.*, 2002), and gas chromatography (GC) (Friedman, 2004; Lu and Foo, 1998), although in this chapter we will focus only on liquid chromatography.

Available HPLC detectors have various limitations. Although low detection limits and good sensitivity are obtained by UV, fluorescence, refractive index, light scattering or electrochemical detectors, the structural information they provide lacks detail. The introduction of methods that combine two or more analytical techniques, such as HPLC-UV coupled with photodiode array detection (HPLC-UV-DAD) and HPLC coupled with mass spectrometry (HPLC-MS), has improved structural elucidation of metabolites (Marston and Hostettmann, 2009).

Nowadays, the best analytical tool to quantify and characterize phenolic compounds is considered to be liquid chromatography coupled with ultraviolet-photodiode array detection (UV-DAD) (Chen *et al.*, 2009; Crozier *et al.*, 1997; Epriliati *et al.*, 2010; Fang *et al.*, 2009; Kerem *et al.*, 2004; Liu *et al.*, 2008; Sakakibara *et al.*,

2003; Sun *et al.*, 2007; Wang *et al.*, 2009) or mass spectrometry (MS) (Cao *et al.*, 2009; Chiva-Blanch *et al.*, 2011; Cimpan and Gocan, 2002; Gruz *et al.*, 2008; Han *et al.*, 2008; Martinez-Huelamo *et al.*, 2012; Sanchez-Rabaneda *et al.*, 2003a; Tsao and Deng, 2004; Tulipani *et al.*, 2012; Urpi-Sarda *et al.*, 2009; Vallverdú-Queralt *et al.*, 2010; Volpi and Bergonzini, 2006).

14.2 Sample preparation
14.2.1 Analyte isolation

Accurate identification and quantification of analytes greatly depends on the extraction step. The lack of a standard extraction procedure, which is due to the variability and complexity of phenolic chemical structures and the matrices in which they are found, has led to the proliferation of multiple extraction techniques and methods (Table 14.1).

Extraction can also ensure a more sensitive determination of phenolic compounds and metabolites found at very low concentrations by eliminating interfering components, especially in biological matrices.

Liquid–liquid (LLE) and solid–liquid extraction (which may be followed by solid-phase extraction (SPE) to purify the extract) are the most widely used techniques. Common extraction solvents are methanol, ethanol, acetone, ethyl acetate, and diethyl ether, containing only a small amount of acid. However, polar phenolic acids such as cinnamic acids cannot be extracted with pure organic solvents, and require alcohol–water or acetone–water mixtures.

Liquid samples are usually centrifuged and/or filtered and then the sample is either directly injected into the separation system or analytes are isolated using LLE or SPE.

Conventional methods such as boiling, heating or refluxing can be used to extract natural phenolic compounds from samples, but polyphenols can be lost due to hydrolysis, ionization and oxidation during the process (Li *et al.*, 2005). In recent years, other techniques have been developed for polyphenol extraction, including ultrasound-assisted, microwave-assisted, supercritical fluid, and high hydrostatic pressure extraction (HHP) (Wang and Weller, 2006).

Supercritical fluid extraction is being increasingly used in food and pharmaceutical industries as it is more environmentally friendly, avoiding the use of large amounts of toxic solvents, as well as being rapid, automatable, and selective (Bleve *et al.*, 2008; Maróstica-Junior *et al.*, 2010). The intrinsic low viscosity and high diffusivity of supercritical CO_2 has permitted faster and more efficient separation, and relatively clean extracts. In addition, the absence of light and air during extraction reduces the degradation of analytes that occur in traditional extraction techniques. Supercritical fluids have solvating powers similar to organic solvents but with higher diffusivity, lower viscosity, and lower surface tension. However, the solvating power of a supercritical fluid needs to be controlled by temperature and pressure, or by adding organic modifiers such as methanol. For example, owing to the polarity of

Table 14.1 Examples of Extraction and Analysis of Polyphenols in Food and Biological Samples

Matrix	Analytes	Extraction	Analysis	LOD (mg/l)	References
Food and Beverage Samples					
Fruits and fruit juices	Phenolic acids, anthocyanins, hydroxybenzoic acids, flavan-3-oles, hydroxycinnamic acids, coumarins, flavanones, flavones, dihydrochalcones, flavonols	SE LLE	HPLC-UV HPLC-ESI-MS	0.03–0.005	Abad-Garcia et al., 2007; Liu et al., 2012; Sakakibara et al., 2003; Xu et al., 2012; Fang et al., 2009
Grapes and grape juices	Anthocyanins, flavanols, flavonols, hydroxycinnamates	SE LLE	HPLC/Q-TOF HPLC-MS/MS HPLC-DAD HPLC-MS	3–0.5	Liang et al., 2011; Muñoz et al., 2008; Xu et al., 2012
Vegetables	Quercetin glycosides, hydroxycinnamic acids, phenolic acids, flavanols, flavonols, flavones	SE SPE	HPLC-UV CE UHPLC-DAD	0.62–0.005	Caridi et al., 2007; Silva et al., 2012
Tea leaves and derived products	Flavanols, hydroxycinnamic acids, phenolic acids, flavones, phenolic terpenes, hydroxy-benzoic acids	SE	HPLC-DAD-ESI-MS/MS UHPLC-UV UHPLC-MS/MS	0.048–0.0301	Aura et al., 2002 ; Spáčil et al., 2010; Wang et al., 2008a
Apples	Flavanols, flavonols, hydroxycinnamates, anthocyanins, dihydrochalcones	SE	HPLC-DAD HPLC-MS	3×10^{-7}– 3×10^{-8}	Alonso-Salces et al., 2005; Vrhovsek et al., 2004
Wine	Flavanols, flavonols, phenolic acids, stilbenes, hydroxycinnamates, hydroxybenzoic acids, procyanidyns, cinnamic acids	LLE MEPS Filtration	HPLC-UV-FLD UHPLC-DAD HPLC-UV-DAD	0.54–0.02 0.2–0.01 0.05–0.003	Bétes-Saura et al., 1996; Gonçalves et al., 2013; Rodri;guez-Delgado et al., 2001
Alcohol-free beer	Flavanols, hydroxycinnamates	SPE	HPLC-UV	0.2–0.01	Garcia et al., 2004

Sample	Analytes	Extraction	Method	LOD/range	References
Beans, soy beans and derived products	Flavanols, phenolic acids, hydroxycinnamates, isoflavones	SE	HPLC-UV-DAD HPLC-ESI-MS UHPLC-UV	<0.5	Griffith and Collison, 2001; Ross et al., 2009; Toro-Funes et al., 2012
Cocoa and chocolate	Catechin and epicatechin, procyanidins	LLE SE	HPLC-FLD UHPLC-MS/MS	$0.002-2 \times 10^{-6}$ $20-9$	Machonis et al., 2012; Ortega et al., 2010
Olive oil	Tyrosols and flavonols	LLE	HPLC-ECD	<4	Capannesi et al., 2000
Water	Phenolic acids, flavonols, hydroxycoummarics	SE	HPLC-DAD	$0.3-0.1$	Liu et al., 2008
Propolis	Flavonoids	LLE	HPLC-MS	<0.0025	Volpi and Bergonzini, 2006
Tomato and derived products	Phenolic acids, flavonols, flavanones, hydroxy-cinnamic acids	LLE	HPLC-MS/MS, HPLC-QTOF, HPLC-Orbitrap	1.7×10^{-5} 3×10^{-7}	Vallverdú-Queralt et al., 2010, 2011a, 2011b; 2011d, e 2012b, c
Biological Samples					
Rat urine and plasma	Isoflavones and its metabolites Puerarin (daidzein-8-C-glucoside) and its metabolites Epicatechin, epigallocatechin and its metabolites	SPE Protein precipitation LLE	HPLC-MS/MS HPLC-ESI-MS/MS HPLC-UV	$0.125-0.025$	Fang et al., 2002; Fu et al., 2008; Prasain et al., 2004a
Serum	Isoflavones and lignans Procyandin B1	SPE LLE	HPLC-ESI-MS/MS HPLC-MS	$<1 \times 10^{-10}$	Grace et al., 2003; Sano et al., 2003
Urine	Phenolic acids, Flavanols, flavonols Flavanols, flavonols, phenolic acids and related metabolites	LLE Protein precipitation SPE	HPLC-ESI-MS/MS UHPLC-ESI-MS/ MS	$1.3 \times 10^{-7}-$ 1×10^{-10}	Magiera et al., 2012; Martinez-Huelamo et al., 2012; Rios et al., 2003
Plasma	Epicatechin and related metabolites Quercetin Flavanols, flavonols, phenolic acids and related metabolites	SPE	HPLC-UV/ Vis-FLD-ECD HPLC-ECD HPLC-ESI-MS/MS	<0.2 $2.9 \times 10^{-7}-$ 5×10^{-9}	Erlund et al., 1999; Martinez-Huelamo et al., 2012; Ottaviani et al., 2012

LLE, liquid–liquid extraction; SE, solid extraction; SPE, solid-phase extraction.

anthocyanins, their extraction by the $SC–CO_2$ method requires high pressures and the presence of methanol or ethanol (Bleve *et al.*, 2008).

Ultrasound-assisted extraction is an inexpensive, simple, and efficient alternative to conventional extraction techniques (Wang *et al.*, 2008b). This method extracts non-volatile and semi-volatile compounds from the matrix. The ultrasonic process facilitates contact between the sample matrix and extraction solvent. Ultrasonication is often carried out to improve phenolic compound extraction from plants; for example, a study with *Folium eucommiae* (Huang *et al.*, 2009) found it to be more efficient than conventional extraction techniques.

Another promising approach for extracting phenolic compounds is microwave-assisted extraction, which was satisfactorily used to analyze gallic acid, protocatechuic acid, chlorogenic acid and caffeic acid in Eucommia ulmodies (Li *et al.*, 2004). Zhang *et al.* extracted polyphenols from Camellia oleifera fruit hull using microwave-assisted extraction (Zhang *et al.*, 2011), finding the optimal conditions to be a liquid:solid ratio of 15.33:1 (ml/g), extraction time of 35 min and extraction temperature of 76°C. The same method has also been used to extract polyphenols from green tea (Nkhili *et al.*, 2009) and found to be more efficient than conventional heating.

Enzymatic release of phenolic compounds is also employed to extract phenolic compounds, for example, from grape pomace (Maier *et al.*, 2008). Another study investigated the ability of three enzymes (Ultraflo L, Viscozyme L, and a-Amylase) to release phenolic compounds from *Ipomoea batatas* L. (sweet potato) stems (Min *et al.*, 2006). Ferulic acid release rate was optimal when Ultraflo L (1.0%) was used, whereas Viscozyme L was the most effective for the release of vanillic acid and vanillin.

Another technique that enhances the extraction of polyphenols is HPP. Studies carried out by Shouqin *et al.* (2005) have demonstrated the benefits of hydrostatic pressure for the extraction of flavanols.

14.2.2 Analyte purification. Solid phase Extraction

SPE is an extraction technique used as a clean-up procedure and a pre-concentration step with crude plant, biological, environmental, food, and pharmaceutical samples (Ho *et al.*, 2012; Kerio *et al.*, 2012; Martinez-Huelamo *et al.*, 2009; Navas, 2012; Olmos-Espejel *et al.*, 2012).

Alkyl-bonded silica or copolymer sorbents are commonly used to extract analytes, reverse-phase sorbents being most chosen for polyphenols. Samples and solvents are usually slightly acidified to prevent ionization of phenolic compounds, which would lead to a weaker analyte retention in the sorbent (Navas, 2012; Vinas *et al.*, 2011).

In a recent study, reverse-phase HLB cartridges were used to extract phenolic compounds and metabolites from the urine and plasma of volunteers who had consumed different types of tomato sauce: without oil, or containing 5% of virgin olive oil or 5% of refined olive oil (Martinez-Huelamo *et al.*, 2012). Urpi-Sardà *et al.* (2009) also used HLB cartridges to extract conjugated phenolics from urine and plasma after regular consumption of cocoa. SPE is essential when working with LC-MS in order to reduce the matrix effect, especially when analytes are found in low concentrations, as occurs in biological samples. However, HLB cartridges have also been used to determine resveratrol and piceid in beer matrices (Chiva-Blanch *et al.*, 2011).

Mix-mode cation/anion exchange reverse-phase sorbents have also been applied to extract phenolic compounds from biological matrices due to their higher capacity to clean up samples. In a study by Medina-Remon *et al.* (2009), HLB, MCX (mix-mode cation-exchange reverse-phase sorbent) and MAX (mix-mode anion-exchange reverse-phase sorbent) were compared in the extraction of 10 representative polyphenols from urine samples, and the best recoveries were obtained with MAX cartridges. Vallverdú-Queralt *et al.* used MAX cartridges to analyze phenolic compounds of different tomato varieties (Vallverdú-Queralt *et al.*, 2011e) and to distinguish between organic and conventional tomatoes (Vallverdú-Queralt *et al.*, 2012a). In a study carried out to analyze colonic microbial metabolites, MCX was used to extract polyphenols from urine and plasma after regular consumption of cocoa (Urpi-Sarda *et al.*, 2009).

14.3 **High performance liquid chromatography (HPLC)**

The type of column used to separate phenolics and their glycosides is almost exclusively a reverse-phase C18-bonded silica column ranging from 100 to 300 mm in length and with an internal diameter of 2–4.6 mm (Merken and Beecher, 2000; Stalikas, 2010; Tsao and Deng, 2004; Tulipani *et al.*, 2012), although occasionally C8 columns are used to separate phenolic acids. Columns are maintained from room temperature to 40°C during the analysis but thermostated columns give more repeatable elution times and greater resolution, and allow the backpressure of the LC column to be reduced at high flow rates.

The use of a binary system is essential for the separation of structurally varied phenolic compounds. Gradient elution is usually performed with a solvent A, including an aqueous acidified polar solvent or water-containing buffer, and a solvent B, which can be an organic solvent such as methanol or acetonitrile, pure or acidified (Merken and Beecher, 2000; Tsao and Deng, 2004). The volume of injection ranges from 1 to 100 μl, depending on the internal diameter of the column used (Merken and Beecher, 2000).

14.3.1 **Ultraviolet detection**

Ultraviolet (UV) is the simplest and most commonly used HPLC detector due to its greater sensitivity, linearity, versatility, and reliability (Wolfender, 2009). The existence of conjugated double and aromatic bonds in phenolic compounds allows them to absorb UV or UV-VIS light (Stalikas, 2010). There are different types of UV detectors: fixed wavelength, multiple wavelength, or photodiode array (DAD)

(Wolfender, 2009), the last one being the most frequently used to detect phenolic compounds.

Polyphenols absorb light at different wavelengths. Flavonoids have two characteristic absorption bands: the first has a maximum in the 240 to 285nm range, corresponding to the A-ring, while the second band has a maximum in the 300 to 550nm range, which is attributed to the substitution pattern and conjugation of the C-ring. Anthocyanins also present two absorption bands, in the regions of 265–275 and 465–560 nm. Flavones, flavonols, and flavonols are detected at 280 and 350 nm. UV spectra of catechins give peaks at 210, 278, and 280 nm. Flavones and flavonols have bands in the ranges of 240–280 nm and 300–380 nm. Flavanones and isoflavones are detected at 280–290 nm and 236–262 nm, respectively (Crozier *et al.*, 1997; Sakakibara *et al.*, 2003). Figure 14.1 shows the UV spectra of representative polyphenols.

UV detection became the preferred detector in LC analysis since it is cheap and robust, especially for food matrices containing high phenol concentrations (Table 14.1). In a study by Bétes-Saura *et al.* (1996) an HPLC coupled with a UV-DAD detector was used to identify and quantify 30 polyphenols in white wines. The column used was a C18 (250×4 mm), with 5 μm particle size. Flow rate was set at 1.5 ml/min and gradient elution was performed with glacial acetic acid in water at pH 2.65 (phase A) and 20% solvent A mixed with 80% acetonitrile (phase B). The chromatogram was monitored simultaneously at three wavelengths: 280, 320, and 365 nm. Benzoic acids, tyrosol, flavan-3-ols, and the oligomeric procyanidins were quantified at 280 nm, cinnamic acids and their tartaric esters at 320 nm and flavonols at 365 nm (Caporaso *et al.*, 2011). The method was validated, providing good precision and linearity and low limits of detection, which varied from 0.003 mg/l for cis-caftaric acid to 0.051 mg/l for tyrosol.

Liu *et al.* (2008) developed a method to determine polyphenols in water by HPLC-DAD. The separation of phenolic compounds was carried out in a C18 column (150×4.6 mm, 5 μm). Gradient elution was performed using acetic acid/water solution (1:99, v/v) as the aqueous mobile phase and methanol as the organic phase. The photodiode array detector operated between 210 and 400 nm. The method was validated, with recoveries between 83 and 95% and limits of detection ranging from 0.1 to 0.3 mg/l. The developed method allowed the identification and quantification of seven polyphenols (chlorogenic acid, esculetin, caffeic acid, scopoletin, rutin, quercetin hydrate, kaempferol) in tobacco-polluted water.

Another study by Lachman *et al.* (2009) used HPLC-DAD to analyze anthocyanidins in red- and purple-fleshed potatoes from 15 cultivars. Anthocyanidins were determined using a reverse-phase column C18 (4×250 mm, 7 μm). Solvent A was aqueous 1% (v/v) phosphoric acid, 10% (v/v) acetic acid, 5% acetonitrile (v/v), and solvent B was 100% HPLC grade acetonitrile at a flow rate of 1 ml/min. The quantification of anthocyanidins was set at $\lambda = 530$ nm. Prior to HPLC analysis, samples were hydrolyzed by acidic hydrolysis. The results showed that individual cultivars differed significantly in the relative proportion of anthocyanidins. However, the most abundant anthocyanidin in red- and purple-fleshed potatoes was petunidin (46.9%),

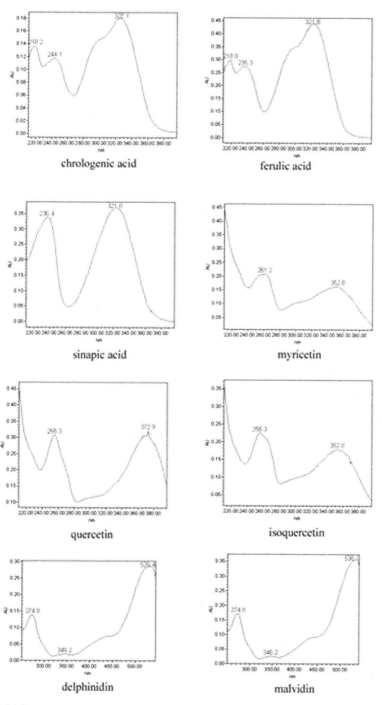

FIGURE 14.1

UV spectra of some polyphenols.

followed by malvidin (22.8%) and pelargonidin (22.1%), cyanidin (5.38%), peonidin (2.74%), and delphinidin (0.15%).

14.3.2 Fluorescence detection

Fluorescence detection is rarely used to analyze polyphenols since only a few exhibit natural fluorescence, including isoflavones without an OH group at position 5 (de Rijke *et al.*, 2002) and flavonoids with an OH group at position 3 (Sengupta and Kasha, 1979), such as catechin and methoxylated flavones (Huck and Bonn, 2001). The analysis of these compounds by HPLC-fluorimetric detection is a more selective and sensitive technique for complex mixtures and provides lower limits of detection (LODs) than UV detection. Moreover, the combination of UV detection and fluorescence makes it possible to distinguish between fluorescent and non-fluorescent co-eluting compounds, and allows a more sensitive detection of the former (Rodriguez-Delgado *et al.*, 2001).

De Quirós *et al.* (2009) proposed a new method for the analysis of flavanol, procyanidin, hydroxycinnamate, flavonol, and stilbene derivatives in white wines based on HPLC-UV-fluorimetric detection. A reverse-phase (250×4.0 mm, $5\,\mu$m) column was used, and the mobile phases consisted of (A) water–acetonitrile–acetic acid, $67:32:1$ v/v/v, and (B) water–acetic acid, $99:1$ v/v at 0.8 ml/min. The identification and quantification of phenolic compounds was achieved by setting the UV-Vis and fluorimetric detectors at selected wavelengths. The fluorescence detector was set at λ_{em} 360 nm and λ_{ex} 278 nm for (+)-catechin, (−)-epicatechin, procyanidin B1 and procyanidin B2, and at λ_{em} 392 nm and λ_{ex} 300 nm for trans-resveratrol. The UV–Vis detector system was set at 280 nm for flavanols, procyanidins, and trans-resveratrol, 320 nm for caftaric acid and 360 nm for flavonols. The method was fully validated, providing great repeatability with an RSD lower than 5%, limits of detection from 0.02 to 0.2 µg/ml, and recoveries of 97.3%.

Another study (Ottaviani *et al.*, 2012) used an HPLC-fluorimetric detection technique to determine epicatechin metabolites in human plasma after a dietary intervention of a dairy-based cocoa drink rich in epicatechin and procyanidins. The separation was achieved with a C18 column (150×4.6 mm, $3\,\mu$m) using 4% (v/v) methanol and 42 mM sodium acetate and an acetonitrile gradient with a flow rate of 0.8 ml/min. The detection of (−)-epicatechin and its related metabolites was achieved following the traces of fluorescence at 276 nm excitation and 316 nm emission and UV absorption at 280 nm. This method demonstrated that (−)-epicatechin-3′-β-D-glucuronide, (−)-epicatechin-3′-sulfate, and 3′-O-methyl-(−)-epicatechin-5/7-sulfate are the predominant (−)-epicatechin-related metabolites in humans, and confirmed the relevance of the stereochemical configuration in the context of flavanol metabolism.

Nevertheless, when working with fluorescence detection, it is necessary to take into account that emission spectra of polyphenols may show pH-dependence or might undergo solvent-dependent dual emission.

To implant fluorescence detection of phenolic compounds that do not exhibit natural fluorescence, derivatization with metal cations has been used. For example, quercetin and kaempferol can form highly fluorescent complexes with metal cations such as Al (III) (Hollman *et al.*, 1996).

14.3.3 Electrochemical detection

Electrochemical detection is another technique that can be used to analyze polyphenols in food and biological matrices, since most flavonoids are electroactive due to the presence of phenolic groups. Electrochemical detection is a selective technique and can be very sensitive for compounds that are oxidized or reduced at low-voltage (Milbury, 2001). Most flavonoids show two values of maximum detector response: the first corresponds to the oxidation of phenolic substituents on the B-ring, while the second might be due to the other, less oxidizable, phenolic groups.

In a study by Aaby *et al.* (2004), the electrochemical behavior of 20 flavonoids and cinnamic acid derivatives was correlated with antioxidant activity measured by FRAP, DPPH, and ORAC assays. The aim of the work was to determine whether the analysis of phenolic compounds by HPLC coupled with a coulometric array detector could be used to predict antioxidant activity assessed by the three tests. It was concluded that the electrochemical response at a relatively low oxidation potential (300 mV) and the cumulative response at medium oxidation potential (400 and 500 mV) showed considerable correlation with antioxidant activities, with the highest correlations being found with FRAP and DPPH assays after short reaction periods.

Another example of polyphenol analysis by electrochemical detection is given by Peng *et al.* (2005). The aim of the study was to determine phloridzin, (−)-epicatechin, chlorogenic acid and myricetin in apple juice and cider by capillary electrophoresis with electrochemical detection. The analytes were separated in 20 min in a 75 cm length capillary at a separation voltage of 18 kV in a 50 mmol/l borate buffer (pH 8.7). The limits of detection ranged from 1×10^{-7} to 5×10^{-7} g/ml for all analytes. The proposed method gave high recoveries of 95–98% and good reproducibility, with an RSD lower than 3.6%.

Capannesi *et al.* (2000) compared different techniques to evaluate the phenolic content of an extra-virgin olive oil with varying storage time and conditions. The techniques used were a disposable screen-printed sensor coupled with differential pulse voltammetry, and a tyrosinase-based biosensor operating in an organic solvent, using an amperometric oxygen probe as the transducer. Electrochemical detection revealed the degradation reaction of large molecules (such as oleuropein derivatives) into smaller ones.

Electrochemical detection has also been applied to biological samples, as described by Jin *et al.* The aim of the study was to validate and apply a method for the quantification of quercetin in human plasma after the ingestion of a commercial canned green tea (Jin *et al.*, 2004). The analysis was performed by an HPLC system coupled with electrochemical detection. A microbore octadecylsilica column (150 × 1.0 mm, 3 μm) was used and the mobile phase was methanol–water (4 : 6, v/v) containing 0.5% phosphoric acid, with a flow rate of 25 μl/min. Quercetin was oxidized at a detection potential of +0.5 V *versus* Ag/AgCl. The method proved highly selective and sensitive with a detection limit of 0.33 pg.

14.3.4 **Mass spectrometry**

Liquid chromatography coupled with mass spectrometry is an efficient method to detect and quantify phenolic compounds in plant extracts and biological fluids. The mass spectrometer ionizes the compounds to generate charged molecules and molecule fragments, measuring their mass-to-charge ratios (Ignat *et al.*, 2011; Marston and Hostettmann, 2009; Stalikas, 2010; Wolfender, 2009). Different sources can be used for compound ionization: fast atom bombardment (FAB), electrospray ionization (ESI), atmospheric pressure chemical ionization (APCI), atmospheric pressure photo-ionization (APPI), and matrix-assisted laser desorption ionization (MALDI). The detection of the compounds can be performed in positive or negative ion mode, the latter being more common in polyphenol analysis (Ignat *et al.*, 2011; Magiera *et al.*, 2012; Prasain *et al.*, 2004b; Schieber *et al.*, 2000; Sporns and Wang, 1998; Stobiecki, 2000). Table 14.2 shows a list of fragment ions of representative polyphenols and their metabolites in negative mode.

Different types of mass analyzers can be used in polyphenol analysis: single quadrupole (MS), triple-quadrupole (MS/MS), ion-trap mass spectrometers (MS^n), time-of-flight (TOF), quadrupole-time-of-flight (QTOF), Fourier transform mass spectrometry (FTMS), and *Orbitrap-based* hybrid mass spectrometers (LTQ-Orbitrap) (Liang *et al.*, 2012; Meda *et al.*, 2011; Mikulic-Petkovsek *et al.*, 2012; Van Der Hooft *et al.*, 2012; Xie *et al.*, 2011).

Quadrupoles consist of four parallel rods connected together, with voltages applied between one pair of rods and the other. Ions with a specific mass-to-charge ratio (*m/z*) will pass through the quadrupole when a particular voltage is applied. This enables quadrupoles to filter the ions en route to the detector. As well as single, triple quadrupole systems are also available, in which the first (Q1) and third quadrupole (Q3) work as filters while Q2 acts as the collision cell. The generic mode for screening in MS systems is the full scan, where a mass spectrum is acquired every few seconds, thus allowing the identification of the protonated or deprotonated molecule and consequently the calculation of the molecular weight of the substance. More sensitive modes of working in quadrupole systems include selected ion monitoring (SIM) in single quadrupole instruments and multiple reaction monitoring (MRM) mode in triple quadrupole instruments. In SIM experiments, the use of a fixed voltage allows the detection of a single *m/z*, whereas in MRM experiments, Q1 filters a precursor ion, and the Q2 is the collision cell, which produces a product ion by collision of the precursor ion with a neutral collision gas. The product ion is transferred into Q3 where only a specific *m/z* is allowed to pass.

Tandem mass spectrometry enables polyphenols to be detected and quantified in complex matrices through MS/MS techniques such as product ion scan, precursor ion scan, and neutral loss scan. A product ion scan mass spectrum contains the fragment ions generated by the collision of the molecular ion. A precursor ion mass spectrum is obtained by limiting the fragment ion to a single ion of interest. Parent ions (molecular ions) are scanned to determine which of them give the target fragment ion. Neutral loss mass spectra show fragment ions with a particular loss of mass, for example, glucoside polyphenols would have a mass loss of 162 μ, which corresponds to a glucoside. Working in MRM mode in combination with precursor

Table 14.2 List of Fragment Ions of some Polyphenols and Related Metabolites Obtained Working in Negative-Ion ESI Mode

Compound	MW	[M-H]⁻	*m/z* ions
3,3/4-Hydroxyphenyl propionic acid glucuronide	342	341	165
3,3/4-Hydroxyphenyl propionic acid sulfate	246	245	165
3/4-Hydroxyphenyl acetic acid glucuronide	328	327	151
3/4-Hydroxyphenyl acetic acid sulfate	232	231	151
3-Hydroxybenzoic	138	137	93
4-Hydroxyhippuric acid	195	194	100
4-Hydroxybenzoic	138	137	93
8-Prenylnaringenin	340	339	219, 175, 119
Apigenin-*C*-hexoside-hexoside	594	593	503, 473, 383, 353
Apigenin-*C*-hexoside-pentoside	564	563	563, 503, 473, 443, 353
Caffeic acid	180	179	135, 107
Caffeic acid-*O*-hexoside 1	342	341	179, 135
Caffeic acid glucuronide	356	355	179
Caffeic acid sulfate	230	259	179
Carboxyacteyl tryptophan	290	289	203, 159, 142
Chlorogenic acid	354	353	191
Coumaric acid glucuronide	340	339	163
Coumaric acid sulfate	244	243	163
Cryptochlorogenic acid	354	353	191, 173, 135
Dihydrocaffeic acid sulfate	262	261	181
Dihydrocaffeic acid glucuronide	358	357	181
Dihydrocaffeic acid, 3,4-dihidroxyphenylpropionic acid	182	181	137
Dihydroxyphenyl acetic acid, homoprotocatechuic acid	168	167	123
Ethyl galate	198	197	169
Ferulic acid	194	193	134
Ferulic acid-*O*-hexoside	356	355	193, 178, 149
Ferulic acid glucuronide	370	369	193
Ferulic acid sulfate	274	273	193
Glutamylphenylalanine	294	293	164, 147, 103
Hydroferulic acid	196	195	136
Homovanillic acid	182	181	137
Hydroferulic acid 3,4-*O*-glucuronide	372	371	195
Hydroferulic acid 3,4-*O*-sulfate	276	275	195
Hydroxyphenyl acetic acid	152	151	107

Continued

Table 14.2 List of Fragment Ions of some Polyphenols and Related Metabolites Obtained Working in Negative-Ion ESI Mode—cont'd

Compound	MW	[M-H]⁻	m/z ions
Hydroxyphenyl propionic acid	166	165	121
Isoferulic acid	194	193	134
Isorhamnetin	316	315	301
Isorhamnetin glucuronide	492	491	315
Isorhamnetin sulfate	396	395	315
Isoxanthohumol	354	353	233, 165, 119
Kaempferol	286	285	251
Kaempferol-3-O-rutinoside	594	593	593, 285
Kaempferol-O-rutinoside-hexoside	756	755	593, 285
m-Coumaric acid	164	163	119
Naringenin	272	271	151, 119
Naringenin 4'-glucuronide	448	447	271
Naringenin 7-glucuronide	448	447	271
Naringenin-7-O-glucoside (prunin)	434	433	433, 271
Neochlorogenic acid	354	353	191, 179, 135
p-Coumaric acid	164	163	119
Phenyl acetic acid	136	135	91
Phenyl acetic glucuronide	312	311	135
Phenyl acetic sulfate	216	215	135
Phloretin-C-diglycoside	598	597	477, 387, 357, 417
Piceid	390	389	227, 185
Quercetin	302	301	301, 151
Quercetin sulfate	382	381	301
Quercetin-3-O-glucuronide	478	477	301
Resveratrol	228	227	185, 143
Rutin	610	609	609, 300
Rutin-O-hexoside-pentoside	904	903	741, 609, 300
Taxifolin	304	303	285
Xanthohumol	354	353	233, 119

ion scan and product ion scan can be helpful in characterizing a particular compound found in complex mixtures. This kind of mass analyzer is the most commonly used.

An ion-trap mass spectrometer (MS^n) consists of a chamber with two electrodes and two end pieces that trap ions with a series of electromagnetic fields. Once the ions are inside, another magnetic field is applied, and only selected ions remain in the chamber. This mass analyzer is useful for structural elucidation purposes, performing multiple stage MS^n (Anari *et al.*, 2004; Wolfender, 2009).

The introduction of high-resolution spectrometers like TOF, QTOF, FTMS, and LTQ-Orbitrap has provided increased resolution and mass accuracy. A TOF mass analyzer consists of an ion source and a detector. The ions are accelerated towards the detector with the same amount of energy through an accelerating potential. Ions with different m/z reach the detector at different times, with lighter ions arriving first due to their greater velocity. This spectrometer permits the analysis of a wide mass range, supplying molecular formula information and precise ion trace extraction.

The QTOF is a hybrid configuration of the TOF spectrometer. The ions are filtered in Q1, the collision takes place in Q2, and finally the product ion is determined by TOF. Compared to a triple quadrupole spectrometer, the QTOF offers greater sensitivity and accuracy when working in full scan mode, and unlike TOF equipment, measures MS/MS. The QTOF spectrometer is useful for the characterization of molecules with a wide range of mass.

FTMS is based on the effect of a magnetic field on an ion rotating in a radiofrequency field. Using the magnetic field, the ions are directed to a chamber where they rotate, describing small orbits with minimum frequency. The application of a radio frequency signal excites the ions to describe spiral orbits with increasing amplitude. When the diameter of the orbit is equal to the distance between the two electrodes, the ions are detected, generating an image of power, which is a direct function of their m/z relationship. This image is integrated by a Fourier transformation and converted to a signal proportional to its intensity. The full spectrum is obtained by scanning a radio frequency field that varies between 8 kHz and 100 MHz. The main advantages of this type of analyzer are its high precision mass measurements (0.001% and above) and almost unlimited resolving power.

The LTQ-Orbitrap, which combines an ion-trap analyzer with FTMS, allows MS and MS^n analysis with an error of less than 2 ppm. LTQ-Orbitrap-MS is a good tool for qualitative analysis, facilitating the structural elucidation of unknown compounds (Peterman *et al.*, 2006).

14.3.4.1 HPLC-MS

A liquid chromatograph coupled to a single quadrupole mass spectrometer allowed the identification of trans-resveratrol and cis-resveratrol (up to 10 μg/l) and trans-piceid and cis-piceid (up to 3 μg/l) in red and white wines. Separation was performed in a C18 (50×2.0 mm i.d., 5 μm) column, using a mobile phase of (A) water (0.5 ml/l acetic acid) and (B) acetone : acetonitrile : acetic acid (70 : 30 : 0.4 ml). Flow-rate was set at 500 μl/min.

A large number of phenolic compounds were successfully identified in propolis samples from different countries using HPLC-MS. The flavonoids were separated in a C18 column (150×4.6 mm, 4 μm) with (A) 0.25% acetic acid and (B) methanol as the mobile phase at a flow rate of 0.5 ml/min (Volpi and Bergonzini, 2006).

Anthocyanidins of 15 grape juice samples, four grape berries, and four grape skins were quantified using LC-MS. Separation was performed in an amide-C18 column (250×4.6 mm, 5 μm) with a mobile phase of (A) 0.4% TFA (v/v) in water and (B) 0.4% TFA (v/v) in acetonitrile at 1 ml/min. The results indicated that anthocyanidin concentration was higher in grape skins than the corresponding berries, and varied among the different grape juice samples (Xu *et al.*, 2012).

14.3.4.2 HPLC-MS/MS

The triple quadrupole mass spectrometer has been widely used because it provides higher selectivity, accuracy, and reproducibility, and better limits of detection and quantification, compared with a single quadrupole mass spectrometer.

A liquid chromatograph coupled to a triple quadrupole mass spectrometer equipped with a Turbo Ionspray source in negative-ion mode was used to study the levels of phenolics in different varieties of tomato (Vallverdú-Queralt et al., 2011e), diced tomatoes (Vallverdú-Queralt et al., 2011c), and tomato sauces (Vallverdú-Queralt A. et al., 2012b). It was also used to evaluate the effects of storage on phenolic compounds (Vallverdú-Queralt et al., 2011a) and the effects of pulsed electric fields on tomato polyphenols (Vallverdú-Queralt et al., 2012c). Separation was performed in a C18 column (50×2.0 mm i.d., 5 μm) with a flow rate of 0.4 ml/min. Mobile phases consisted of (A) 0.1% formic acid in Milli-Q water and (B) 0.1% formic acid in acetonitrile. These conditions varied slightly depending on the product being analyzed. First, the presence of polyphenolic compounds was tested by MS/MS experiments of precursor ion scan, neutral loss scan, and product ion scan. The main objective of precursor ion scan experiments is to identify compounds belonging to a group of substances. In neutral loss experiments, the loss of 162 μ or 176 μ is used to confirm the loss of glucose, or galactose and glucuronides, respectively. Finally, product ion scan allows the identification of aglycones by comparison of their MS/MS spectra with those corresponding to the standards after typical fragmentations. MS/MS experiments were carried out by collision-activated dissociation (CAD) of selected precursor ions in the collision cell of the triple quadrupole mass spectrometer and the mass was determined by the second analyzer of the instrument. MS/MS has been used to analyze phenolic compounds in cocoa (Sanchez-Rabaneda et al., 2003b), fennel (Parejo et al., 2004), and artichoke (Sanchez-Rabaneda et al., 2003a). Data was collected in MRM mode for quantification purposes, tracking the transition of the specific parent and product ions for each compound.

HPLC-MS/MS techniques can also be applied to determine phenolic compounds as potential taxonomical markers in food and plant samples (Andres-Lacueva C. et al., 2002; De la Presa-Owens et al., 1995; Romero-Perez et al., 1996; Russo et al., 1998; Singleton and Trousdale, 1983; Vallverdú-Queralt et al., 2011e). For instance, phenolic and hydroxycinnamic acids, flavonoids, total polyphenols, and hydrophilic antioxidant capacity can be used as chemotaxonomic tomato markers to distinguish between tomato varieties (Vallverdú-Queralt et al., 2011e). The polyphenolic profile determined by HPLC revealed a similarity within grape varieties and differences between varieties. Similarly, varieties of white musts (De la Presa-Owens et al., 1995), wines (Romero-Perez et al., 1996), and sparkling wines (Parejo et al., 2004) have been shown to have different phenolic profiles.

Apart from food analysis, HPLC-MS/MS is widely used for biological fluid analysis due to its high sensitivity. In this way, an HPLC-MS/MS was used to identify and quantify phenolic compounds and metabolites from different tomato sauces in human plasma and urine in an intervention study (Tulipani et al., 2012). There were three interventions in this prospective randomized, cross-over study: tomato sauce elaborated without oil, and with the addition of 5% virgin olive oil or refined olive oil. Chromatographic separation was achieved in a C18 (50×2.0 mm, $5\,\mu$m) column, with a gradient elution of 0.1% aqueous formic acid and 0.1% formic acid in acetonitrile, and a flow-rate of 0.6 ml/min. Naringenin, ferulic acid, caffeic acid, and their corresponding glucuronide metabolites were detected in urine after the ingestion of the tomato sauces, while only two of the six urinary phenolic metabolites were identified in plasma as can be seen in Figure 14.2. Polyphenol levels of between 300 and 727 nmol/l have been detected using HPLC-MS/MS with a triple quadrupole instrument, thus showing the high sensitivity and selectivity of this system in the analysis of polyphenols in biological samples with a simple SPE extraction and clean-up process.

Using a triple quadrupole mass spectrometer, Aura *et al.* (2002) demonstrated that fecal microflora can deconjugate rutin, isoquercitrin, and quercetine glucuronides *in vitro* due to the presence of β,D-glucosidase, α,L-rhamnosidase, and β, D-glucuronidase. Fecal samples were freeze-dried before analysis. Then, polyphenol metabolites underwent liquid–liquid extraction using methanol/water (90 : 10, v/v) and were concentrated with a rotary evaporator. Samples were filtered and injected into the HPLC system. Chromatographic separation was performed using a reverse-phase column (100×1 mm). Mobile phases consisted of (A) 10 mmol/l ammonium acetate in water with 0.2% (v/v) acetic acid and (B) 10 mmol/l

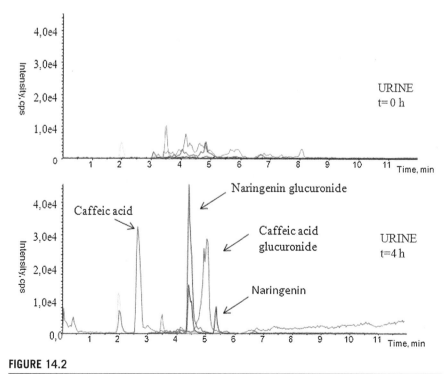

FIGURE 14.2

MRM chromatograms of urine sample at baseline (t = 0 h) and at 4 h after tomato sauce consumption.

ammonium acetate in methanol with 0.2% (v/v) acetic acid. Methanol with 0.1% (v/v) ammonium hydroxide was added as a post-column solvent (30 μl/min) to promote the desprotonation process of phenols previous to the ESI source. The measurements were performed in negative-ionization mode, and analyses were conducted in MRM. One or two fragment ions from the product ion spectra of the metabolites were used to identify the metabolites. This study showed that deconjugation and conversion of isoquercitrin and quercetin glucuronides to hydroxyphenylacetic acid occurs very rapidly in *in vitro* colonic fermentation. In contrast, rutin is deglycosylated at a slower rate, suggesting that rutin would be hydrolyzed at a slower rate than the other substrates. Therefore, the resulting quercetin aglycone appeared only transiently before further metabolism.

14.3.4.3 HPLC-HRMS

High-resolution mass spectrometry (HRMS) is used for qualitative analysis. A widely used technique is liquid chromatography/electrospray ionization–time-of-flight–mass spectrometry (HPLC-ESI-QTOF). QTOF technologies allow exact mass measurements of both MS and MS/MS ions. QTOF-MS has been used to determine phenolic compounds in gazpachos, ketchups, and tomato juices (Vallverdú-Queralt A. *et al.*, 2011b), resulting in the identification for the first time of apigenin-*C*-hexoside-hexoside and apigenin-*C*-hexoside-pentoside in tomato-based products. These compounds were distinguished by the presence of the ion [M-H-60]⁻ (Figure 14.3A and B), following the method of Han *et al.* (2008) which involves liquid chromatography coupled with electrospray ionization mass spectrometry. Vallverdú-Queralt *et al.* (2011b) reported for the first time the presence of protocatechuic acid-*O*-hexoside,

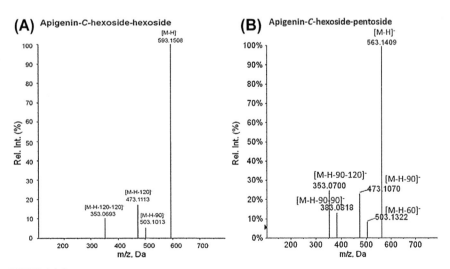

FIGURE 14.3

(A) Mass spectrum of apigenin-*C*-hexoside-hexoside. (B) Mass spectrum of apigenin-*C*-hexoside-pentoside.

caffeic acid-O-dihexoside, apigenin-C-hexoside-hexoside and apigenin-C-hexoside-pentoside in tomato-based products.

Metabolomics, a combination of analytical and statistical techniques, facilitates sample differentiation by quantitatively and qualitatively measuring the dynamic range of metabolites. With the recent developments in plant metabolomic techniques, it is possible to detect several metabolites simultaneously and reliably compare samples for differences and similarities in a semi-automated and untargeted manner. Metabolomics is predicted to play a crucial role in "bridging the phenotype–genotype gap" and in achieving complete genome sequence annotation and the understanding of gene function (Hall, 2006).

Metabolomics has also been used in the quality control of medicinal plants (Kim *et al.*, 2011). An ultraperformance liquid chromatography-quadrupole time-of-flight mass spectrometry (UPLC-QTOF MS)-based metabolomic technique was applied for the metabolite profiling of 60 *Panax ginseng* samples of 1 to 6year-old plants. After submitting the data for classification by various metabolite selection methods, the results showed variations according to the age of the samples, especially for those of 4 to 6year-old plants. Thus, a UPLC-QTOF MS-based metabolomics approach was able to quickly and accurately distinguish between *P. ginseng* samples according to their cultivation period.

HPLC-ESI-QTOF has also been used as a non-targeted strategy to differentiate between organic and conventional ketchups (Vallverdú-Queralt *et al.*, 2011d). Interpretation of the observed MS/MS spectra in comparison with results found in the literature was the main tool for putative identification of metabolites. The compounds found in significantly higher ($p < 0.05$) amounts in organic than in conventional ketchups were: caffeoylquinic and dicaffeoylquinic acids, caffeic and caffeic acid hexosides, kaempferol-3-O-rutinoside, ferulic-O-hexoside, naringenin-7-O-glucoside, naringenin, rutin, and quercetin. Examination of the chromatograms in TOF-MS mode also suggested the presence of glutamyl phenylalanine (m/z 293) and N-malonyltryptophan (m/z 289).

Recently, LTQ-Orbitrap-MS has been proposed as one of the most suitable strategies for qualitative analysis, since it routinely delivers the highest resolution and mass accuracy, which are necessary to reduce analysis times and increase confidence in results. Due to its ability to eliminate interference in the initial mass selection stage and to the specificity of MS/MS measurements, this spectrometer facilitates qualitative analysis of nontarget compounds. Elemental composition assignment and exact mass measurements are essential for molecule characterization. The structural elucidation of unknown compounds is easily accomplished by using accurate mass measurement of the product ions formed in MS^n experiments.

Thus, accurate mass experiments have yielded the elemental composition of polyphenol compounds, with MS^n fragment ions providing additional structural confirmation. The LTQ Orbitrap provides accurate mass MS and MS^n spectral data on a chromatographic time-scale with scan cycles of 1 s (at R = 60,000) or less. Up to five to eight sequential fragmentation spectra can be obtained, depending on the concentration and ionization efficiency of the compound. Using multiple-stage mass spectra it is possible to generate spectral trees of the compounds (Sheldon *et al.*, 2009). Van der Hooft *et al.* (2011) validated and applied an accurate mass MS^n spectral tree approach to 121 polyphenolic compounds of different chemical flavonoid subclasses, including isomeric forms. The study focused on the possibility of discriminating between positional and stereoisomeric forms. Accurate mass spectra of polyphenols were obtained using an LTQ-Orbitrap hybrid mass spectrometer in negative and positive ionization mode. The accurate MS^n fragmentation spectra enabled isomeric compounds to be differentiated. Spectral trees of 119 polyphenols (except catechin and epicatechin) showed unique fragments and differences in relative intensities of fragment ions. Thus, spectral trees constitute a potent tool for the identification of phenolic compounds or their metabolites. This tool could be applied to generate an MS^n metabolite database based on MS^n fragmentation and exact mass measurement.

Another study used liquid chromatography coupled with an LTQ Orbitrap to analyze polyphenols in tomato samples (Vallverdú-Queralt *et al.*, 2010). A C18 column (50×2.0 mm i.d, 5 μm) was used to separate the compounds. Gradient elution was performed with water/0.1% formic acid and acetonitrile/0.1% formic acid at a constant flow rate of 0.4 ml/min. A total of 38 compounds were identified in the tomato samples with very good mass accuracy (< 2 mDa). The spectra generated for cinnamic and benzoic acids showed the deprotonated molecule $[M-H]^-$ and some fragments. The typical loss of CO_2 was observed for gallic, protocatechuic, caffeic, and ferulic acids, giving $[M-H-44]^-$ as a characteristic ion, and loss of a methyl group $[M-H-15]^-$ was observed for ferulic acid. Flavonol aglycones such as quercetin gave the deprotonated molecule $[M-H]^-$ as a characteristic ion and ions corresponding to retro-Diels Alder fragmentation in the C-ring involving 1,3 scission, as described by other authors (Gruz *et al.*, 2008). The LTQ-Orbitrap was crucial for the structural determination of kaempferol-*O*-rutinoside-hexoside and rutin-*O*-hexoside-pentoside, which were not discernible under lower-resolution conditions.

Phloridzin-*C*-diglycoside (*m/z* 759) was only identified in the LTQ-Orbitrap due to the lack of sensitivity of the triple quadrupole. Figure 14.4 shows the MS^2 of *m/z* 759 of phloridzin-*C*-diglycoside, displaying losses of 90 u and 120 u from *m/z* 759 and 639, respectively, which confirmed the presence of two hexose units. Moreover, loss of H_2O was observed in the product ion spectra of *m/z* 759, showing an ion at *m/z* 741, which displayed a loss of 120 u (*m/z* 621). Losses of 90 and 120 u are characteristic fragment ions in the MS/MS mode of *C*-glycosides (Parejo *et al.*, 2004; Sanchez-Rabaneda *et al.*, 2003a).

In another study, 53 *O*-glycosyl-*C*-glycosyl flavones with *O*-glycosylation on phenolic hydroxyl or on the *C*-glycosyl residue, or a combination of both forms, were studied by liquid chromatography-UV diode array detection-electrospray ionization mass spectrometry ion trap in the negative mode. The study of the relative abundance of the main ions from the MS preferential fragmentation on -MS^2 and/or -MS^3 events allowed the differentiation of the *O*-glycosylation position,

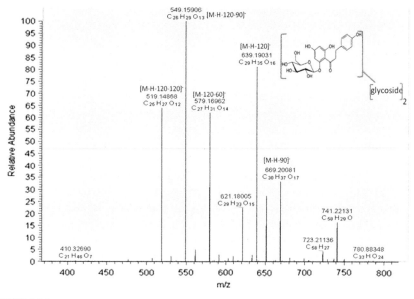

FIGURE 14.4

Identification of phloridzin-*C*-diglycoside in tomato samples. The MS^2 in product ion scan of *m/z* 759 shows the characteristic fragment ions of a *C*-diglycoside polyphenol.

either on a phenolic hydroxyl or on the sugar moiety of *C*-glycosylation (Ferreres *et al.*, 2007).

De Paepe *et al.* (2013) developed and validated a method to identify phenolic compounds in apple extracts using UHPLC coupled with an Orbitrap. An accurate mass spectrometry technique allowed the identification of 39 phenolic compounds in apples, including flavonoids, proanthocyanidins, and phenolic acids.

14.4 Ultra-high-performance liquid chromatography (UHPLC)

Liquid chromatographic performance has been improved by the introduction of UHPLC. To improve chromatographic separation, new columns with a very small particle packing size (1.7 μm) have been developed. Column efficiency is inversely proportional to particle size as the Van Deemter equations prove, so columns with 1.7μm particles provide higher resolution and better efficiency than the conventional ones (Novakova *et al.*, 2006). Improved mobile phase systems can operate at high backpressures (15000 psi), thus enhancing mobile phase viscosity and the capacity to dissolve analytes (Epriliati *et al.*, 2010). Due to these high pressures, new hardware in LC technology has been developed. As a result of the combination of columns and high pressure, UHPLC has enhanced sensitivity and peak resolution, and reduced both analysis time and costs (Gruz *et al.*, 2008; Leandro *et al.*, 2006; Ortega *et al.*, 2010; Wu *et al.*, 2008).

14.4.1 UHPLC-UV

An ultra high pressure liquid chromatography technique was used to develop a new method for analyzing conjugated isoflavones in commercial soy milks (Toro-Funes *et al.*, 2012) using UV detection. This approach allowed the determination of 12 isoflavones in less than 8 min in a single run. The method was fully validated, with limits of detection lower than 0.05 mg/l and a limit of quantification below 0.2 mg/l. Chromatographic separation of analytes was achieved using a C18 column (50×2.1 mm, 1.7 μm). Solvent A was ultrapure water with 0.1% formic acid, and solvent B was acetonitrile with 0.1% formic acid, and the flow rate was set at 0.6 ml/min. The quantification of isoflavones was performed at 262 nm.

Gonçalves *et al.* (2013) also developed a new method to analyze hydroxy-benzoics and hydroxycinnamics acids in wine samples using UHPLC coupled to a photodiode array detector in 11 min. The method was validated and showed limits of detection of 0.01–0.2 mg/l and a limit of quantification of 0.03–0.7 mg/l. A C18 (100×2.1 mm, 1.8 μm) column was used, and the gradient elution was performed with 0.1% formic acid and methanol as mobile phases at 0.25 ml/min. Before UHPLC analysis, phenolic compounds were extracted using microextraction by packed sorbent optimized for hydroxybenzoic and hydroxycinnamic acids.

Another study used UHPLC coupled with a UV detector to identify and quantify 58 polypyhenols in sage tea (Zimmermann *et al.*, 2011) within 28 min. Separation of the phenolic compounds was carried out using a reverse-phase column (150×2.1 mm, 1.7 μm). 0.1% formic acid was used as the aqueous mobile phase, and acetonitrile containing 0.1% formic acid as the organic mobile phase, at a flow rate of 0.4 ml/min. Wavelengths of the UV detector were set for quantification at 273, 320, and 360 nm. The method was applied to characterize 16 commercial brands of sage tea.

14.4.2 UHPLC-MS/MS

Coupling UHPLC with electrospray ionization tandem mass spectrometry offers a strong alternative to conventional HPLC-MS/MS in terms of analysis time, costs, and improved resolution and sensitivity.

Using UHPLC-MS/MS, 17 phenolic acids were quantified in white wine, grapefruit juice and green tea infusion within 10 min. Separation was performed in a reverse-phase column C8 (2.1×150 mm, 1.7 μm) with a mobile phase of aqueous 7.5 mM HCOOH (A) and acetonitrile (B) at a flow rate of 250 μl/min (Gruz *et al.*, 2008). The validated method offered good precision and accuracy, and limits of detection of 0.15–15 pmol/injection. A UHPLC-MS/MS system is also suitable for routine analysis in laboratories.

A UHPLC-MS/MS was used to analyze procyandins and alkaloids (caffeine and theobromine) in samples of cocoa nibs (Ortega *et al.*, 2010). A high-strength silica separation column (100×2.1 mm i.d., 1.8-mm particle size) was used with (A) water/acetic acid (99.8/0.2, v/v) and (B) acetonitrile at a flow rate of 0.4 ml/min. This method allowed catechin and epicatechin to be quantified separately, which was not possible with HPLC-UV due to the coelution of the compounds. Also, the total analysis time for a cocoa phenolic extract was reduced from 80 min by HPLC to 12.50 min by UHPLC, and limits of detection of the procyanidin were enhanced from 0.009–0.02 mg/ml with HPLC-MS/MS to 0.007–0.01 mg/ml with UHPLC-MS/MS.

Twenty-six phenolic compounds, including 15 isoflavonoids, five flavones, four flavanones, a coumestan, and a coumarine of plant material, were analyzed in 17 min by UHPLC–ESI–MS/MS. The 26 compounds were separated with a C8 column (2.1 × 150 mm, 1.7 mm), using a mobile phase of methanol (A) and 10 mM aqueous formic acid (B) 100% A at a flow rate of 0.2 ml/min. The validated method achieved good accuracy and precision, with limits of detection ranging between 0.0001 and 10 pmol/injection. Solvent consumption and time were reduced compared to conventional HPLC systems (approximately 30–70 min) (Prokudina *et al.*, 2012).

14.5 **Conclusions**

Polyphenol characterization is a difficult task due to the large number of phenolic compounds present in nature and the few available commercial standards.

Efficient identification and quantification involves pretreating the samples to avoid interference. Food and beverage samples are normally analyzed after liquid–liquid or solid–liquid extraction, but biological samples such as plasma, urine, or serum usually require solid-phase extraction, since they contain far lower levels of compounds than food samples. Solid-phase extraction procedures are also used to clean up the samples by eliminating interfering compounds, thus avoiding the matrix effect.

Polyphenols can be identified or quantified with various analytical techniques, but the most commonly used are liquid chromatography coupled with UV detection, and liquid chromatography coupled with mass spectrometry.

The most widespread technique for analyzing polyphenols in food and plant samples is HPLC-UV, which also constitutes an excellent option for routine analysis in analytical laboratories and the food industry. In contrast, when working with biological samples, a more sensitive technique like mass spectrometry is required due to the complexity of the matrix and because mass spectrometry offers lower limits of detection than HPLC-UV (Table 14.1).

References

Aaby, K., Hvattum, E., Skrede, G., 2004. Analysis of flavonoids and other phenolic compounds using high-performance liquid chromatography with coulometric array detection: relationship to antioxidant activity. J. Agri. Food Chem. 52, 4595–4603.

Abad-García, B., Berrueta, L.A., López-Márquez, D.M., Crespo-Ferrer, I., Gallo, B., Vicente, F., 2007. Optimization and validation of a methodology based on solvent extraction and liquid chromatography for the simultaneous determination of several polyphenolic families in fruit juices. J. Chromatogr. A 1154, 87–96.

Alonso-Salces, R.M., Barranco, A., Corta, E., Berrueta, L.A., Gallo, B., Vicente, F., 2005. A validated solid–liquid extraction method for the HPLC determination of polyphenols in apple tissues. Comparison with pressurised liquid extraction. Talanta 65, 654–662.

Anari, M.R., Sanchez, R.I., Bakhtiar, R., Franklin, R.B., Baillie, T.A., 2004. Integration of knowledge-based metabolic predictions with liquid chromatography data-dependent tandem mass spectrometry for drug metabolism studies: application to studies on the biotransformation of indinavir. Anal. Chem. 76, 823–832.

Andres-Lacueva, C., Ibern-Gomez, M., Lamuela-Raventos, R.M., Buxaderas, S., de la Torre-Boronat, M., 2002. Cinnamates and resveratrol content for sparkling wine characterization. Am. J. Enol. Viticulture 53, 147–150.

Aura, M.A., O'Leary, K.A., Williamson, G., Ojala, M., Bailey, M., Puupponene-Pimia, R., et al., 2002. Quercetin Derivatives Are Deconjugated and Converted to Hydroxyphenylacetic Acids but Not Methylated by Human Fecal Flora In Vitro. J. Agri. Food Chem. 50, 1725–1730.

Baydar, N.G., Özkan, G., Sağdiç, O., 2004. Total phenolic contents and antibacterial activities of grape (*Vitis vinifera L.*) extracts. Food Control 15, 335–339.

Bétes-Saura, C., Andres-Lacueva, C., LamuelaRaventos, R.M., 1996. Phenolics in White Free Run Juices and Wines from Penedès by High-Performance Liquid Chromatography: Changes during Vinification. J. Agri. Food Chem. 44, 3040–3046.

Bleve, M., Ciurlia, L., Erroi, E., Lionetto, G., Longo, L., Rescio, L., et al., 2008. An innovative method for the purification of anthocyanins from grape skin extracts by using liquid and sub-critical carbon dioxide. Sep. Purif. Technol. 64, 192–197.

Cao, X., Wang, C., Pei, H., Sun, B., 2009. Separation and identification of polyphenols in apple pomace by high-speed counter-current chromatography and high-performance liquid chromatography coupled with mass spectrometry. J. Chromatogr. A 1216, 4268–4274.

Capannesi, C., Palchetti, I., Mascini, M., Parenti, A., 2000. Electrochemical sensor and biosensor for polyphenols detection in olive oils. Food Chem. 71, 553–562.

Caporaso, J.G., Lauber, C.L., Walters, W.A., Berg-Lyons, D., Lozupone, C.A., Turnbaugh, P.J., et al., 2011. Global patterns of 16S rRNA diversity at a depth of millions of sequences per sample. Proc. Natl. Acad. Sci. U. S. A. 108 (Suppl. 1), 4516–4522.

Caridi, D., Trenerry, V.C., Rochfort, S., Duong, S., Laugher, D., Jones, R., 2007. Profiling and quantifying quercetin glucosides in onion (*Allium cepa* L.) varieties using capillary zone electrophoresis and high performance liquid chromatography. Food Chem. 105, 691–699.

Chen, Q., Zhao, J., Chaitep, S., Guo, Z., 2009. Simultaneous analysis of main catechins contents in green tea (*Camellia sinensis* (L.)) by Fourier transform near infrared reflectance (FT-NIR) spectroscopy. Food Chem. 113, 1272–1277.

Chiva-Blanch, G., Urpi-Sarda, M., Rotches-Ribalta, M., Zamora-Ros, R., Llorach, R., Lamuela-Raventos, R.M., et al., 2011. Determination of resveratrol and piceid in beer matrices by solid-phase extraction and liquid chromatography-tandem mass spectrometry. J. Chromatogr. A 1218, 698–705.

Cimpan, G., Gocan, S., 2002. Analysis of medicinal plants by HPLC: Recent approaches. J. Liquid Chromatogr. Relat. Technol. 25, 2225–2292.

Crozier, A., Lean, M.E.J., McDonald, M.S., Black, C., 1997. Quantitative Analysis of the Flavonoid Content of Commercial Tomatoes, Onions, Lettuce, and Celery. J. Agri. Food Chem. 45, 590–595.

De la Presa-Owens, C., Lamuela-Raventos, R.M., Buxaderas, S., De la Torre-Boronat, M.C., 1995. Differentiation and Grouping Characteristics of Varietal Grape Musts from Penedes Region. Am. J. Enol. Viticulture 46, 283–291.

De Paepe, D., Servaes, K., Noten, B., Diels, L., De Loose, M., Van Droogenbroeck, B., et al., 2013. An improved mass spectrometric method for identification and quantification of phenolic compounds in apple fruits. Food Chem. 136, 368–375.

de Quirós, A.R., Lage-Yusty, M.A., López-Hernández, J., 2009. HPLC-analysis of polyphenolic compounds in Spanish white wines and determination of their antioxidant activity by radical scavenging assay. Food Res. Int. 42, 1018–1022.

de Rijke, E., Joshi, H.C., Sanderse, H.R., Ariese, F., Brinkman, U.A.T., Gooijer, C., 2002. Natively fluorescent isoflavones exhibiting anomalous Stokes' shifts. Anal. Chim. Acta 468, 3–11.

Epriliati, I., Kerven, G., D'Arcy, B., Gidley, M.J., 2010. Chromatographic analysis of diverse fruit components using HPLC and UPLC. Anal. Methods 1606–1613.

Erlund, I., Alfthan, G., Siren, H., Ariniemi, K., Aro, A., 1999. Validated method for the quantitation of quercetin from human plasma using high-performance liquid chromatography with electrochemical detection. J. Chromatogr. B. Biomed. Sci. Appl. 727, 179–189.

Fang, N., Yu, S., Badger, T.M., 2002. Characterization of isoflavones and their conjugates in female rat urine using LC/MS/MS. J. Agri. Food Chem. 50, 2700–2707.

Fang, Z., Zhang, Y., Lü, Y., Ma, G., Chen, J., Liu, D., et al., 2009. Phenolic compounds and antioxidant capacities of bayberry juices. Food Chem. 113, 884–888.

Ferreres, F., Gil-Izquierdo, A., Andrade, P.B., Valentao, P., Tomas-Barberan, F.A., 2007. Characterization of C-glycosyl flavones O-glycosylated by liquid chromatography-tandem mass spectrometry. J. Chromatogr. A 1161, 214–223.

Friedman, M., 2004. Analysis of biologically active compounds in potatoes (*Solanum tuberosum*), tomatoes (*Lycopersicon esculentum*), and jimson weed (*Datura stramonium*) seeds. J. Chromatogr. A 1054, 143–155.

Fu, T., Liang, J., Han, G., Lv, L., Li, N., 2008. Simultaneous determination of the major active components of tea polyphenols in rat plasma by a simple and specific HPLC assay. J. Chromatogr. B 875, 363–367.

Garcia, A.A., Grande, B.C., Gandara, J.S., 2004. Development of a rapid method based on solid-phase extraction and liquid chromatography with ultraviolet absorbance detection for the determination of polyphenols in alcohol-free beers. J. Chromatogr. A 1054, 175–180.

Gonçalves, J., Silva, C.L., Castilho, P.C., Câmara, J.S., 2013. An attractive, sensitive and high-throughput strategy based on microextraction by packed sorbent followed by UHPLC-PDA analysis for quantification of hydroxybenzoic and hydroxycinnamic acids in wines. Microchem. J. 106, 129–138.

Grace, P.B., Taylor, J.I., Botting, N.P., Fryatt, T., Oldfield, M.F., Al-Maharik, N., et al., 2003. Quantification of isoflavones and lignans in serum using isotope dilution liquid chromatography/tandem mass spectrometry. Rapid Commun. Mass Spectrom. 17, 1350–1357.

Griffith, A.P., Collison, M.W., 2001. Improved methods for the extraction and analysis of isoflavones from soy-containing foods and nutritional supplements by reversed-phase high-performance liquid chromatography and liquid chromatography-mass spectrometry. J. Chromatogr. A 913, 397–413.

Gruz, J., Novák, O., Strnad, M., 2008. Rapid analysis of phenolic acids in beverages by UPLC–MS/MS. Food Chem. 111, 789–794.

Hall, R.D., 2006. Plant metabolomics: from holistic hope, to hype, to hot topic. New Phytologist 169, 453–468.

Han, J., Ye, M., Qiao, X., Xu, M., Wang, B.R., Guo, D.A., 2008. Characterization of phenolic compounds in the Chinese herbal drug Artemisia annua by liquid chromatography coupled to electrospray ionization mass spectrometry. J. Pharm. Biomed. Anal. 47, 516–525.

Herrero-Martinez, J.M., Sanmartin, M., Roses, M., Bosch, E., Rafols, C., 2005. Determination of dissociation constants of flavonoids by capillary electrophoresis. Electrophoresis 26, 1886–1895.

Ho, Y.B., Zakaria, M.P., Latif, P.A., Saari, N., 2012. Simultaneous determination of veterinary antibiotics and hormone in broiler manure, soil and manure compost by liquid chromatography-tandem mass spectrometry. J. Chromatogr. A 1262, 160–168.

Hollman, P.C.H., van Trijp, J.M.P., Buysman, M.N.C.P., 1996. Fluorescence Detection of Flavonols in HPLC by Postcolumn Chelation with Aluminum. Anal. Chem. 68, 3511.

Huang, W., Xue, A., Niu, H., Jia, Z., Wang, J., 2009. Optimised ultrasonic-assisted extraction of flavonoids from Folium eucommiae and evaluation of antioxidant activity in multi-test systems in vitro. Food Chem. 114, 1147–1154.

Huck, C.W., Bonn, G.K., 2001. Evaluation of detection methods for the reversed-phase HPLC determination of 3-,4-,5-trimethoxyflavone in different phytopharmaceutical products and in human serum. Phytochem. Anal. 12, 104–109.

Ignat, I., Volf, I., Popa, V.I., 2011. A critical review of methods for characterisation of polyphenolic compounds in fruits and vegetables. Food Chem. 126, 1821–1835.

Jin, D., Hakamata, H., Takahashi, K., Kotani, A., Kusu, F., 2004. Determination of quercetin in human plasma after ingestion of commercial canned green tea by semi-micro HPLC with electrochemical detection. Biomed. Chromatogr. 18, 662–666.

Kamangerpour, A., Ashraf-Khorassani, M., Taylor, L., McNair, H., Chorida, L., 2002. Supercritical Fluid Chromatography of polyphenolic compounds in grape seed extract. Chromatographia 55, 417–421.

Kerem, Z., Bravdo, B.A., Shoseyov, O., Tugendhaft, Y., 2004. Rapid liquid chromatography-ultraviolet determination of organic acids and phenolic compounds in red wine and must. J. Chromatogr. A 1052, 211–215.

Kerio, L.C., Wachira, F.N., Wanyoko, J.K., Rotich, M.K., 2012. Characterization of anthocyanins in Kenyan teas: Extraction and identification. Food Chem. 131, 31–38.

Kim, N., Kim, K., Choi, B.Y., Lee, D., Shin, Y.S., Bang, K.H., et al., 2011. Metabolomic approach for age discrimination of Panax ginseng using UPLC-Q-Tof MS. J. Agri. Food Chem. 59, 10435–10441.

Lachman, J., Hamouz, K., Šulc, M., Orsák, M., Pivec, V., Hejtmánková, A., et al., 2009. Cultivar differences of total anthocyanins and anthocyanidins in red and purple-fleshed potatoes and their relation to antioxidant activity. Food Chem. 114, 836–843.

Leandro, C.C., Hancock, P., Fussell, R.J., Keely, B.J., 2006. Comparison of ultra-performance liquid chromatography and high-performance liquid chromatography for the determination of priority pesticides in baby foods by tandem quadrupole mass spectrometry. J. Chromatogr. A 1103, 94–101.

Li, H., Chen, B., Nie, L., Yao, S., 2004. Solvent effects on focused microwave assisted extraction of polyphenolic acids from Eucommia ulmodies. Phytochem. Anal. 15, 306–312.

Li, H., Chen, B., Yao, S., 2005. Application of ultrasonic technique for extracting chlorogenic acid from *Eucommia ulmodies* Oliv. (*E. ulmodies*). Ultrason. Sonochem. 12, 295–300.

Liang, Z., Yang, Y., Cheng, L., Zhong, G., 2012. Polyphenolic composition and content in the ripe berries of wild Vitis species. Food Chem. 132, 730–738.

Liang, Z., Sang, M., Fan, P., Wu, B., Wang, L., Duan, W., et al., 2011. Changes of Polyphenols, Sugars, and Organic Acid in 5 Vitis Genotypes during Berry Ripening. J. Food Sci. 76, C1231–C1238.

Liu, Q., Cai, W., Shao, X., 2008. Determination of seven polyphenols in water by high performance liquid chromatography combined with preconcentration. Talanta 77, 679–683.

Liu, G.L., Guo, H.H., Sun, Y.M., 2012. Optimization of the Extraction of Anthocyanins from the Fruit Skin of *Rhodomyrtus tomentosa* (Ait.) Hassk and Identification of Anthocyanins in the Extract Using High-Performance Liquid Chromatography–Electrospray Ionization–Mass Spectrometry (HPLC-ESI-MS). Int. J. Mol. Sci. 13, 6292–6302.

Lu, Y., Foo, Y., 1998. Constitution of some chemical components of apple seed. Food Chem. 61, 29–33.

Machonis, P.R., Jones, M.A., Schaneberg, B.T., Kwik-Uribe, C.L., 2012. Method for the determination of catechin and epicatechin enantiomers in cocoa-based ingredients and products by high-performance liquid chromatography: single-laboratory validation. J. AOAC Int. 95, 500–507.

Magiera, S., Baranowska, I., Kusa, J., 2012. Development and validation of UHPLC-ESI-MS/MS method for the determination of selected cardiovascular drugs, polyphenols and their metabolites in human urine. Talanta 89, 47–56.

Maier, T., Goppert, A., Kammerer, D.R., Schieber, A., Carle, R., 2008. Optimization of a process for enzyme-assisted pigment extraction from grape (*Vitis vinifera* L.) pomace. Eur. Food Res. Technol. 267–275.

Maróstica-Junior, M., Vieira-Leite, A., Vicente-Dragano, N., 2010. Supercritical Fluid Extraction and Stabilization of Phenolic Compounds From Natural Sources—Review (Supercritical Extraction and Stabilization of Phenolic Compounds). Open Chem. Eng. J. 4, 51–60.

Marston, A., Hostettmann, K., 2009. Natural product analysis over the last decades. Planta Med. 75, 672–682.

Martinez-Huelamo, M., Jimenez-Gamez, E., Hermo, M.P., Barron, D., Barbosa, J., 2009. Determination of penicillins in milk using LC-UV, LC-MS and LC-MS/MS. J. Sep. Sci. 32, 2385–2393.

Martinez-Huelamo, M., Tulipani, S., Torrado, X., Estruch, R., Lamuela-Raventos, R.M., 2012. Validation of a new LC-MS/MS method for the detection and quantification of phenolic metabolites from tomato sauce in biological samples. J. Agri. Food Chem. 60, 4542–4549.

Meda, R.N.T., Vlase, L., Lamien-Meda, A., Lamien, C.E., Muntean, D., Tiperciuc, B., et al., 2011. Identification and quantification of phenolic compounds from *Balanites aegyptiaca* (L) Del (Balanitaceae) galls and leaves by HPLC-MS. Nat. Prod. Res. 25, 93–99.

Medina-Remon, A., Barrionuevo-Gonzalez, A., Zamora-Ros, R., Andres-Lacueva, C., Estruch, R., Martinez-Gonzalez, M.A., et al., 2009. Rapid Folin–Ciocalteu method using microtiter 96-well plate cartridges for solid phase extraction to assess urinary total phenolic compounds, as a biomarker of total polyphenols intake. Anal. Chim. Acta 634, 54–60.

Merken, H.M., Beecher, G.R., 2000. Measurement of food flavonoids by high-performance liquid chromatography: A review. J. Agri. Food Chem. 48, 577–599.

Metcalf, R.L., 1987. Plant volatile and insect atractant. Crit. Rev. Plant Sci. 5, 251–301.

Mikulic-Petkovsek, M., Slatnar, A., Stampar, F., Veberic, R., 2012. HPLC–MSn identification and quantification of flavonol glycosides in 28 wild and cultivated berry species. Food Chem. 135, 2138–2146.

Milbury, P.E., 2001. Analysis of complex mixtures of flavonoids and polyphenols by high-performance liquid chromatography electrochemical detection methods. Methods Enzymol. 335, 15–26.

Min, J.Y., Kang, S.M., Park, D.J., Kim, Y.D., Jung, H.N., Yang, J.K., 2006. Enzymatic release of ferulic acid from *Ipomea batatas* L. (sweet potato) stem. Biotechnol. Bioprocess Eng. 11.

Muñoz, S., Mestres, M., Busto, O., Guasch, J., 2008. Determination of some flavan-3-ols and anthocyanins in red grape seed and skin extracts by HPLC-DAD: Validation study and response comparison of different standards. Anal. Chim. Acta 628, 104–110.

Navas, M.J., 2012. Analysis and Antioxidant Capacity of Anthocyanin Pigments. Part IV: Extraction of Anthocyanins. Crit. Rev. Anal. Chem. 42, 313.

Nkhili, E., Tomao, V., El Hajji, H., El Boustani, E.S., Chemat, F., Dangles, O., 2009. Microwave-assisted water extraction of green tea polyphenols. Phytochem. Anal. 20, 408–415.

Novakova, L., Solichova, D., Solich, P., 2006. Advantages of ultra performance liquid chromatography over high-performance liquid chromatography: comparison of different analytical approaches during analysis of diclofenac gel. J. Sep. Sci. 29, 2433–2443.

Olmos-Espejel, J.J., Garcia de Llasera, M.P., Velasco-Cruz, M., 2012. Extraction and analysis of polycyclic aromatic hydrocarbons and benzo[a]pyrene metabolites in microalgae cultures by off-line/on-line methodology based on matrix solid-phase dispersion, solid-phase extraction and high-performance liquid chromatography. J. Chromatogr. A 1262, 138–147.

Ortega, N., Romero, M., Macià, A., Reguant, J., Anglès, N., Morelló, J., et al., 2010. Comparative study of UPLC–MS/MS and HPLC–MS/MS to determine procyanidins and alkaloids in cocoa samples. J. Food Composition Anal. 23, 298–305.

Ottaviani, J.I., Momma, T.Y., Kuhnle, G.K., Keen, C.L., Schroeter, H., 2012. Structurally related (−)-epicatechin metabolites in humans: assessment using de novo chemically synthesized authentic standards. Free Radic. Biol. Med. 52, 1403–1412.

Palenzuela, B., Arce, L., Macho, A., Munoz, E., Rios, A., Valcarcel, M., 2004. Bioguided extraction of polyphenols from grape marc by using an alternative supercritical-fluid extraction method based on a liquid solvent trap. Anal. Bioanal. Chem. 378, 2021–2027.

Palma, M., Taylor, L.T., 1999. Extraction of polyphenolic compounds from grape seeds with near critical carbon dioxide. J. Chromatogr. A 849, 117–124.

Parejo, I., Jauregui, O., Sanchez-Rabaneda, F., Viladomat, F., Bastida, J., Codina, C., 2004. Separation and characterization of phenolic compounds in fennel (*Foeniculum vulgare*) using liquid chromatography-negative electrospray ionization tandem mass spectrometry. J. Agri. Food Chem. 52, 3679–3687.

Peng, Y., Liu, F., Peng, Y., Ye, J., 2005. Determination of polyphenols in apple juice and cider by capillary electrophoresis with electrochemical detection. Food Chem. 92, 169–175.

Peterman, S.M., Duczak Jr., N., Kalgutkar, A.S., Lame, M.E., Soglia, J.R., 2006. Application of a linear ion trap/orbitrap mass spectrometer in metabolite characterization studies: examination of the human liver microsomal metabolism of the non-tricyclic anti-depressant nefazodone using data-dependent accurate mass measurements. J. Am. Soc. Mass Spectrom. 17, 363–375.

Prasain, J.K., Jones, K., Brissie, N., Moore, R., Wyss, J.M., Barnes, S., 2004a. Identification of puerarin and its metabolites in rats by liquid chromatography-tandem mass spectrometry. J. Agri. Food Chem. 52, 3708–3712.

Prasain, J.K., Wang, C.C., Barnes, S., 2004b. Mass spectrometric methods for the determination of flavonoids in biological samples. Free Radic. Biol. Med. 37, 1324–1350.

Prokudina, E.A., Havlíček, L., Al-Maharik, N., Lapčík, O., Strnad, M., Gruz, J., 2012. Rapid UPLC–ESI–MS/MS method for the analysis of isoflavonoids and other phenylpropanoids. J. Food Composition Anal. 26, 36–42.

Rios, L.Y., Gonthier, M.P., Remesy, C., Mila, I., Lapierre, C., Lazarus, S.A., et al., 2003. Chocolate intake increases urinary excretion of polyphenol-derived phenolic acids in healthy human subjects. Am. J. Clin. Nutr. 77, 912–918.

Rodríguez-Delgado, M.A., Malovaná, S., Pérez, J.P., Borges, T., García Montelongo, F.J., 2001. Separation of phenolic compounds by high-performance liquid chromatography with absorbance and fluorimetric detection. J. Chromatogr. A 912, 249–257.

Romero-Perez, A., Lamuela-Raventos, R.M., Buxaderas, S., de laTorre-Boronat, R.C., 1996. Resveratrol and piceid as varietal markers of white wines. J. Agri. Food Chem. 44, 1975–1978.

Ross, K.A., Beta, T., Arntfield, S.D., 2009. A comparative study on the phenolic acids identified and quantified in dry beans using HPLC as affected by different extraction and hydrolysis methods. Food Chem. 113, 336–344.

Russo, M., Galletti, G.C., Bocchini, P., Carnacini, A., 1998. Essential oil chemical composition of wild populations of Italian oregano spice (*Origanum vulgare* ssp. *hirtum* (Link) *Ietswaart*): A preliminary evaluation of their use in chemotaxonomy by cluster analysis. J. Agri. Food Chem. 46, 3741–3746.

Sakakibara, H., Honda, Y., Nakagawa, S., Ashida, H., Kanazawa, K., 2003. Simultaneous determination of all polyphenols in vegetables, fruits, and teas. J. Agri. Food Chem. 51, 571–581.

Sanchez-Rabaneda, F., Jauregui, O., Casals, I., Andres-Lacueva, C., Izquierdo-Pulido, M., Lamuela-Raventos, R.M., 2003a. Liquid chromatographic/electrospray ionization tandem mass spectrometric study of the phenolic composition of cocoa (*Theobroma cacao*). J. Mass Spectrom. 38, 35–42.

Sanchez-Rabaneda, F., Jauregui, O., Casals, I., Andres-Lacueva, C., Izquierdo-Pulido, M., Lamuela-Raventos, R.M., 2003b. Liquid chromatographic/electrospray ionization tandem mass spectrometric study of the phenolic composition of cocoa (*Theobroma cacao*). J. Mass Spectrom. 38, 35–42.

Sano, A., Yamakoshi, J., Tokutake, S., Tobe, K., Kubota, Y., Kikuchi, M., 2003. Procyanidin B1 is detected in human serum after intake of proanthocyanidin-rich grape seed extract. Biosci. Biotechnol. Biochem. 67, 1140–1143.

Schieber, A., Ullrich, W., Carle, R., 2000. Characterization of polyphenols in mango puree concentrate by HPLC with diode array and mass spectrometric detection. Innovative Food Sci. Emerg. Technol. 1, 161–166.

Seabra, R.M., Andrade, P.B., Valentão, P., Fernandes, E., Carvalho, F., Bastos, M.L., 2006. In: Fingerman, M., Nagabhushanam, R. (Eds.), Biomaterial from Aquatic and Terrestrial organisms. Science Publishers, Enfield, NH, USA.

Sengupta, P.K., Kasha, M., 1979. Excited state proton-transfer spectroscopy of 3-hytdroxyflavone and quercetin. Chem. Phys. Lett. 68, 382–385.

Sheldon, M.T., Mistrik, R., Croley, T.R., 2009. Determination of Ion Structures in Structurally Related Compounds Using Precursor Ion Fingerprinting. J. Am. Soc. Mass Spectrom. 20, 370–376.

Shouqin, Z., Jun, X., Changzheng, W., 2005. High hydrostatic pressure extraction of flavonoids from propolis. J. Chem. Technol. Biotechnol. 80, 50–54.

Silva, C.L., Haesen, N., Câmara, J.S., 2012. A new and improved strategy combining a dispersive-solid phase extraction-based multiclass method with ultra high pressure liquid chromatography for analysis of low molecular weight polyphenols in vegetables. J. Chromatogr. A 1260, 154–163.

Singleton, V.L., Trousdale, E., 1983. White Wine Phenolics: Varietal and Processing Differences As Shown by HPLC. Am. J. Enol. Viticulture 34, 27–34.

Slimestad, R., Fossen, T., Verheul, M.J., 2008. The flavonoids of tomatoes. J. Agri. Food Chem. 56, 2436–2441.

Spáčil, Z., Nováková, L., Solich, P., 2010. Comparison of positive and negative ion detection of tea catechins using tandem mass spectrometry and ultra high performance liquid chromatography. Food Chem. 123, 535–541.

Sporns, P., Wang, J., 1998. Exploring new frontiers in food analysis using MALDI-MS. Food Res. Int. 31, 181–189.

Stalikas, C.D., 2010. Phenolic acids and flavonoids: occurrence and analytical methods. Methods Mol. Biol. (Clifton, N.J.) 610, 65–90.

Stobiecki, M., 2000. Application of mass spectrometry for identification and structural studies of flavonoid glycosides. Phytochemistry 54, 237–256.

Sun, J., Liang, F., Bin, Y., Li, P., Duan, C., 2007. Screening non-colored phenolics in red wines using liquid chromatography/ultraviolet and mass spectrometry/mass spectrometry libraries. Molecules (Basel, Switzerland) 12, 679–693.

Toro-Funes, N., Odriozola-Serrano, I., Bosch-Fusté, J., Latorre-Moratalla, M.L., Veciana-Nogués, M.T., Izquierdo-Pulido, M., et al., 2012. Fast simultaneous determination of free and conjugated isoflavones in soy milk by UHPLC–UV. Food Chem. 135, 2832–2838.

Tsao, R., Deng, Z., 2004. Separation procedures for naturally occurring antioxidant phytochemicals. J. Chromatogr. B. Anal. Technol. Biomed. Life Sci. 812, 85–99.

Tulipani, S., Martinez Huelamo, M., Rotches Ribalta, M., Estruch, R., Ferrer, E.E., Andres-Lacueva, C., et al., 2012. Oil matrix effects on plasma exposure and urinary excretion of phenolic compounds from tomato sauces: Evidence from a human pilot study. Food Chem. 130, 581–590.

Urpi-Sarda, M., Monagas, M., Khan, N., Llorach, R., Lamuela-Raventos, R.M., Jauregui, O., et al., 2009. Targeted metabolic profiling of phenolics in urine and plasma after regular consumption of cocoa by liquid chromatography-tandem mass spectrometry. J. Chromatogr. A 1216, 7258–7267.

Vallverdú-Queralt, A., Jauregui, O., Medina-Remon, A., Andres-Lacueva, C., Lamuela-Raventos, R.M., 2010. Improved characterization of tomato polyphenols using liquid chromatography/electrospray ionization linear ion trap quadrupole Orbitrap mass spectrometry and liquid chromatography/electrospray ionization tandem mass spectrometry. Rapid Commun. Mass Spectrom. 24, 2986–2992.

Vallverdú-Queralt, A., Arranz, S., Medina-Remon, A., Casals-Ribes, I., Lamuela-Raventos, R.M., 2011a. Changes in phenolic content of tomato products during storage. J. Agri. Food Chem. 59, 9358–9365.

Vallverdú-Queralt, A., Jáuregui, O., Di Lecce, G., Andrés-Lacueva, C., Lamuela-Raventós, R.M., 2011b. Screening of the polyphenol content of tomato-based products through accurate-mass spectrometry (HPLC–ESI-QTOF). Food Chem. 129, 877–883.

Vallverdú-Queralt, A., Medina-Remón, A., Andres-Lacueva, C., Lamuela-Raventos, R.M., 2011c. Changes in phenolic profile and antioxidant activity during production of diced tomatoes. Food Chem. 126, 1700–1707.

Vallverdú-Queralt, A., Medina-Remon, A., Casals-Ribes, I., Amat, M., Lamuela-Raventos, R.M., 2011d. A metabolomic approach differentiates between conventional and organic ketchups. J. Agri. Food Chem. 59, 11703–11710.

Vallverdú-Queralt, A., Medina-Remon, A., Martinez-Huelamo, M., Jauregui, O., Andres-Lacueva, C., Lamuela-Raventos, R.M., 2011e. Phenolic profile and hydrophilic antioxidant capacity as chemotaxonomic markers of tomato varieties. J. Agri. Food Chem. 59, 3994–4001.

Vallverdú-Queralt, A., Jauregui, O., Medina-Remon, A., Lamuela-Raventos, R.M., 2012a. Evaluation of a method to characterize the phenolic profile of organic and conventional tomatoes. J. Agri. Food Chem. 60, 3373–3380.

Vallverdú-Queralt, A., Medina-Remon, A., Casals-Ribes, I., Andres-Lacueva, C., Waterhouse, A.L., Lamuela-Raventos, R.M., 2012b. Effect of tomato industrial processing on phenolic profile and hydrophilic antioxidant capacity. Food Sci. Technol. 47, 154–160.

Vallverdú-Queralt, A., Oms-Oliu, G., Odriozola-Serrano, I., Lamuela-Raventos, R.M., Martin-Belloso, O., Elez-Martinez, P., 2012c. Effects of pulsed electric fields on the bioactive compound content and antioxidant capacity of tomato fruit. J. Agri. Food Chem. 60, 3126–3134.

Van der Hooft, J.J., Vervoort, J., Bino, R.J., Beekwilder, J., de Vos, R.C., 2011. Polyphenol identification based on systematic and robust high-resolution accurate mass spectrometry fragmentation. Anal. Chem. 83, 409–416.

Van Der Hooft, J.J.J., Akermi, M., Ünlü, F.Y., Mihaleva, V., Roldan, V.G., Bino, R.J., et al., 2012. Structural annotation and elucidation of conjugated phenolic compounds in black, green, and white tea extracts. J. Agri. Food Chem. 60, 8841–8850.

Vinas, P., Martinez-Castillo, N., Campillo, N., Hernandez-Cordoba, M., 2011. Directly suspended droplet microextraction with in injection-port derivatization coupled to gas chromatography–mass spectrometry for the analysis of polyphenols in herbal infusions, fruits and functional foods. J. Chromatogr. A 1218, 639–646.

Volpi, N., Bergonzini, G., 2006. Analysis of flavonoids from propolis by on-line HPLC-electrospray mass spectrometry. J. Pharma. Biomed. Anal. 42, 354–361.

Vrhovsek, U., Rigo, A., Tonon, D., Mattivi, F., 2004. Quantitation of polyphenols in different apple varieties. J. Agri. Food Chem. 52, 6532–6538.

Wang, L., Weller, C.L., 2006. Recent advances in extraction of nutraceuticals from plants. Trends Food Sci. Technol. 17, 300–312.

Wang, D., Lu, J., Miao, A., Xie, Z., Yang, D., 2008a. HPLC-DAD-ESI-MS/MS analysis of polyphenols and purine alkaloids in leaves of 22 tea cultivars in China. J. Food Composition Anal. 21, 361–369.

Wang, J., Sun, B., Cao, Y., Tian, Y., Li, X., 2008b. Optimisation of ultrasound-assisted extraction of phenolic compounds from wheat bran. Food Chem. 106, 804–810.

Wang, Z., Hsu, C., Yin, M., 2009. Antioxidative characteristics of aqueous and ethanol extracts of glossy privet fruit. Food Chem. 112, 914–918.

Wolfender, J.L., 2009. HPLC in natural product analysis: the detection issue. Planta Med. 75, 719–734.

Wu, T., Wang, C., Wang, X., Xiao, H., Ma, Q., Zhang, Q., 2008. Comparison of UPLC and HPLC for Analysis of 12 Phthalates. Chromatographia 68, 803–806.

Xie, J., Zhang, Y., Kong, D., Rexit, M., 2011. Rapid identification and determination of 11 polyphenols in Herba lycopi by HPLC–MS/MS with multiple reactions monitoring mode (MRM). J. Food Composition Anal. 24, 1069–1072.

Xu, Y., Simon, J.E., Ferruzzi, M.G., Ho, L., Pasinetti, G.M., Wu, Q., 2012. Quantification of anthocyanidins in the grapes and grape juice products with acid assisted hydrolysis using LC/MS. J. Funct. Foods 4, 710–717.

Yanagida, A., Shoji, A., Shibusawa, Y., Shindo, H., Tagashira, M., Ikeda, M., et al., 2006. Analytical separation of tea catechins and food-related polyphenols by high-speed counter-current chromatography. J. Chromatogr. A 1112, 195–201.

Zhang, L., Wang, Y., Wu, D., Xu, M., Chen, J., 2011. Microwave-assisted extraction of polyphenols from *Camellia oleifera* fruit hull. Molecules (Basel, Switzerland) 16, 4428–4437.

Zimmermann, B.F., Walch, S.G., Tinzoh, L.N., Stühlinger, W., Lachenmeier, D.W., 2011. Rapid UHPLC determination of polyphenols in aqueous infusions of *Salvia officinalis* L. (sage tea). J. Chromatogr. B 879, 2459–2464.

Characterization and Quantification of Polyphenols in Fruits

15

Fabian Weber, Nadine Schulze-Kaysers, Andreas Schieber

Institute of Nutritional and Food Sciences, University of Bonn, Bonn, Germany

CHAPTER OUTLINE HEAD

15.1 Introduction

Phenolic compounds represent a chemically extremely diverse subclass of secondary metabolites that occur ubiquitously in plants and therefore constitute an integral part of the human diet. They are involved in the plant's defense system and serve as natural antioxidants, signal molecules, UV screens, and antimicrobial agents (Quideau *et al.*, 2011). Because of their protective role, they are located primarily in the outer layers and seeds of the plant tissue. They can broadly be classified into simple phenolic acids, flavonoids, xanthones, stilbenes, and lignans. More complex polyphenols include condensed and hydrolysable tannins, which strongly interact with proteins ("tanning"). The phenolic acids can be subdivided into hydroxybenzoic acids (e.g., gallic, protocatechuic, salicylic, syringic, and *p*-hydroxybenzoic acids) and hydroxycinnamic acids such as caffeic, *p*-coumaric, synapic, and ferulic acids. Within the flavonoids (C_6–C_3–C_6 backbone), there are six main subgroups, which are the flavanols, flavonols, flavones, flavanones, anthocyanins, and isoflavones, the latter compounds differing in the attachment of the B-ring to position 3 of the C-nucleus. Among the above-mentioned subgroups, only the anthocyanins are intensely colored, with absorption maxima at around 520 nm. The other flavonoids are either colorless or slightly yellow. Glycosylation of the flavonoid nuclei with various sugars such as glucose, galactose, rhamnose, xylose, and arabinose enhances their water solubility and, especially in the case of anthocyanins, leads to stabilization of the aglycone. Additional variation occurs through hydroxylation and methoxylation of the flavonoid core, and through acylation of the sugar moieties with aromatic or aliphatic acids. Xanthones show a

Polyphenols in Plants. http://dx.doi.org/10.1016/B978-0-12-397934-6.00015-2

293

C_6–C_1–C_6 structure and have been found, for example, in mango and mangosteen (Obolskiy *et al.*, 2009). Stilbenes consist of a C_6–C_2–C_6 backbone and are produced in several plants as stress metabolites (phytoalexins). In the human diet, they are found mainly in grapes, e.g., resveratrol and its glucoside polydatin (or piceid), and peanuts (Burns *et al.*, 2002). The basic structure of lignans consists of two phenylpropane (C_6–C_3) units. Flaxseed is by far the most abundant source of lignans, whereas in fruits only very small quantities are found (Touré and Xueming, 2010). Although phenolic compounds, as a class, are ubiquitously found in plants, some members are biosynthesized only by few plants and can therefore be used for chemotaxonomic purposes.

Previously, phenolic compounds as well as other secondary plant metabolites were considered antinutrients because of the detrimental properties of some of their members. For example, tannins may have inhibitory effects on digestive enzymes and thus impair the absorption of nutrients. As a consequence, undesired phenolic compounds were removed from plants by breeding or selection (Treutter, 2010) and from foods by appropriate technological treatment, also because some phenolics adversely affect sensory properties. However, epidemiological studies and numerous investigations suggest that polyphenols may impart health benefits in humans (Fraga *et al.*, 2010). This paradigm shift has led to an intense interest of food chemists and technologists, nutritionists, plant scientists, biochemists, pharmacists, and members of allied disciplines, in polyphenols. It is not surprising that, in view of this situation, the development of methods for the characterization and quantification of phenolic compounds in fruits, vegetables, plant-derived processed food, and functional foods and dietary supplements has also significantly increased, as evidenced by numerous relevant review articles published during the past few years (Ajila *et al.*, 2011; Arapitsas, 2012; Bueno *et al.*, 2012a, b; Dai and Mumper, 2010; Haminiuk, *et al.*, 2012; Hümmer and Schreier, 2008).

Analyzing phenolic components in fruits and other plant-derived matrices may serve various purposes in addition to merely extending our knowledge of the polyphenol profile of plants for inclusion in databases. The structural diversity of this class of compounds has always attracted, and challenged, analytical chemists; and even in a time when the analytical portfolio is continuously expanding, polyphenols are not characterized and quantified as easily as many other compounds. In particular tannins, both condensed and hydrolyzable, require sophisticated techniques for unambiguous structure elucidation. Determining the structure of polyphenols is important also for establishing structure–bioactivity relationships and for the understanding of their sensory impact. Furthermore, there is increasing interest in characterizing the effects of processing on the stability of polyphenols, degradation pathways, and interactions of phenolics with other matrix constituents, for example proteins (Quideau *et al.*, 2011). Since some phenolic compounds occur only in a limited number of plants, they may be used as markers for authentication purposes. In this context, the question has been addressed whether the authenticity of organic products may be determined based on the contents of secondary plant metabolites (Roose *et al.*, 2010).

15.2 **Sample preparation**

Exhaustive extraction, usually followed by purification, is essential for the analysis of polyphenols from fruits or other plant materials. Due to the wide range of compounds, many different methods have been published but only a few claim to be universally applicable (Sakakibara *et al.*, 2003). Therefore, the choice of sample preparation methods is of utmost importance to prevent discrimination of certain compounds on the one hand and to obtain reliable results on the other hand. Besides simple extraction with different solvents varying in polarity to account for the solubility of the analytes, or the continuous extraction with a Soxhlet apparatus, there are several more advanced or sophisticated methods for the extraction of polyphenols from plant material. To support simple solvent extraction, either microwave irradiation or ultrasound may be applied. Both methods are advantageous mainly in terms of time. Although many studies aim to examine the ultrasonic- or microwave-assisted extraction (UAE or MAE, respectively) as methods for the industrial production of polyphenol-rich extracts, there is no reason why these methods should not be considered as sample preparation methods for subsequent analysis of the polyphenols. Both methods provide the possibility to adjust several parameters for the optimization of the extraction method. In addition to the type of solvent, the time of extraction, and the temperature, the power of microwave irradiation or ultrasound is the main parameter that needs to be optimized. Several studies applied the response surface methodology (RSM) to determine the optimal extraction parameters.

Li *et al.* (2011) reported an extraction yield of over 90% polyphenols from grape seeds in only 4.6 min by MAE, and pointed out that simple solvent extraction consumes about 200 min to achieve similar results. Conversely, Liu *et al.* (2010) achieved a total extraction of 23.5% (41.2% polyphenols) from hawthorn after 13 min of microwave irradiation. This huge difference is obviously due to the two completely different plant materials but it shows the importance of a thorough investigation of the extraction procedure for each type of sample. RSM was also applied by several authors for the evaluation of the ultrasonic-supported extraction of polyphenols, for example from pomegranate peel (Tabaraki *et al.*, 2012), orange (Khan *et al.*, 2010), and black chokeberry (Galvan d'Alessandro *et al.*, 2012).

A rather new method that has been applied for extraction, mainly for the analysis of contaminants or toxins, is the accelerated solvent extraction (ASE), also called pressurized solvent extraction (PSE) or pressurized liquid extraction (PLE), which applies high pressure (approximately 100 bar) in order to decrease extraction time and solvent consumption (Richter *et al.*, 1996). Carabias-Martínez *et al.* (2005) summarized numerous studies applying ASE for the extraction of various analytes from different foodstuffs including polyphenols from fruits. According to Carabias-Martínez *et al.* (2005), ASE is applied for two reasons; either for the automation of the extraction in order to increase reliability and reproducibility or for the development of specialized strategies for the selective extraction of target compounds. However, automation is not limited to the extraction itself. Papagiannopoulos *et al.* (2002) reported the online coupling of ASE, solid phase extraction (SPE), and HPLC analysis.

After the extraction of the plant material, further purification steps are typically performed either to reduce matrix interferences or to selectively enrich minor compounds. Among the different purification strategies, SPE is the most common way to obtain certain polyphenolic subclasses as a fraction. In addition, also liquid–liquid extraction or more advanced solid-supported liquid–liquid extraction (SS-LLE) may be used for fractionating polyphenols of different polarity. The latter method was investigated by Nave *et al.* (2007) for the analysis of non-anthocyanin phenols from wine, but it should also be applicable for the purification of any fruit extract.

As already mentioned, the possibility of automation in the sample preparation is of great interest, and thus numerous commercial devices for SPE are available. As there are many different solid-phase materials that may be used for SPE, this technique is widely employable for many different types of polyphenols. Krämer-Schafhalter *et al.* (1998) compared 16 different materials for the SPE of anthocyanins from *Aronia melanocarpa* and showed that, beside the common C18 reversed phase, non-ionic acrylic ester resins like Amberlite XAD 7 are suitable for the purification of the very polar anthocyanins. They have become a widely accepted adsorbent especially for anthocyanins and anthocyanin-derived compounds. The third important group of SPE materials are dextran gels like Toyopearl or Sephadex, which are mainly used for the fractionation of polyphenols by size or molecular weight, e.g., for the separation of anthocyanins and proanthocyanidins (Cuevas-Rodríguez *et al.*, 2010). Numerous studies applying different SPE techniques for the purification and fractionation of polyphenols exist. Lee *et al.* (2004) used SPE for the separation of polar anthocyanins from non-anthocyanin phenols such as hydroxycinnamates or flavonol glycosides from different *Vaccinium* species. A combination of column chromatography and SPE may be applied for a more complex separation of different polyphenol classes (Svedstroem *et al.*, 2006). A comprehensive overview of the various methods of extraction and purification was given by Garcia-Salas *et al.* (2010).

15.3 Analytical methods

The numerous methods for determination of polyphenols can be classified into three groups: simple sum parameters, chromatographic techniques, and chemometric methods. The quantification of total polyphenols or single subclasses is carried out mainly by photometric measurements, either with or without previous derivatization. For the quantification of individual compounds, a chromatographic separation needs to be performed because no selective derivatization methods exist. As chromatographic methods are usually time-consuming and relatively expensive, chemometric methods have increasingly been developed during the past years.

The most common method for the quantification of total polyphenols is based on the reaction with the Folin–Ciocalteu reagent. Many different methods with slight modifications have been published, the method of Singleton and Rossi (1965) being one of the most frequently cited. Although this method is prone to several interferences and its results are more or less inaccurate, it has become the standard method for the fast and simple determination of the total polyphenol content. Newer, more advanced methods have been published (Schoonen and Sales, 2002).

Some methods have been developed for the determination of several phenolic subclasses, in particular the very heterogeneous oligomeric or polymeric proanthocyanidins, especially since chromatographic methods are of limited use in this field. The determination of proanthocyanidins may be performed after acid-catalyzed cleavage or derivatization (Bate-Smith Assay, Vanillin Assay), or through their ability to precipitate other polymers (Adams–Harbertson Assay, MCP Assay). The Bate-Smith assay makes use of the fact that proanthocyanidins, as their name suggest, release anthocyanidins under acidic conditions, which may then be measured with a spectrophotometer (Bate-Smith, 1975). Catechin or epicatechin moieties in the proanthocyanidins can react with aromatic aldehydes like vanillin, due to their dihydroxy-substituted B-ring, and form red pigments (Swain and Hillis, 1959). By replacing vanillin with 4-(dimethylamino)cinnamaldehyde, the absorption maximum shifts to over 600 nm and thus the interference with anthocyanins is reduced (Cheynier and Fulcrand, 2003). The interference of procyanidins and monomeric anthocyanins or flavanols may also be avoided by precipitation of the higher molecular compounds with proteins or polysaccharides. Based on the assay of Hagerman and Butler (1978), a protein precipitation-based assay was developed by Adams and Harbertson, which combines the separation of low- and high-molecular compounds by precipitation with bovine serum albumin, the bleaching reaction of anthocyanins with SO_2, and the reaction of phenols with ferric chloride, to determine monomeric anthocyanins, small polymeric pigments, large polymeric pigments and iron- reactive tannins in one complex assay (Harbertson et al., 2003). However, the iron- reactive part of the assay was criticized by some authors to be invalid and to suffer from dilution effects (Brooks et al., 2008; Jensen et al., 2008). By replacing the protein with a polysaccharide such as methyl cellulose, the interferences with the precipitating agent during measurement at 280 nm may be overcome. The Methyl Cellulose Precipitation Assay (MCP Assay) published by Sarneckis et al. (2006) is a fast method for the quantification of condensed tannins (as catechin or epicatechin equivalents) and may even be performed in a 96-well plate format (Mercurio et al., 2007). However, it is still not clear which components are precipitated with MC and which are not, and thus the method needs to be adjusted and has been revised several times.

Not only monomeric polyphenols are released under acidic conditions but also the sugar moieties from glycosides may be cleaved. The original glycosyl-glucose assay by Williams et al. (1995) was primarily developed for the determination of glycosidically bound aroma precursors but may also be applied for the determination of anthocyanins or other flavonoid glucosides.

The most important information that has to be revealed when dealing with oligomeric or polymeric polyphenols is the molecular size, which depends on the degree of polymerization and the monomeric composition. The acid-catalyzed cleavage releases the extension units as carbocations, which may be intercepted with nucleophilic agents and the corresponding adducts analyzed. The ratio between the adducts (the extension units) and the neutrally released terminal units yields the mean degree of polymerization (mdp). Dependent on the nucleophile used, the method is called thiolysis or phloroglucinolysis. Benzyl mercaptan is used for its strong nucleophilic character (Matthews *et al.*, 1997), but it is also a very strong odorant and therefore phloroglucinol was applied (Kennedy and Jones, 2001). The application of cysteamine provides the opportunity to analyze the reaction products with ion chromatography (Torres and Selga, 2003). The major drawback of thiolysis or phloroglucinolysis is the heterogenic composition of the proanthocyanidins, which results in side reactions and discriminations. The different interflavanoid bonds are cleaved with considerably different kinetics, whereas A-type proanthocyanidins are not cleaved at all. This needs to be considered when the mdp is calculated.

Among the various separation techniques for polyphenols, HPLC, and more recently UHPLC, are indisputably the most widespread and most important methods to separate and quantify distinct compounds. The analysis of polyphenols by HPLC is mainly conducted by separation on a reversed-phase column (C18) and detection with a diode array detector (DAD) at the specific wavelength. The use of DADs is nowadays the standard detection method, but phenolics can be assigned only by their retention time and their specific absorption spectra, as opposed to a more comprehensive identification by mass spectrometry. Additionally, different types of compounds can be quantified simultaneously by extraction of the corresponding wavelength. Flavanols are most commonly quantified at 280 nm, hydroxycinnamic acids at 320 nm, flavanols at 360 nm, and anthocyanins at 520 nm (Singleton, 1988). A binary solvent system with acidified water and acetonitrile or methanol in gradient elution mode has most commonly been employed. The elution order of polyphenols, while using conventional C18 stationary phases, is determined by the polarity of the analytes and hence is identical regardless of the column used. For example, the elution pattern for the 3-*O*-monoglucosides of the anthocyanins in grapes and wine is always delphinidin < cyanidin < petunidin < peonidin < malvidin. Diglucosides will elute earlier and acylated derivatives will elute later than the monoglucosides (Santos-Buelga *et al.*, 2003). Pelargonidin-3-*O*-glucoside, which is only rarely found in grapes but constitutes the predominant anthocyanin in strawberries, would elute between the monoglucosides of cyanidin and petunidin. Further developments of newer and more specific column materials offer new ways for the determination of polyphenols. The most substantial development is the reduction of particle sizes and the introduction of working pressures above 600 bars namely the development of UHPLC, which considerably reduced the analysis times. This topic and other special LC applications like two-dimensional HPLC have recently been reviewed by Kalili and de Villiers (2011).

Many different methods have been published for the analysis of each polyphenolic subclass. For the analysis of anthocyanins, the addition of acid to the eluent is crucial to obtain sharp peaks and avoid low resolution due to structural conversions during the chromatographic run (Melander *et al.*, 1984). Flavanols and proanthocyanidins require a good separation or extensive sample pretreatment, due to their less specific detection wavelength. While low-molecular proanthocyanidins may be separated using conventional RP18 stationary phases, high-molecular, polymeric proanthocyanidins are poorly separated and result in a chromatographic hump, as the number of possible structures increases dramatically with the degree of polymerization (dp). Therefore, the application of normal-phase columns may be a useful way to separate proanthocyanidins up to decamers (Hammerstone *et al.*, 1999). Nevertheless, this method also fails to separate distinct compounds of the same dp. The characterization of polymeric proanthocyanidins regarding their size and their dp, respectively, may also be accomplished by size exclusion HPLC (Yanagida *et al.*, 2002). A comprehensive treatise on the analysis of proanthocyanidins has been published by Hümmer and Schreier (2008).

Coupling of HPLC/UHPLC systems with a mass spectrometer has become the standard method for the identification and quantification not only of minor compounds in complex matrices. Among the different ionization methods, electrospray ionization (ESI) is the most versatile, while matrix-assisted laser desorption ionization (MALDI) is mainly used for high-molecular procanthocyanidins and atmospheric pressure chemical ionization (APCI) is applied only for less polar compounds. Time of Flight (TOF) MS allows the determination of the exact mass of the analyte, ion traps (IT) allow multiple fragmentation, which is useful for tentative structure elucidation, and triple quadrupols (QqQ) are particularly suitable for quantification of analytes.

While mass spectrometry is used mainly for the identification of fruit-derived polyphenols, its application may extend also to the quantification of phenolic compounds. The main phenolic compounds in red and white grape pomace were identified by IT MS by their MS^2 fragmentation patterns (Kammerer *et al.*, 2004). Cavaliere *et al.* (2008) applied QqQ MS for the quantification of 33 phenols including some resveratrol oligomers in grape skins and some proanthocyanidins in grape seeds by the combined ion current profile for the three most abundant ions of each compound. An extensive overview of the many different MS methods used in grape and wine analysis was given by Flamini (2003).

Proper identification of the phenolic profile is crucial for the authentication of fruit-derived products or for chemotaxonomic studies. For example, phloridzin has long been used as a taxonomic marker for apples (Schieber *et al.*, 2001; Tomás-Barberán *et al.*, 1993) but was more recently detected also in strawberries (Hilt *et al.*, 2003), which demonstrates the importance of the identification also of minor compounds for the detection of adulterations (Alonso-Salces *et al.*, 2004). By application of QqQ MS, these authors detected five isorhamnetin glycosides, two hydroxyphloretin glycosides and quercetin in apple peel and noted their possible use for authentication. Isorhamnetin glycosides had been detected in apple fruits of the cultivar 'Brettacher' for the first time by Schieber *et al.* (2002). An advanced technique for the identification of polyphenols is multiple reaction monitoring (MRM), which was used for the characterization of anthocyanins in red onion, strawberry, and cherry (Steimer and Sjüberg, 2011) and for the determination of the complete phenolic profile of strawberry with over 60 compounds (Del Bubba *et al.*, 2012). The application of MALDI-TOF MS for the analysis of different oligomeric polyphenols from grapes, cranberry, sorghum, and pomegranate was shown by Reed *et al.* (2005) and reviewed by Monagas *et al.* (2010).

A wide range of LC applications with diode array and mass spectrometric detection for different compounds and matrices was given by Kalili and de Villiers (2011) and Ignat *et al.* (2011). According to Ignat *et al.* (2011), a versatile but scarcely applied method for the analysis of polyphenols and especially anthocyanins is capillary electrophoresis (CE). A more detailed review of the different electrophoretic methods and a list with applications was given by Valls *et al.* (2009).

Another chromatographic application that is mainly used for the isolation of polyphenols is countercurrent chromatography (CCC). This liquid–liquid partition technique has proven its versatility for the fractionation of crude plant extracts and for the purification of individual compounds. In view of the limited availability of reference compounds, CCC has become an important tool to obtain standards both for analytical purposes and bioassays. The application of CCC has extensively been reviewed (Marston and Hostettmann, 2006; Valls *et al.*, 2009). It is worth mentioning that CCC may be coupled with MS in order to separate and identify phenolic compounds simultaneously, for example for the characterization of the phenolic profile of sea buckthorn (Gutzeit *et al.*, 2007) and of anacardic acids in cashew (Jerz *et al.*, 2012).

Although several other analytical tools for the characterization of phenolic compounds, such as gas chromatography (GC) (Muñoz-González *et al.*, 2012) or high-performance thin layer chromatography (HPTLC) (Skalicka-Woźniak *et al.*, 2009), may be applied, liquid chromatography, either conventional (HPLC) or with elevated pressure (UHPLC), remains the technique of choice for the analysis of polyphenols.

The number of identified fruit-derived polyphenols has enormously increased in the last decades, while the demand for quick and reliable methods of analysis has increased simultaneously. The need for methods to determine the polyphenolic profiles, e.g., for authentication purposes, may be fulfilled by the development of faster, more sensitive chromatographic methods with advanced resolutions on one side but also by spectroscopic methods like NMR or IR spectroscopy on the other side. These non-selective methods enable the fast profiling of the phenolic fraction with minimal sample preparation, whereas the chromatographic methods require extensive purification, especially for the determination of minor compounds. Currently, NMR and IR spectroscopy are mainly used for the quantification of major compounds in wine and fruit juices for the detection of adulterations or the determination of the product origin, but the application of these techniques will increase during the next years.

References

Ajila, C.M., Brar, S.K., Verma, M., Tyagi, R.D., Godbout, S., Valéro, J.R., 2011. Extraction and analysis of polyphenols: recent trends. Crit. Rev. Biotechnol. 31, 227–249.

Alonso-Salces, R.M., Ndjoko, K., Queiroz, E.F., Ioset, J.R., Hostettmann, K., Berrueta, L.A., et al., 2004. On-line characterisation of apple polyphenols by liquid chromatography coupled with mass spectrometry and ultraviolet absorbance detection. J. Chromatogr. A. 1046, 89–100.

Arapitsas, P., 2012. Hydrolyzable tannin analysis in food. Food Chem. 135, 1708–1717.

Bate-Smith, E.C., 1975. Phytochemistry of proanthocyanidins. Phytochemistry 14, 1107–1113.

Brooks, L., McCloskey, L., McKesson, D., Sylvan, M., 2008. Adams–Harbertson protein precipitation-based wine tannin method found invalid. J. AOAC. Int. 91, 1090–1094.

Bueno, J.M., Ramos-Escudero, F., Sáez-Plaza, P., Muñoz, A.M., José Navas, M., Asuero, A.G., 2012a. Analysis and antioxidant capacity of anthocyanin pigments. Part I: General considerations concerning polyphenols and flavonoids. Crit. Rev. Anal. Chem. 42, 102–125.

Bueno, J.M., Sáez-Plaza, P., Ramos-Escudero, F., Jiménez, A.M., Fett, R., Asuero, A.G., 2012b. Analysis and antioxidant capacity of anthocyanin pigments. Part II: Chemical structure, color, and intake of anthocyanins. Crit. Rev. Anal. Chem. 42, 126–151.

Burns, J., Yokota, T., Ashihara, H., Lean, M.E.J., Crozier, A., 2002. Plant foods and herbal sources of resveratrol. J. Agric. Food Chem. 50, 3337–3340.

Carabias-Martínez, R., Rodríguez-Gonzalo, E., Revilla-Ruiz, P., Hernández-Méndez, J., 2005. Pressurized liquid extraction in the analysis of food and biological samples. J. Chromatogr. A. 1089, 1–17.

Cavaliere, C., Foglia, P., Gubbiotti, R., Sacchetti, P., Samperi, R., Laganà, A., 2008. Rapid-resolution liquid chromatography/mass spectrometry for determination and quantitation of polyphenols in grape berries. Rapid Commun. Mass Spectrom. 22, 3089–3099.

Cheynier, V., Fulcrand, H., 2003. Analysis of polymeric proanthocyanidins and complex polyphenols. In: Santos Buelga, C., Williamson, G. (Eds.), Methods in Polyphenol Analysis. Royal Society of Chemistry, Cambridge, pp. 284–313.

Cuevas-Rodríguez, E.O., Yousef, G.G., García-Saucedo, P.A., López-Medina, J., Paredes-López, O., Lila, M.A., 2010. Characterization of anthocyanins and proanthocyanidins in wild and domesticated mexican blackberries (*Rubus* spp.). J. Agric. Food Chem. 58, 7458–7464.

Dai, J., Mumper, R.J., 2010. Plant phenolics: Extraction, analysis and their antioxidant and anticancer properties. Molecules 15, 7313–7352.

Del Bubba, M., Checchini, L., Chiuminatto, U., Doumett, S., Fibbi, D., Giordani, E., 2012. Liquid chromatographic/electrospray ionization tandem mass spectrometric study of polyphenolic composition of four cultivars of *Fragaria vesca* L. berries and their comparative evaluation. J. Mass. Spectrom. 47, 1207–1220.

Flamini, R., 2003. Mass spectrometry in grape and wine chemistry. Part I: Polyphenols. Mass. Spectrom. Rev. 22, 218–250.

Fraga, C.G., Galleano, M., Verstraeten, S.V., Oteiza, P.I., 2010. Basic biochemical mechanisms behind the health benefits of polyphenols. Mol. Aspects Med. 31, 435–445.

Galvan d'Alessandro, L., Kriaa, K., Nikov, I., Dimitrov, K., 2012. Ultrasound assisted extraction of polyphenols from black chokeberry. Sep. Purif. Technol. 93, 42–47.

Garcia-Salas, P., Morales-Soto, A., Segura-Carretero, A., Fernández-Gutiérrez, A., 2010. Phenolic-compound-extraction systems for fruit and vegetable samples. Molecules 15, 8813–8826.

Gutzeit, D., Winterhalter, P., Jerz, G., 2007. Application of preparative high-speed counter-current chromatography/electrospray ionization mass spectrometry for a fast screening and fractionation of polyphenols. J. Chromatogr. A. 1172, 40–46.

Hagerman, A.E., Butler, L.G., 1978. Protein precipitation method for the quantitative determination of tannins. J. Agric. Food Chem. 26, 809–812.

Haminiuk, C.W.I., Maciel, G.M., Plata-Oviedo, M.S.V., Peralta, R.M., 2012. Phenolic compounds in fruits – an overview. Int. J. Food Sci. Tech. 47, 2023–2044.

Hammerstone, J.F., Lazarus, S.A., Mitchell, A.E., Rucker, R., Schmitz, H.H., 1999. Identification of procyanidins in cocoa (*Theobroma cacao*) and chocolate using high-performance liquid chromatography/mass spectrometry. J. Agric. Food Chem. 47, 490–496.

Harbertson, J.F., Picciotto, E.A., Adams, D.O., 2003. Measurement of polymeric pigments in grape berry extracts and wines using a protein precipitation assay combined with bisulfite bleaching. Am. J. Enol. Viticulture 54, 301–306.

Hilt, P., Schieber, A., Yildirim, C., Arnold, G., Klaiber, I., Conrad, J., et al., 2003. Detection of phloridzin in strawberries (*Fragaria x ananassa* Duch.) by HPLC–PDA–MS/MS and NMR spectroscopy. J. Agric. Food Chem. 51, 2896–2899.

Hümmer, W., Schreier, P., 2008. Analysis of proanthocyanidins. Mol. Nutr. Food Res. 52, 1381–1398.

Ignat, I., Volf, I., Popa, V.I., 2011. A critical review of methods for characterisation of polyphenolic compounds in fruits and vegetables. Food Chem. 126, 1821–1835.

Jensen, J.S., Werge, H.H.M., Egebo, M., Meyer, A.S., 2008. Effect of wine dilution on the reliability of tannin analysis by protein precipitation. Am. J. Enol. Viticulture 59, 103–105.

Jerz, G., Murillo-Velásquez, J.A., Skrjabin, I., Gök, R., Winterhalter, P., 2012. Anacardic acid profiling in cashew nuts by direct coupling of preparative high-speed countercurrent chromatography and mass spectrometry (prep HSCCC-ESI-/APCI-MS/MS). In: Toth, S. (Ed.), Recent Advances in the Analysis of Food and Flavors. American Chemical Society, Washington DC, pp. 145–165.

Kalili, K.M., de Villiers, A., 2011. Recent developments in the HPLC separation of phenolic compounds. J. Sep. Sci. 34, 854–876.

Kammerer, D., Claus, A., Carle, R., Schieber, A., 2004. Polyphenol screening of pomace from red and white grape varieties (*Vitis vinifera* L.) by HPLC-DAD-MS/MS. J. Agric. Food Chem. 52, 4360–4367.

Kennedy, J.A., Jones, G.P., 2001. Analysis of proanthocyanidin cleavage products following acid-catalysis in the presence of excess phloroglucinol. J. Agric. Food Chem. 49, 1740–1746.

Khan, M.K., Abert-Vian, M., Fabiano-Tixier, A.-S., Dangles, O., Chemat, F., 2010. Ultrasound-assisted extraction of polyphenols (flavanone glycosides) from orange (*Citrus sinensis* L.) peel. Food Chem. 119, 851–858.

Krämer-Schafhalter, A., Fuchs, H., Pfannhauser, W., 1998. Solid-phase extraction (SPE)—A comparison of 16 materials for the purification of anthocyanins from *Aronia melanocarpa* var. Nero. J. Sci. Food Agric. 78, 435–440.

Lee, J., Finn, C.E., Wrolstad, R.E., 2004. Comparison of anthocyanin pigment and other phenolic compounds of *Vaccinium membranaceum* and *Vaccinium ovatum* native to the pacific northwest of north america. J. Agric. Food Chem. 52, 7039–7044.

Li, Y., Skouroumounis, G.K., Elsey, G.M., Taylor, D.K., 2011. Microwave-assistance provides very rapid and efficient extraction of grape seed polyphenols. Food Chem. 129, 570–576.

Liu, J.-L., Yuan, J.-F., Zhang, Z.-Q., 2010. Microwave-assisted extraction optimised with response surface methodology and antioxidant activity of polyphenols from hawthorn (*Crataegus pinnatifida* Bge.) fruit. Int. J. Food Sci. Tech. 45, 2400–2406.

Matthews, S., Mila, I., Scalbert, A., Pollet, B., Lapierre, C., Hervé du Penhoat, C.L.M., et al., 1997. Method for estimation of proanthocyanidins based on their acid depolymerization in the presence of nucleophiles. J. Agric. Food Chem. 45, 1195–1201.

Marston, A., Hostettmann, K., 2006. Developments in the application of counter-current chromatography to plant analysis. J. of Chromatogr. A 1112, 181–194.

Melander, W.R., Lin, H.J., Jacobson, J., Horvath, C., 1984. Dynamic effect of secondary equilibria in reversed-phase chromatography. J. Phys. Chem. A 88, 4527–4536.

Mercurio, M.D., Dambergs, R.G., Herderich, M.J., Smith, P.A., 2007. High throughput analysis of red wine and grape phenolics—adaptation and validation of methyl cellulose precipitable tannin assay and modified Somers color assay to a rapid 96 well plate format. J. Agric. Food. Chem. 55, 4651–4657.

Monagas, M., Quintanilla-López, J.E., Gómez-Cordovés, C., Bartolomé, B., Lebrón-Aguilar, R., 2010. MALDI-TOF MS analysis of plant proanthocyanidins. J. Pharm. Biomed. Anal. 51, 358–372.

Muñoz-González, C., Moreno-Arribas, M.V., Rodríguez-Bencomo, J.J., Cueva, C., Martín Álvarez, P.J., et al., 2012. Feasibility and application of liquid–liquid extraction combined with gas chromatography–mass spectrometry for the analysis of phenolic acids from grape polyphenols degraded by human faecal microbiota. Food Chem. 133, 526–535.

Nave, F., Cabrita, M.J., Teixeira, d. C.C., 2007. Use of solid-supported liquid–liquid extraction in the analysis of polyphenols in wine. J. Chromatogr. A. 1169, 23–30.

Obolskiy, D., Pischel, I., Siriwatanametanon, N., Heinrich, M., 2009. *Garcinia mangostana* L.: A phytochemical and pharmacological review. Phytother. Res. 23, 1047–1065.

Papagiannopoulos, M., Zimmermann, B., Mellenthin, A., Krappe, M., Maio, G., Galensa, R., 2002. Online coupling of pressurized liquid extraction, solid-phase extraction and high-performance liquid chromatography for automated analysis of proanthocyanidins in malt. J. Chromatogr. A. 958, 9–16.

Quideau, S., Deffieux, D., Douat-Casassus, C., Pouységu, L., 2011. Plant polyphenols: Chemical properties, biological activities, and synthesis. Angew. Chem. Int. Ed. 50, 586–621.

Reed, J.D., Krueger, C.G., Vestling, M.M., 2005. MALDI-TOF mass spectrometry of oligomeric food polyphenols. Phytochemistry 66, 2248–2263.

Richter, B.E., Jones, B.A., Ezzell, J.L., Porter, N.L., Avdalovic, N., Pohl, C., 1996. Accelerated solvent extraction: A technique for sample preparation. Anal. Chem. 68, 1033–1039.

Roose, M., Kahl, J., Körner, K., Ploeger, A., 2010. Can the authenticity of products be proved by plant substances? Biol. Agri. Horticulture 27, 129–138.

Sakakibara, H., Honda, Y., Nakagawa, S., Ashida, H., Kanazawa, K., 2003. Simultaneous determination of all polyphenols in vegetables, fruits, and teas. J. Agric. Food Chem. 51, 571–581.

Santos–Buelga, C., García–Viguera, C., Tomás–Barberán, F.A., 2003. On–line identification of flavonoids by HPLC coupled to diode array detection. In: Santos–Buelga, C., Williamson, G. (Eds.), Methods in Polyphenol Analysis. Royal Society of Chemistry, Cambridge, pp. 92–127.

Sarneckis, C.J., Dambergs, R.G., Jones, P., Mercurio, M., Herderich, M.J., Smith, P.A., 2006. Quantification of condensed tannins by precipitation with methyl cellulose: Development and validation of an optimised tool for grape and wine analysis. Aust. J. Grape Wine Res. 12, 39–49.

Schieber, A., Keller, P., Carle, R., 2001. Determination of phenolic acids and flavonoids of apple and pear by high-performance liquid chromatography. J. Chromatogr. A. 910, 265–273.

Schieber, A., Keller, P., Streker, P., Klaiber, I., Carle, R., 2002. Detection of isorhamnetin glycosides in extracts of apples (*Malus domestica* cv. "Brettacher") by HPLC-PDA and HPLC-APCI-MS/MS. Phytochem. Anal. 13, 87–94.

Schoonen, J., Sales, G., 2002. Determination of polyphenols in wines by reaction with 4-aminoantipyrine and photometric flow-injection analysis. Anal. Bioanal. Chem. 372, 822–828.

Singleton, V.L., Rossi Jr., J.A., 1965. Colorimetry of total phenolics with phosphomolybdic-phosphotungstic acid reagents. Am. J. Enol. Viticulture 16, 144–158.

Singleton, V.L., 1988. Wine phenols. In: Linskens, H.F., Jackson, J.F. (Eds.), Wine Analysis. Springer, Berlin, pp. 173–218.

Skalicka-Woźniak, K., Hajnos, M., Głowniak, K., 2009. High-performance thin-layer chromatography combined with densitometry for quantitative analysis of chlorogenic acid in fruits of *Peucedanum alsaticum* L.. J. Planar Chrom—Modern TLC 22, 297–300.

Steimer, S., Sjöberg, P.J.R., 2011. Anthocyanin characterization utilizing liquid chromatography combined with advanced mass spectrometric detection. J. Agric. Food Chem. 59, 2988–2996.

Svedstroem, U., Vuorela, H., Kostiainen, R., Laakso, I., Hiltunen, R., 2006. Fractionation of polyphenols in hawthorn into polymeric procyanidins, phenolic acids and flavonoids prior to high-performance liquid chromatographic analysis. J. Chromatogr. A. 1112, 103–111.

Swain, T., Hillis, W.E., 1959. The phenolic constituents of *Prunus domestica* L—The quantitative analysis of phenolic constituents. J. Sci. Food Agric. 10, 63–68.

Tabaraki, R., Heidarizadi, E., Benvidi, A., 2012. Optimization of ultrasonic-assisted extraction of pomegranate (*Punica granatum* L.) peel antioxidants by response surface methodology. Sep. Purif. Technol. 98, 16–23.

Tomás-Barberán, F.A., García-Viguera, C., Nieto, J.L., Ferreres, F., Tomás-Lorente, F., 1993. Dihydrochalcones from apple juices and jams. Food Chem. 46, 33–36.

Torres, J., Selga, A., 2003. Procyanidin size and composition by thiolysis with cysteamine hydrochloride and chromatography. Chromatographia 57, 441–445.

Touré, A., Xueming, X., 2010. Flaxseed lignans: Source, biosynthesis, metabolism, antioxidant activity, bio-active components, and health benefits. Compr. Rev. Food Sci. Food Saf. 9, 261–269.

Treutter, D., 2010. Managing phenol contents in crop plants by phytochemical farming and breeding—Visions and constraints. Inter. J. Mol. Sci. 11, 807–857.

Valls, J., Millán, S., Martí, M.P., Borràs, E., Arola, L., 2009. Advanced separation methods of food anthocyanins, isoflavones and flavanols. J. Chromatogr. A. 1216, 7143–7172.

Williams, P.J., Cynkar, W., Francis, I.L., Gray, J.D., Iland, P.G., Coombe, B.G., 1995. Quantification of glycosides in grapes, juices, and wines through a determination of glycosyl glucose. J. Agric. Food Chem. 43, 121–128.

Yanagida, A., Shoji, T., Kanda, T., 2002. Characterization of polymerized polyphenols by size-exclusion HPLC. Biosci. Biotechnol. Biochem. 66, 1972–1975.

Determination of Polyphenols, Flavonoids, and Antioxidant Capacity in Dry Seeds

16

Shmuel Galili, Ran Hovav

Agricultural Research Organization, The Volcani Center, Bet Dagan, Israel

CHAPTER OUTLINE HEAD

16.1 Introduction

Dry seeds are a good source of protein, dietary fibers, minerals, and bioactive polyphenols—such as anthocyanins, flavonoids, isoflavones, phenolic acids, and lignans—which exhibit high levels of antioxidant activity (Geil and Anderson, 1994; Lin and

Polyphenols in Plants. http://dx.doi.org/10.1016/B978-0-12-397934-6.00016-4

Lai, 2006; Segev *et al.*, 2010). This antioxidant capacity has been linked to a reduced risk of developing chronic diseases such as cancer, diabetes, obesity, and cardiovascular diseases (Adams and Standridge, 2006; Aparicio-Fernandez *et al.*, 2006, 2008; Azevedo *et al.*, 2003; Cardador-Martínez *et al.*, 2002, 2006; Heimler *et al.*, 2005). These polyphenol compounds also exhibit antinutritional effects, mainly reduction in protein digestibility (Yasmin *et al.*, 2008). Therefore, some seeds, mainly legumes, must be processed before consumption, for example by soaking and cooking, a procedure that also improves their flavor and palatability (Chau *et al.*, 1997). However, some of these processes have been found to decrease the levels of polyphenols and other compounds with antioxidant capacity, as well as antioxidant activity in the final product (Han and Baik, 2008; Segev *et al.*, 2010, 2011, 2012; Xu and Chang, 2008). To determine the levels and composition of polyphenols and the antioxidant capacity of different seeds and seed parts following different treatments, polyphenols must be extracted from the seeds, quantified, and analyzed by colorimetric methods (Xu and Chang, 2007) or by analytical methods such as liquid, gas or other types of chromatography.

16.2 Extraction of polyphenols from seeds

Polyphenols are extracted from seeds using water, organic solvents, or their combination. The selection of extraction method and solvent is critical for estimations of the concentration of any compound in foods. Thus extraction protocols and analytical techniques for the quantification of these compounds need to be standardized for each seed species.

16.2.1 Sample preparation

Sample preparation is relevant to efficiency, accuracy, and reproducibility, as it can be responsible for up to 30% of the error in analytical measurements (Alonso-Salces *et al.*, 2001; Robbins, 2003). It is understood that collected samples must represent the actual pool. As a general rule, dry seed samples are directly subjected to milling or grinding. Processed seeds must be air-, heat-, vacuum- or freeze-dried before grinding and extraction to avoid degradation of native polyphenols, because high moisture or water content induces enzyme activity (Stalikas, 2007), and reduces grinding efficiency, particularly of the seed coats. Heating and exposure to light and oxygen may also affect polyphenol composition, and therefore high-temperature drying should be avoided. Freeze-drying seems to be the optimal method due to conservation of most of the total polyphenols and their composition upon extraction (Abascal *et al.*, 2005). However, drying processes, including freeze-drying, can have undesirable effects on the ingredient profiles of plant samples, thus caution should be taken when studying the medicinal properties of seeds. For example, freeze-drying, similar to other drying methods, reduced antioxidant activity in raspberry fruits compared to their fresh counterparts due to the reduction in both glycoside and aglycone polyphenols (Mejia-Meza *et al.*, 2010). Freeze-dried samples of wine grape pomace retained higher total phenolic content and antioxidant activity (Tseng and Zhao, 2012), whereas their levels were lower in pumpkin flour (Que *et al.*, 2008) compared to other drying methods.

Several seed types also contain non-phenolic substances that might influence extraction efficiency. Thus, additional steps may be required to remove these unwanted components; for example, oil seeds must be defatted before extraction. This is done mainly by removing the seed's oil with an organic solvent such as petroleum ether in a Soxhlet apparatus (Kyari, 2008), or mechanically in a Carver press, prior to polyphenol extraction (Nepote *et al.*, 2002; Yvonne *et al.*, 2007). Peanuts defatted by both methods gave significantly higher amounts of isoflavones and trans-resveratrol compared to non-defatted peanuts (Yvonne *et al.*, 2007).

To solvate a solid particle, there must be contact between the two. A smaller particle size is preferable to a larger one, because more surface area is in contact with the solvent. Thus, to increase extraction efficiency, the seeds are broken down into fine particles using a seed grinder followed by sieving through a sieve with less than 1 mm pore size. Sieve pore size has a large influence on extraction efficiency. Mukhopadhyay *et al.* (2006) found a *ca.* three-fold increase in the extraction efficiency of phenolic acid from black cohosh with a particle size decrease from greater than 2.00 mm to less than 0.25 mm. This is because surface area per unit mass of the plant material increases as the particle size decreases. Similar observations have been made for the extraction of polyphenols from spent coffee (Pinelo *et al.*, 2007) and sumac (Kossah *et al.*, 2010). No significant differences were found between particle sizes of 0.5 and 1 mm, but smaller particles resulted in processing difficulties such as dust, heat generation during grinding, and blocked filters during extraction (Kossah *et al.*, 2010). Nevertheless, Nepote *et al.* (2005) found no difference between 0–1-mm, 1–2-mm, 2–10-mm and non-crushed peanut skin extracted by soaking for 1 h in 70% ethanol. However, following maceration for 24 h of the same peanut skin, only particles of 0–1 mm exhibited significantly lower values of total phenolic compounds extracted. This could be explained by the compactness of the solid, resulting in a smaller contact surface and a higher amount of retained solvent relative to the largest particle sizes.

16.2.2 **The extraction process**

Extraction is the first step in the isolation of phenolic compounds, and it is therefore critically important to select an efficient extraction procedure and conditions that will optimize extraction yield and maintain the stability of phenolic compounds. The most common method of extracting free polyphenols from seeds involves polar organic solvents, which destroy the cell membranes and dissolve the polyphenols. This is due to ease of use, efficiency, and wide applicability. Total—free and bound—polyphenols can be extracted by alkali solution followed by acidification, defatting, and solubilization (Krygier *et al.*, 1982). These two fractions (free and bound polyphenols) can be separated by a two-step extraction, with organic solvent followed by alkali extraction (Dewanto *et al.*, 2002; Han and Baik, 2008). Extraction efficiency is related to the interactions between matrix and solute and/or the solute's diffusion path through the sample matrix. Strong interactions and a long diffusion path can be overcome by reducing particle size, as mentioned above, or by increasing the extraction temperature and time. Weak interactions, on the other hand, can be overcome by using more extractions steps. Thus, several factors can influence the efficiency and accuracy of polyphenol extraction: solvent type, extraction duration, extraction temperature, sample-to-solvent, and solvent-to-water ratios, and the seed's polyphenol composition which may influence its solubility in the extraction solvent. The effect of each factor on extraction efficiency is not always obvious and cannot be predicted. Thus, it is very important to optimize the extraction conditions so as to maximize the extraction efficiency for each seed type.

16.2.3 Solvent type

Elementary aspects in the choice of extraction solvent are the solubility and diffusivity of the desired solute in the extraction solvent. Cost, safety, toxicity, and sustainability must also be taken into account, mainly in commercial extractions. Free seed polyphenols can be extracted by different solvents: water, ethanol, methanol, acetone, or ethyl acetate, often with different proportions of water, are commonly used to extract antioxidants from seeds (Xu and Chang 2007). Ethanol and water are generally used as the solvents when extracting polyphenols for human consumption because they are less toxic than other organic solvents, such as methanol and acetone (Chew *et al.*, 2011). For anthocyanin-rich seeds, small amounts (in most cases up to 5%) of organic acids, such as formic acid, acetic acid, citric acid, tartaric acid, and phosphoric acid, should be added to the organic solvent. These acidic solvents denature the cell membranes, and dissolve and stabilize the anthocyanins (Dai and Mumper, 2010). It is important not to use excessive amounts of acid as it can hydrolyze the molecules that are naturally conjugated to anthocyanin and other polyphenolic compounds (Dai and Mumper, 2010). Different solvent compositions may affect the amount of polyphenols, flavonoids, anthocyanins, and tannins extracted from the seed sample. For example, 50% acetone is the best solvent for polyphenol and flavonoid extraction from chickpea (Segev *et al.*, 2010), peanut skin (Shem-Tov *et al.*, 2012), and quinoa (Brand *et al.*, 2012). Acidic 70% acetone is best for extraction of polyphenols, flavonoids, and tannins from black bean, lentil, black soybean, and red kidney bean (Xu and Chang 2007).

16.2.4 Solvent-to-water ratio

Extraction solvents are combinations of organic solvent (0–100%) and water (100%–solvent proportion). The organic solvent dissolves the cell membranes and non-polar polyphenols, and the water dissolves the polar polyphenols. Water also causes the plant material to swell thereby allowing the solvent to penetrate more easily into the solid matrix and increasing polyphenol extractability (Gertenbach, 2001). Thus, the choice of solvent-to-water ratio depends on polyphenol composition. The solubility of polyphenols in water is determined by their polarity. Plant materials may contain phenolics that vary from simple to highly polymerized substances (e.g., tannins) in different quantities. Moreover, phenolics may also be associated with other plant components, such as carbohydrates and proteins. Simple polyphenols such as flavanols (e.g., catechin or epicatechin) and anthocyanins (e.g., delphinidin and cyanidin), both aglycones in glycosylated form, are known to be present in many seeds; they are polar compounds, and are therefore soluble in water (Mukhopadhyay *et al.*, 2006). In contrast, highly polymerized polyphenols such as condensed tannins and those associated with carbohydrates and proteins are less soluble in water, and a high solvent-to-water ratio will be more efficient for their extraction (Mukhopadhyay *et al.*, 2006; Xu *et al.*, 2007).

16.2.5 **Solvent-to-solid ratio**

The solvent-to-solid ratio is in most cases positively correlated with extraction yield—a higher solvent-to-solid ratio increases the total amount of solids extracted, regardless of the solvent used (Al-Farsi and Lee, 2008). This is consistent with mass-transfer lows: in which solvent-to-solid ratio is used influence the concentration gradient between the solid and the bulk of the liquid (Pinelo *et al.*, 2005). The solvent-to-solid ratio can be increased by increasing solvent volume or decreasing sample size. Increasing solvent volume may influence the economic feasibility of the extraction, dilute the solute, and lengthen the extraction procedure. A reduction in sample size may cause sampling error. Many extraction protocols use solvent-to-solid ratios from 10 to 30. Lower ratios may cause saturation problems, whereas higher ratios may require a solute concentration step by evaporating of the solvent.

16.2.6 **Number of extraction steps**

Extraction efficiency increases with the use of multiple extraction cycles with the same or different solvents. For example, it is more efficient to carry out four extractions in 1 ml than one extraction in 4 ml. Up to five, but in most cases, two to three extraction steps are sufficient for a quantitative determination of seed polyphenols (Liu *et al.*, 2013; Nepote *et al.*, 2005; Shi *et al.*, 2003). More extractions are not recommended due to the time expended, the consumption of solvent, and compound dilution. Multistep extractions can also minimize polyphenol oxidation and decomposition caused by high temperature or long extraction duration (Al-Farsi *et al.*, 2008).

16.2.7 **Extraction duration**

The recovery of phenolic compounds from plant materials is also influenced by the duration of the extraction, which may affect polyphenol solubilization. Long extraction times can increase the chances of phenolics oxidation which will decrease the yield and change the conformation of the extracted polyphenols (Dai and Mumper,

2010). Thus, a shorter extraction time that does not affect extraction yield is preferred. Al-Farsi *et al.* (2008) found only a slight or no increase in phenolics extracted from date seeds using water and 50% acetone at increasing extraction times from 0.5 to 4h. Best extraction yields for polyphenols of grape seeds were obtained after an extraction time of 1.5h. A longer extraction time decreased the total phenolics extracted, possibly due to loss of some oxidized phenolic compounds that might polymerize into insoluble compounds (Shi *et al.*, 2003). Nepote *et al.* (2005) showed that polyphenol extraction from peanut skin via maceration requires 60min to extract the same quantity of total phenolic compounds extracted by the shaking method in 10min.

16.2.8 Extraction temperature

Several studies have found that elevating the extraction temperature from ambient to 90°C shows a positive correlation with extraction yield (Chew *et al.*, 2011; Mukhopadhyay *et al.*, 2006; Sardsaengjun and Jutiviboonsuk, 2010; Shi *et al.*, 2003). However, temperatures higher than 100°C might cause a significant reduction in extraction yield (Mukhopadhyay *et al.*, 2006; Sardsaengjun and Jutiviboonsuk, 2010). High extraction temperatures improve extraction efficiency of phenolic compounds by softening the plant tissues, weakening the cell wall (Juntachote *et al.*, 2006; Mukhopadhyay *et al.*, 2006; Sardsaengjun and Jutiviboonsuk, 2010; Shi *et al.*, 2003; Spigno *et al.*, 2007), increasing polyphenol solubility and increasing diffusion and mass-transfer rates of the extracted compound (Cacace and Mazza, 2003; Kim *et al.*, 2007; Vongsangnak *et al.*, 2004). High temperature also reduces solvent viscosity which increases the solvent's penetration into the sample matrices, further improving the extraction rate. However, a very high temperature may degrade polyphenolic compounds. Extraction temperature should take into account the boiling point of the extraction solvent, as temperatures above this point might change the solvent-to-water ratio due to evaporation of solvent from the aqueous-solvent solution (Shi *et al.*, 2003). However, many phenolic compounds, like anthocyanins, are easily hydrolyzed and oxidized. Thus, extraction and concentration are typically conducted at temperatures ranging from 20 to 50°C (Jackman *et al.*, 1987), because high temperatures may cause rapid degradation of the anthocyanin (Patrasa *et al.*, 2010). In addition, polyphenol hydroxylation and oxidation might cause a reduction in the antioxidant capacity of the crude extract (Chew *et al.*, 2011; Spigno *et al.*, 2007; Vongsangnak *et al.*, 2004).

16.3 Additional extraction techniques for seed polyphenols

Although solid–liquid extraction is widely employed for phenolics extraction from seeds, this method can encounter several problems, such as long extraction time, utilization of high solvent volumes and in some cases, poor solute recovery. Thus, other extraction techniques have been applied to recover antioxidant phenolic compounds, including: ultrasound-assisted extraction, microwave-assisted extraction, supercritical fluid extraction, and pressurized liquid extraction. These extraction techniques reduce extraction time and the amount of extraction solvent needed, demonstrate high reproducibility and can be adapted to small or large scales for use in the laboratory and industry, respectively (Chemat *et al.*, 2011).

16.3.1 **Ultrasound-assisted extraction**

Ultrasounds are sound waves of very high frequency (2 MHz or greater) which are propagated via compression and rarefaction, and require a medium (e.g., tissue) in which to travel (en.wikipedia.org/wiki/Ultrasound). Ultrasound-assisted extraction makes use of high-intensity ultrasonic energy created by the implosion of cavitation bubbles. The bubbles' collapse can produce physical, chemical, and mechanical effects. When this energy reaches the surface of the seed material through the extraction solvent, it is transformed into mechanical energy that is equivalent to several thousand atmospheres of pressure (Liu *et al.*, 2008). The high pressure breaks the seed particles, destroys cell membranes, improves penetration of solvent into the sample matrix and increases the contact surface area between the solid and liquid phases resulting in the release of polyphenol compounds to the extraction solvent (Chemat *et al.*, 2011; Rostagno *et al.*, 2003). Ultrasound-assisted extraction is a simple, environmentally friendly and efficient alternative to conventional extraction techniques (Ghafoor *et al.*, 2009). The method's main advantages are simplicity of use and low instrumental requirements (Chemat *et al.*, 2011; Liu *et al.*, 2008; Zhang *et al.*, 2009). Ultrasonic devices include an ultrasonic bath—mainly used for small-scale extractions—or an ultrasonic probe system for large-scale industrial extraction (Luque-Garciá and Luque de Castro, 2003; Vinatoru, 2001; Wang and Weller, 2006; Zhang *et al.*, 2009). However, ultrasound-assisted extraction still utilizes more extraction solvent and longer extraction times than other novel extraction techniques (Chemat *et al.*, 2011). Nevertheless, Rostagno *et al.* (2003) showed that ultrasound-assisted extractions are up to 15% more efficient for soybean isoflavone extraction than the regular mixing–stirring extraction. Optimal extraction conditions were 50% ethanol, and 60°C for 20 min. About 90% and 98% of total isoflavones were obtained after the first and second extraction cycles, respectively, with no significant differences between and ultrasonic bath and probe. Similar optimum conditions were found for polyphenol extraction from wheat bran (Wang and Clements, 2008), litchi seeds (Chen *et al.*, 2011) and grape seeds (Ghafoor *et al.*, 2009).

16.3.2 **Microwave-assisted extraction**

Microwaves are radio waves with lengths ranging from 1 m to 1 mm, or equivalently, with frequencies ranging from 300 MHz (0.3 GHz) to 300 GHz (http://en.wikipedia .org/wiki/Microwave). These wavelengths heat up polar molecules, such as water, ethanol, and methanol, by a dual mechanism of ionic conduction and dipole rotation (Smith and Carpentier, 1983). Microwave-assisted extraction has garnered increasing interest as it enables the efficient use of microwave energy to extract valuable compounds from solid samples (for a review, see Tatke and Jaiswal, 2011). The idea of using microwave energy to extract polyphenols from plant seeds stems from the fact that dry seeds contain about 10–15% water. The water molecules, located inside the seeds' cells, absorb the microwave energy and heat up, resulting in high pressure buildup. The high pressure destroys the cell wall and increases the flow of polyphenols out of the cells. The heat of the plant tissue enhances the penetration of the extraction solvent into the seed particles and the diffusion of the polyphenols from the plant cells to the extraction solvent (Tatke and Jaiswal, 2011). Thus, extraction time and solvent volume are reduced and extraction efficiencies improved relative to conventional extraction techniques (Chemat *et al.*, 2011).

There are two types of commercially available microwave extraction systems: closed and open vessels. The use of closed vessels results in higher pressure, which allows the extraction temperature of the solvent to be increased above its boiling point, thereby enhancing extraction efficiency and shortening the extraction time; however, this can also damage the extracted solute. The use of high pressure and temperature requires an additional cooling step and thus limits the amount of sample use. The use of open vessels is much safer, requires less expensive equipment, can operate with large samples, and additional regent can be added during the extraction. However, it requires longer extraction times (Tatke *et al.*, 2011). The use of microwave-assisted extraction to extract phenols from flaxseeds was up to 1.7-fold more efficient than traditional extraction (Beejmohun *et al.*, 2007). Those authors found that microwave power determines extraction yield mainly for the shorter extraction times. Compared with other extraction methods of polyphenols from grape seeds, microwave-assisted extraction provided comparable or better extraction yields with shorter extraction times. However, as observed with flaxseeds, changing the power applied to the extraction has basically no effect on the sufficiency of the extraction; inspection of the applied power profile during extraction revealed that the power must be strictly reduced to maintain a constant temperature in the reaction cell. Similar observations were made when extracting anthocyanins from Chinese purple corn cob (Yang and Zhai, 2010) and with peanut skins (Ballard *et al.*, 2010).

16.3.3 Supercritical fluid extraction

Among the various extraction techniques used on analytical and preparative scales for separating desired components from solid or liquid matrices, supercritical fluid extraction is one of the most commonly used (Herrero *et al.*, 2010). This technique makes use of critical conditions of temperature and pressure beyond which the liquid and vapor phases of the extraction solvent are indistinguishable and the extraction solvent becomes a supercritical fluid. Under these conditions, the extraction solvent undergoes remarkable physical changes including: liquid-like densities, reduction in surface tension, and gas-like viscosity, compressibility, and diffusivity. Although several solvents, including nitrous oxide, ethane, propane, n-pentane, ammonia, fluoroform, sulfur hexafluoride, and water have been used as extraction solvents, the most commonly used supercritical fluid extraction solvent for polyphenols is carbon dioxide (Zougagh *et al.*, 2004). This is due to its low critical temperature and pressure, low cost, availability in pure form, relative non-toxicity and non-flammability, and its easy removal at the end of the extraction (Chiu *et al.*, 2002; Herrero *et al.*, 2010; Liza *et al.*, 2010; Rawa-Adkonis *et al.*, 2003). In some cases, co-solvents or modifiers, such as methanol or ethanol, are added to the carbon dioxide to enhance its polarity (Castro-Vargasa *et al.*, 2010; Yilmaz *et al.*, 2011). The main advantages of supercritical fluid extraction are: higher solute-solubilization capacity, lower amount of solvents used, and easy precipitation and purification of the solutes after extraction, resulting in a pure, high-quality product (Seabra *et al.*, 2012; Zougagh *et al.*, 2004). The main disadvantage of supercritical fluid extraction is the need for sophisticated and expensive high-pressure equipment and technology (Zougagh *et al.*, 2004). Supercritical fluid extraction has been utilized for the extraction of polyphenols from seeds of guava (Castro-Vargasa *et al.*, 2010; Zougagh *et al.*, 2004), date (Liu *et al.*, 2013), tamarind (Tsuda *et al.*, 1995), cocoa (Sarmento *et al.*, 2008) and grape (Murga *et al.*, 2000; Palma *et al.*, 1999; Yilmaz *et al.*, 2011). All of these utilized carbon dioxide as the extraction solvent with or without ethanol, methanol or ethyl acetate as co-solvents. The co-solvents improved the extraction yields of polyphenols by increasing their solubility in the extraction solvent (Murga *et al.*, 2000). Castro-Vargasa *et al.* (2010) found that, with guava seeds, carbon dioxide with ethanol results in higher extraction efficiency of polyphenols than pure carbon dioxide or ethyl acetate, due to the higher polarity of ethanol as a co-solvent. Extraction efficiency was positively correlated to both pressure (10–30 mPa) and temperature (40–60°C). However, the highest yields of total polyphenols were obtained at 60°C and 10 MPa. Yilmaz *et al.* (2011) found 30°C, 300 bars and 20% ethanol to be the optimum conditions for extraction of most polyphenols examined. Pressure and amount of ethanol were the most important parameters influencing extraction yield via elevation of carbon dioxide solubility and polarity, respectively. Similar observations were made by Castro-Vargasa *et al.* (2010) and Murga *et al.* (2000).

16.3.4 **Pressurized liquid extraction**

Pressurized liquid extraction, also called accelerated solvent extraction, is a technique that utilizes liquid solvents at elevated temperature and pressure (Benthin *et al.*, 1999; Dai and Mumper, 2010). Pressurized fluid extraction differed from Soxhlet extraction, mainly by utilization of solvents near their supercritical region where they have high extraction properties. This enables high solubility and high diffusion rate of the solute at high temperature (25–200°C); high pressure (3–20 MPa), however, keeps the extraction solvent in a liquid state which enables good penetration of the solvent into the sample (Giergielewicz-Mozajska *et al.*, 2001). Thus, an extraction process is achieved that requires small amounts of solvent (15–40 ml) and shortened extraction times (15–20 min) (for a review, see Mustafa and Turner, 2011). Transferring the extraction conditions used in conventional methods to pressurized liquid extraction usually does not require a change in solvent. Often, the organic solvent or combination of solvents utilized in the existing Soxhlet method can simply be adopted by the pressurized liquid extraction method.

Moreover, by adjusting temperature and pressure, additional solvents that were not effective with conventional methods can be used. Increasing the temperature of water results in a decrease in its viscosity and surface tension, making it a suitable solvent for the extraction of both polar and non-polar organic compounds (Hassas-Roudsari *et al.*, 2009). Two main approaches are taken for pressurized liquid extraction: static, in which the extraction process consists of one or several extraction cycles with replacement of the solvent between cycles, and dynamic, in which fresh extraction solvent is continuously pumped through the sample vessel (Mustafa and Turner, 2011). In recent years, pressurized liquid extraction has been successfully applied to the extraction of phenolic compounds from different plant seeds such as pine (Liazid *et al.*, 2010), grape (Palma *et al.*, 2002), barley (Bonoli *et al.*, 2004), and tara (Seabra *et al.*, 2012).

16.4 Methods for quantification of total polyphenols, total flavonoids and antioxidant activity in seeds

16.4.1 Quantification of total polyphenols, flavonoids, and anthocyanins

Polyphenols are extracted from seeds using water or organic solvent. The amount of total polyphenols can be quantified by colorimetric methods utilizing standard curves derived from known amounts of purified polyphenol molecules such as gallic acid or catechin (Xu *et al.*, 2007). Specific polyphenol compounds are determined by analytical methods such as HPLC and GC and quantified by comparing with known amounts of internal standards.

Total seed polyphenols are usually determined by colorimetric methods with Folin-Ciocalteu reagent (Folin and Ciocalteu, 1927; Singleton *et al.*, 1999). This reagent is formed from a mixture of phosphotungstic acid ($H_3PW_{12}O_{40}$) and phosphomolybdic acid ($H_3PMo_{12}O_{40}$) which, after oxidation of the phenols, are reduced to blue oxides of tungsten (W_8O_{23}) and molybdene (Mo_8O_{23}), respectively. This reaction, which occurs under alkaline conditions, is carried out with sodium carbonate. Under these conditions, the electron is easily removed from the phenol molecule (Singleton *et al.*, 1999). The resultant blue coloration has a maximum absorption in the region of 760 nm, and is proportional to the total quantity of phenolic compounds originally present; this quantity is usually expressed as gallic acid or catechin equivalents (George *et al.*, 2005). The linear range of the calibration curve is 50–1000 µg/ml (r = 0.99) (Xu *et al.*, 2007).

Total seed flavonoids are also generally determined by colorimetric methods, utilizing $NaNO_2$ followed by $AlCl_3*6H_2O$ and NaOH, which results in a bright-orange color (Segev *et al.*, 2010). The absorbance is measured immediately at 510 nm using a UV-visible spectrophotometer. The results are calculated and expressed as micrograms of (+)-catechin equivalents. The linear range of the calibration curve is 10 to 1000 µg/ml (r = 0.99) (Xu *et al.*, 2007).

The simplest assay for the quantification of total anthocyanins is direct measurement of the absorption at 490–550 nm. However, this method also measures polymerized degradation products which might lead to an overestimation of anthocyanin content (Dai and Mumper, 2010). Thus, a better method for total monomeric anthocyanins is the use of a pH-differential method (Fuleki and Francis, 1968; Wrolstada *et al.*, 2005). This method is based on the fact that anthocyanins, but no other pigments, absorb light at 530 nm at pH 1.0 but not at pH 4.5 (Wrolstada *et al.*, 2005). Anthocyanin samples are diluted with aqueous pH 1.0 and 4.5 buffers followed by determination of the absorption at 530 nm. Total monomeric anthocyanin content can be calculated from the difference in absorbance between the two buffer solutions using the following equation (Wrolstada *et al.*, 2005):

$$\text{Monomeric anthocyanin pigment } (\text{mg/l}) = (A \times MW \times DF \times 1{,}000)/(\varepsilon \times 1)$$

where $A = (A_{530} - A_{700}$ at pH 1.0$) - (A_{530} - A_{700}$ at pH 4.5$)$, MW = anthocyanin's molecular weight*; DF = dilution factor; ε = molar extinction coefficient*; l = standard path length (1 cm). *If measuring an unknown anthocyanin, a MW of 449.2 and ε of 26,900 for cyanidin-3-glucoside can be adopted (Truong *et al.*, 2012).

16.4.2 Quantification of antioxidant activity

The main benefit of plant polyphenols is their antioxidant activity, which can protect biological systems against the potentially damaging effects derived from reactive oxygen and nitrogen species (ROS and RNS, respectively) processed. Several methods, which differ in reaction mechanisms and conditions and oxidant species, have been developed and tested (for a review, see Karadag *et al.*, 2009). Of these, mainly three assays are utilized to determine antioxidant activity in seeds: ferric reducing ability of plasma (FRAP) (Benzie and Strain, 1996), oxygen radical absorbance capacity (ORAC) (Cao *et al.*, 1993) and 2,2-diphenyl-1-picrylhydrazyl (DPPH) radical scavenging.

16.4.2.1 Ferric reducing ability of plasma (FRAP)

The FRAP assay is a simple, rapid, and inexpensive method for measuring antioxidant activity. It can be performed using automated, semi-automated, or manual methods (Prior *et al.*, 2005). This assay is based on the reduction of ferric (Fe^{3+}) to ferrous (Fe^{2+}) ions at low pH which causes the formation of a colored ferrous-tripyridyltriazine complex. FRAP values are obtained by comparing the change in absorbance at 593 nm in a test reaction mixture with that in mixtures containing ferrous ions at known concentrations. The changes in absorbance are linear over a wide concentration range with antioxidant mixtures, including seeds (Benzie and Strain, 1996). In most cases, FRAP values are expressed as μmol/ml trolox equivalents or as μmol/ml Fe^{2+} calculated from a standard curve. However, the relative activity of trolox is about twice that of Fe^{2+} (Benzie *et al.*, 1996), and this should be taken into account when comparing antioxidant data from different studies. The FRAP assay cannot detect antioxidant activity derived from compounds that act by radical quenching, particularly thiols and proteins (Karadag *et al.*, 2009; Ou *et al.*, 2002), giving this assay an additional advantage in determining the antioxidant activity of seed polyphenols. High correlation values have been found between FRAP antioxidant activity and total polyphenol and flavonoid contents in different seeds from legumes (Xu *et al.*, 2007), chickpea (Segev *et al.*, 2010), peanut (Shem-Tov *et al.*, 2012), quinoa (Brand *et al.*, 2012), and grape (Thaipong *et al.*, 2006).

16.4.2.2 2,2-Diphenyl-1-picrylhydrazyl (DPPH) radical scavenging

The DPPH molecule is a stable free radical which has a deep-violet color, characterized by absorption at 515–520 nm. The DPPH radical-scavenging assay is based on the reduction of DPPH when mixed with an antioxidant such as polyphenol, which leads to loss of its violet color and a reduction in its absorption at 520 nm (Blois, 1958; Molyneux, 2004). The percentage of DPPH free-radical scavenging is calculated using the following equation:

$$\text{DPPH scavenging effect } (\%) = [(A_0 - A_1)/A_0] \times 100$$

where A_0 and A_1 are the absorbance values at the end of the reaction in the DPPH solution with or without the seed extract sample, respectively (Jothy *et al.*, 2011). DPPH scavenging values are expressed as μmol trolox equivalent/ml. To improve the interpretation of the results of this assay, another parameter is added: the efficient concentration value (EC_{50}), which is the amount of substrate tested that causes 50% loss of DPPH absorption (Bondet *et al.*, 1997; Brand-Williams *et al.*, 1995). Each sample is measured at several concentrations and the EC_{50} values (in mg/ml) are calculated using the dose-inhibition curve in the linear range by plotting the extract concentration *versus* the corresponding scavenging effect (Jothy *et al.*, 2011). Both DPPH scavenging and EC_{50} values are negatively correlated to the concentration of the substrate; thus, the higher the antioxidant levels, the lower the DPPH scavenging and EC_{50} values. As found for the FRAP assay, high correlations were obtained between DPPH scavenging and total polyphenols for legume seeds (Xu *et al.*, 2007), nuts, and oil seeds (Sreeramulu and Raghunath, 2011) and seeds of Malaysian tropical fruits (Norshazila *et al.*, 2010), but not for lupine seeds (Wang and Clements, 2008).

16.4.2.3 Oxygen radical absorbance capacity (ORAC)

The ORAC assay is based on inhibition of the reactive species' activity by an antioxidant which results in loss of phycoerythrin fluorescence (Cao *et al.*, 1993). In contrast to the FRAP and DPPH assays, the ORAC method combines both inhibition time and degree of inhibition into a single quantity (Cao and Prior, 1999). This assay measures the kinetics of the decrease in fluorescence for each sample compared to a blank, by plotting fluorescence emission *versus* time (Figure 16.1). The level of antioxidant (in μmol trolox equivalent/ml) in the seeds is calculated from the area between the sample and blank curves utilizing the following equations (Huang *et al.*, 2002):

$$1.\ \text{Antioxidant units } (AU) = 0.5 + \sum_{i=1}^{35} F_i/F_0$$

where F_i and F_0 are the fluorescence emissions at times i and 0, respectively.

$$2.\ \text{Relative ORAC unit} = (AU_{sample} - AU_{blank}/(AU_{Trolox} - AU_{blank}) * \text{trolox molarity.}$$

The ORAC assay has several limitations, including long and labor-intensive sample preparation, particularly for analyses of large numbers of samples. This

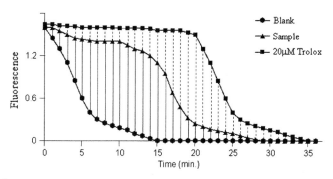

FIGURE 16.1

Representative ORAC plot. ORAC values expressed as the net area under the curve. AU_{sample} is the area between the sample and blank curves (solid lines), and AU_{Trolox} is the area between the trolox and blank curves (solid and dashed lines).

limitation has been overcome, however, by the development of a high-throughput ORAC assay (Huang *et al.*, 2002) using fluorescein as the fluorescent probe (Ou *et al.*, 2001). Zhang *et al.* (2011) analyzed 60 different soybean lines and found significantly higher correlations between ORAC units and FRAP and DPPH assays, as well as total polyphenols, indicating that these methods are consistent in their estimation of antioxidant activity.

References

Abascal, K., Ganora, L., Yarnell, E., 2005. The effect of freeze-drying and its implications for botanical medicine: a review. Phytother. Res. 19, 655–660.

Adams, S.M., Standridge, J.B., 2006. What should we eat? Evidence from observational studies. South Med. J. 99, 744–748.

Al-Farsi, M.A., Lee, C.Y., 2008. Optimization of phenolics and dietary fibre extraction from date seeds. Food Chem. 108, 977–985.

Alonso-Salces, R.M., Korta, E., Barranco, A., Berrueta, A., Gallo, B., Vicente, F., 2001. Pressurized liquidextraction for the determination of polyphenols in apple. J. Chromatogr. A 933, 37–43.

Aparicio-Fernandez, X., Garcia-Gasca, T., Yousef, G.G., Lila, M.A., Gonzalez, d. M.,E., Loarca-Pina, G., 2006. Chemopreventive activity of polyphenolics from black Jamapa bean (*Phaseolus vulgaris* L.) on HeLa and HaCaT cells. J. Agric. Food Chem. 54, 2116–2122.

Aparicio-Fernández, X., Reynoso-Camacho, R., Castaño-Tostado, E., García-Gasca, T., González de Mejía, E., Guzmán-Maldonado, S., et al., 2008. Antiradical capacity and induction of apoptosis on HeLa cells by a *Phaseolus vulgaris* extract. Plant Food Hum. Nutr. 63, 35–40.

Azevedo, L., Gomes, J.C., Stringheta, P.C., Gontijo, A.M.M.C., Padovani, C.R., Ribeiro, L.R., et al., 2003. Black bean (*Phaseolus vulgaris* L.) as protective agent against DNA damage in mice. Food Chem. Toxicol. 41, 1671–1676.

Ballard, T.S., Mallikarjunan, P., Zhou, K., O'Keefe, S., 2010. Microwave-assisted extraction of phenolic antioxidantcompounds from peanut skins. Food Chem. 120, 1185–1192.

Beejmohun, V., Fliniaux, O., Grand, É., Lamblin, F., Bensaddek, L., Christen, P., Kovensky, J., Fliniaux, M.-A., Mesnard, F., 2007. Microwave-assisted extraction of the main phenolic compounds in flaxseed. Phytochem. Anal. 18, 275–282.

Benthin, B., Danz, H., Hamburger, M., 1999. Pressurized liquid extraction of medicinal plants. J. Chromatogr. A 837, 211–219.

Benzie, I.F.F., Strain, J.J., 1996. The ferric reducing ability of plasma (FRAP) as a measure of antioxidant power: the FRAP assay. Anal. Biochem. 239, 70–76.

Blois, M.S., 1958. Antioxidant determinations by the use of a stable free radical. Nature 181, 1199–1200.

Bondet, V., Brand-Williams, W., Berset, C., 1997. Kinetics and mechanisms of antioxidant activity using the DPPH• free radical method. Lebensmittel-Wissenschaft und-Technologie/Food Sci. Tech. 30, 609–615.

Bonoli, M., Marconi, E., Caboni, M.F., 2004. Free and bound phenolic compounds in barley (*Hordeum vulgare* L.) flours. evaluation of the extraction capability of different solvent mixtures and pressurized liquid methods by micellar electrokinetic chromatography and spectrophotometry. J. Chromatogr. A 1057, 1–12.

Brand-Williams, W., Cuvelier, M.E., Berset, C., 1995. Use of a free radical method to evaluate antioxidant activity. Lebensmittel-Wissenschaft und-Technologie/Food Sci. Technol. 28, 25–30.

Brand, Y., Galili, L., Badani, H., Hovav, R., Galili, S., 2012. Total phenolic content and antioxidant activity of red and yellow quinoa (*Chenopodium quinoa* Willd.) seeds as affected by baking and cooking conditions. J. Food Nutr. Sci. 3, 1150–1155.

Cacace, J.E., Mazza, G., 2003. Mass transfer process during extraction of phenolic compounds from milled berries. J. Food Eng. 59, 379–389.

Cao, G.H., Alessio, H.M., Cutler, R.G., 1993. Oxygen-radical absorbance capacity assay for antioxidants. Free. Radic. Biol. Med. 14, 303–311.

Cao, G.H., Prior, R.L., 1999. Measurement of oxygen radical absorbance capacity in biological samples. Oxidants and antioxidants. Meth. Enzymol. 299, 50–62.

Cardador-Martínez, A., Albores, A., Bah, M., Calderón-Salinas, V., Castaño-Tostado, E., Guevara-González, R., et al., 2006. Relationship among antimutagenic, antioxidant and enzymatic activities of methanolic extract from common beans (*Phaseolus vulgaris* L). Plan Food Hum. Nutr 61, 161–168.

Cardador-Martinez, A., Castano-Tostado, E., Loarca-Pina, G., 2002. Antimutagenic activity of natural phenolic compounds presentin the common bean (*Phaseolus vulgaris*) against aflatoxin B1. Food Addit. Contam. 19, 62–69.

Castro-Vargasa, H.I., Rodríguez-Varelab, L.I., Ferreirac, S.R.S., Parada-Alfonso, F., 2010. Extraction of phenolic fraction from guava seeds (*Psidium guajava* L) using supercritical carbon dioxide and co-solvents. J. Supercritical Fluids 51, 319–324.

Chau, C.F., Cheung, P.C., Wong, Y.S., 1997. Effect of cooking on content of amino acids and antinutrients in three Chinese indigenous legume seeds. J. Sci. Food Agri. 75, 447–452.

Chemat, F., Huma, Z.-e., Khan, M.K., 2011. Applications of ultrasound in food technology: Processing, preservation and extraction. Ultrason. Sonochem. 18, 813–835.

Chen, Y., Luo, H., Gao, A., Zhu, M., 2011. Ultrasound-assisted extraction of polysaccharides from litchi (*Litchi chinensis* Sonn.) seed by response surface methodology and their structural characteristics. Innovative Food Sci. Emerg. Technol. 12, 305–309.

Chew, K.K., Khoo, M.Z., Ng, S.Y., Thoo, Y.Y., Wan Aida, W.M., Ho, C.W., 2011. Effect of ethanol concentration, extraction time and extraction temperature on the recovery of phenolic compounds and antioxidant capacity of *Orthosiphon stamineus* extracts. Intern. Food Res. J 18, 1427–1435.

Chiu, K.L., Cheng, Y.C., Chen, J.H., Chang, C., Yang, P.W., 2002. Supercritical fluids extraction of Ginkgo ginkgollaes and flavonoids. J. Supercrit Fluids 24, 77–87.

Dai, J., Mumper, R.J., 2010. Plant phenolics: extraction, analysis and their antioxidant and anticancer properties. Molecules 15, 7313–7352.

Dewanto, V., Wu, X., Liu, R.H., 2002. Processed sweet corn has higher antioxidant activity. J. Agric. Food Chem. 50, 4959–4964.

Folin, O., Ciocalteu, V., 1927. On tyrosine and tryptophane determinations inproteins. J. Biol. Chem. 73, 627–650.

Fuleki, T., Francis, F.J., 1968. Quantitative methods for anthocyanins. J. Food Sci. 33, 78–83.

Geil, P.B., Anderson, J.W., 1994. Nutrition and health implications of dry beans: a review. J. Am. Coll. Nutr. 13, 549–558.

George, S., Brat, P., Alter, P., Amiot, M.J., 2005. Rapid determination of polyphenols and vitamin C in plant-derived products. J. Agri. Food Chem. 53, 1370–1373.

Gertenbach, D.D., 2001. Solid–liquid extraction technologies for manufacturing nutraceuticals from botanicals. In: Functional Foods, Biochemical and Processing Aspects. CRC Press, Boca Raton, Florida, pp. 331–366.

Ghafoor, K., Choi, Y.H., Jeon, J.Y., Jo, I.H., 2009. Optimization of ultrasound-assisted extraction of phenolic compounds, antioxidants and anthocyanins from grape (*Vitis vinifera*) seeds. J. Agric. Food Chem. 57, 4988–4994.

Giergielewicz-Mozajska, H., Dabrowski, L., Namiesnik, J., 2001. Accelerated solvent extraction (ASE) in the analysis of environmental solid samples—some aspects of theory and practice. Crit. Rev. Anal. Chem. 31, 149–165.

Han, H., Baik, B.-K., 2008. Antioxidant activity and phenolic content of lentils (*Lens culinaris*), chickpeas (*Cicer arietinum* L.), peas (*Pisum sativum* L.) and soybeans (*Glycine max*), and their quantitative changes during processing. Int. J. Food Sci. Technol. 43, 1971–1978.

Hassas-Roudsari, M., Chang, P.R., Pegg, R.B., Tyler, R.T., 2009. Antioxidant capacity of bioactives extracted from canola meal by subcritical water, ethanolic and hotwater extraction. Food Chem. 114, 717–726.

Heimler, D., Vignolini, P., Dini, M.G., Romani, A., 2005. Rapid tests to assess the antioxidant activity of *Phaseolus vulgaris* L. dry beans. J. Agric. Food Chem. 53, 3053–3056.

Herrero, M., Mendiola, J.A., Cifuentes, A., Ibáñez, E., 2010. Supercritical fluid extraction: Recent advances and applications. J. Chromatogr. A. 1217, 2495–2511.

Huang, D., Ou, B., Hampsch-Woodill, M., Flanagan, J., Prior, R.L., 2002. High-throughput assay of oxygen radical absorbance capacity (ORAC) using a multichannel liquid handling system coupled with a microplate fluorescence reader in 96-well format. J. Agric. Food Chem. 50, 4437–4444.

Jackman, R.L., Yada, R.Y., Tung, M.A., Speers, R.A., 1987. Anthocyanins as food colorant— a review. J. Food Biochem. 11, 201–247.

Jothy, S.L., Zuraini, Z., Sasidharan, S., 2011. Phytochemicals screening, DPPH free radical scavenging and xanthine oxidase inhibitiory activities of *Cassia fistula* seeds extract. J. Med. Plants Res. 5, 1941–1947.

Juntachote, T., Berghofer, E., Bauer, F., Siebenhandl, S., 2006. The application of response surface methodology to the production of phenolic extracts of lemon grass, galangal, holy basil and rosemary. Int. J. Food Sci. Technol. 41, 121–133.

Karadag, A., Ozcelik, B., Saner, S., 2009. Review of methods to determine antioxidant capacities. Food Anal. Methods 2, 41–60.

Kim, S.J., Murthy, H.N., Hahn, E.J., Lee, H.L., Paek, K.Y., 2007. Parameters affecting the extraction of ginsenosides from the adventitious roots of ginseng (*Panax ginseng* C.A. Meyer). Sep. Purif. Technol. 56, 401–406.

Kossah, R., Nsabimana, C., Zhang, H., Chen, W., 2010. Optimization of extraction of polyphenols from syrian sumac (*Rhus coriaria* L.) and chinese sumac (*Rhus typhina* L) fruits. Res. J. Phtochem. 4, 146–153.

Krygier, K., Sosulski, F., Hogge, L., 1982. Free, esterified, and insoluble bound phenolic acids. 1. Extraction and purification procedure. J. Agric. Food. Chem. 30, 330–334.

Kyari, M.Z., 2008. Extraction and characterization of seed oils. Int. Agrophysics 22, 139–142.

Liazid, A., Schwarz, M., Varela, R.M., Palma, M., Guillén, D.A., Brigui, J., et al., 2010. Evaluation of various extraction techniques for obtaining bioactive extracts from pineseeds. Food and Bioproducts Processing 88, 247–252.

Lin, P.-Y., Lai, H.-M., 2006. Bioactive compounds in legumes and their germinated products. J. Agric. Food Chem. 54, 3807–3814.

Liu, E.-H., Qi, L.-W., Cao, J., Li, P., Li, C.Y., Peng, Y.-B., 2008. Advances of modern chromatographic and electrophoretic methods in separation and analysis of flavonoids Molecules 13, 2521–2544.

Liu, H., Jiao, Z., Liu, J., Zhang, C., Zheng, X., Lai, S., et al., 2013. Optimization of supercritical fluid extraction ofphenolics from date seeds and characterization of its antioxidant activity. Food Anal. Methods 6, 781–788.

Liza, M.S., Abdul Rahman, R., Mandana, B., Jinap, S., Rahmat, A., Zaidul, I.S.M., et al., 2010. Supercritical carbon dioxide extraction of bioactive flavonoid from *strobilanthes crispus* (Pecah Kaca). Food Bioprod. Process 88, 319–326.

Luque-Garciá, J.L., Luque de Castro, M.D., 2003. Ultrasound: A powerful tool for leaching. Trends Anal. Chem. 22, 41–47.

Mejia-Meza, E.I., Yáñez, J.A., Remsberg, C.M., Takemoto, J.K., Davies, N.M., Rasco, B., et al., 2010. Effect of dehydration onraspberries: polyphenol and anthocyanin retention, antioxidant capacity, and antiadipogenic activity. J. Food. Sci. 75, H5–H12.

Molyneux, P., 2004. The use of the stable free radical diphenylpicrylhydrazyl (DPPH) for estimating antioxidant activity. Songklanakarin. J. Sci. Technol. 26, 211–219.

Mukhopadhyay, S., Luthria, D.L., Robbins, R.J., 2006. Optimization of extraction process for phenolic acids from black cohosh (*Cimicifuga racemosa*) by pressurized liquid extraction. J. Sci. Food Agric. 86, 156–162.

Murga, R., Ruiz, R., Beltran, S., Cabezas, J.L., 2000. Extraction of natural complex phenols and tannins from grape seeds by using supercritical mixtures of carbon dioxide and alcohol. J. Agric. Food Chem. 48, 3408–3412.

Mustafa, A., Turner, C., 2011. Pressurized liquid extraction as a green approach in food and herbal plants extraction: A review. Anal. Chim. Acta. 703, 8–18.

Nepote, V., Grosso, N.R., Guzmán, C.A., 2005. Optimization of extraction of phenolic antioxidants from peanut skins. J. Sci. Food Agric 85, 33–38.

Nepote, V., Nelson, R., Grosso, N.R., Guzman, C.A., 2002. Extraction of antioxidant components from peanut skins. Grasas y Aceites 53, 391–395.

Norshazila, S., Syed Zahir, I., Mustapha Suleiman, K., Aisyah, M.R., Kamarul Rahim, K., 2010. Antioxidant levels and activities of selected seeds of malaysian tropical fruits. Mal. J. Nutr. 16, 149–159.

Ou, B., Hampsch-Woodill, M., Flanagan, J., Deemer, E.K., Prior, R.L., Huang, D., 2002. Novel fluorometric assay for hydroxyl radical prevention capacity using fluorescein as the probe. J. Agric. Food Chem. 50, 2772–2777.

Ou, B., Hampsch-Woodill, M., Prior, R.L., 2001. Development and validation of animproved oxygen radical oxygen assay using fluorescein as the fluorescent probe. J. Agric. Food Chem. 49, 4619–4626.

Palma, M., Pineiro, Z., Barroso, C.G., 2002. In-line pressurized-fluid extraction-solidphase extraction for determining phenolic compounds in grapes. J. Chromatogr. A. 968, 1–6.

Palma, M., Taylor, L.T., Varela, R.M., Cutler, S.J., Cutler, H.G., 1999. Fractional extraction of compounds from grape seeds by supercritical fluid extraction and analysis for antimicrobial and agrochemical activities. J. Agric. Food Chem. 47, 5044–5048.

Patrasa, A., Bruntona, N.P., O'Donnellb, C., Tiwari, B.K., 2010. Effect of thermal processing on anthocyanin stability in foods; mechanisms and kinetics of degradation. Trends Food Sci. Technol. 21, 3–11.

Pinelo, M., Rubilar, M., Jerez, M., Sineiro, J., Nunez, M.J., 2005. Effect of solvent, temperature, and solvent-to-solid ratio on the total phenolic content and antiradical activity of extracts from different components of grape pomace. J. Agric. Food Chem. 53, 2111–2117.

Pinelo, M., Tress, A.G., Pedersen, M., Arnous, A., Meyer, A.S., 2007. Effect of cellulases, solvent type and particle size distribution on the extraction of chlorogenic acid and other phenols from spent coffee grounds. Am. J. Food Technol. 2, 641–651.

Prior, R.L., Wu, X., Schaich, K., 2005. Standardized methods for the determination of antioxidant capacity and phenolics in foods and dietary supplements. J. Agric. Food Chem. 53, 4290–4302.

Que, F., Mao, L., Fang, X., Wu, T., 2008. Comparison of hot air-drying and freeze-drying on the physicochemical properties and antioxidant activities of pumpkin (*Cucurbita moschata* Duch.) flours. Intern. J. Food Sci. Technol. 43, 1195–1201.

Rawa-Adkonis, M., Wolska, L., Namiésnik, J., 2003. Modern techniques of extraction of organic analytes from environmental matrices. Crit. Rev. Anal.Chem. 33, 199–248.

Robbins, R.J., 2003. Phenolic acids in foods: an overview of analytical methodology. J. Agric. Food Chem. 51, 2866–2887.

Rostagno, M.A., Palma, M., Barroso, C.G., 2003. Ultrasound-assisted extraction of soy isoflavones. J. Chromatogr. A. 1012, 119–128.

Sardsaengjun, C., Jutiviboonsuk, A., 2010. Effect of temperature and duration time on polyphenols extract of *Areca catechu* Linn. seeds. Thai Pharmaceutical Health Sci. J 5, 14–17.

Sarmento, L.A.V., Machadob, R.A.F., Petrus, J.C.C., Tamanini, T.R., Bolzan, A., 2008. Extraction of polyphenols from cocoa seeds and concentration through polymeric membranes. J. Supercritical Fluids 45, 64–69.

Seabra, I.J., Braga, M.E.M., de Sousa, H.C., 2012. Statistical mixture design investigation of CO_2–Ethanol–H_2O pressurized solvent extractions from tara seed coat. J. Supercritical Fluids 64, 9–18.

Segev, A., Badani, H., Galili, L., Hovav, R., Kapulnik, Y., Shomer, I., et al., 2011. Total phenolic content and antioxidant activity of chickpea (*Cicer arietinum* L.) as affected by soaking and cooking conditions. J. Food Nutr. Sci. 2, 724–730.

Segev, A., Badani, H., Kapulnik, Y., Shomer, I., Oren-Shamir, M., Galili, S., 2010. Determination of polyphenols, flavonoids, and antioxidant capacity in colored chickpea (*Cicer arietinum* L.). J. Food Sci. 75, S115–S119.

Segev, A., Galili, L., Hovav, R., Kapulnik, Y., Shomer, I., Galili, S., 2012. Effects of baking, roasting and frying on total polyphenols and antioxidant activity in colored chickpea seeds. J. Food Nutr. Sci. 3, 369–376.

Shem-Tov, Y., Galili, S., Badani, H., Segev, A., Hedvat, I., Hovav, R., 2012. Determination of total polyphenol, flavonoid and anthocyanin contents and antioxidant capacities of skins from peanut (*Arachis hypogaea*) lines with different skin colors. J. Food. Biochem. 36, 301–308.

Shi, J., Yu, J., Pohorly, J., Young, J.C., Bryan, M., Wu, Y., 2003. Optimization of the extraction of polyphenols from grape seed meal by aqueous ethanol solution. Food Agri. Environ. 1, 42–47.

Singleton, V.l., Orthofer, R., Lamuela-Raventos, R.M., 1999. Analysis of total phenols and other oxidation substrates and antioxidants by means of Folin–Ciocalteu reagent. Methods Enzymol. 299, 152–178.

Smith, B.L., Carpentier, M.H., 1983. The Microwave Engineering Handbook. Chapman and Hall, London.

Spigno, G., Tramelli, L., Faveri, D.M.D., 2007. Effects of extraction time, temperature and solvent onconcentration and antioxidant activity of grape marc phenolics. J. Food Eng. 81, 200–208.

Sreeramulu, D., Raghunath, M., 2011. Antioxidant and phenolic content of nuts, oil seeds, milk and milk products commonly consumed in India. Food Nutr. Sci. 2, 422–427.

Stalikas, C.D., 2007. Extraction, separation, and detection methods for phenolic acids and flavonoids. J. Sep. Sci. 30, 3268–3295.

Tatke, P., Jaiswal, Y., 2011. An overview of microwave assisted extraction and its applicatuions in herbal drug research. Res. J. Med. Plant 5, 21–31.

Thaipong, K., Boonprakob, U., Crosby, K., Cisneros-Zevallos, L., Byrne, D.H., 2006. Comparison of ABTS, DPPH, FRAP, and ORAC assays for estimating antioxidant activity from guava fruit extracts. J. Food Composition Anal. 19, 669–675.

Truong, V.D., Hu, Z., Thompson, R.L., Yencho, G.C., Pecota, K.V., 2012. Pressurized liquid extraction and quantification of anthocyanins in purple-fleshed sweet potato genotypes. J. Food Comp. Anal. 26, 69–103.

Tseng, A., Zhao, Y., 2012. Effect of different drying methods and storage time on the retention of bioactive compounds and antibacterial activity of wine grape pomace (Pinot Noir and Merlot). J. Food Sci. 77, H192–H201.

Tsuda, T., Mizuno, K., Ohshima, K., Kawakishi, S., Osawa, S., 1995. SC-CO2 extraction of antioxidative components from tamarind (*Tamarinus indica* L.) seed coat. J. Agric. Food. Chem. 43, 2803–2806.

Vinatoru, M., 2001. An overview of the ultrasonically assisted extraction of bioactive principles from herbs. Ultrason. Sonochem. 8, 303–313.

Vongsangnak, W., Gua, J., Chauvatcharin, S., Zhong, J.J., 2004. Towards efficient extraction of notoginseng saponins from cultured cells of *Panax notoginseng*. Biochem. Eng. J. 18, 115–120.

Wang, L., Weller, C.L., 2006. Recent advances in extraction of nutraceuticals from plants. Trends Food Sci. Technol. 17, 300–312.

Wang, S., Clements, J., 2008. Antioxidant activities of lupin seeds. In: Palta, J.A., Berger, J.B. (Eds.), Lupins for Health and Wealth' Proceedings of the 12th International Lupin Conference.. International Lupin Association, Canterbury, New Zealand, Fremantle, Western Australia, pp. 546–549.

Wrolstada, R.E., Dursta, R.W., Lee, J., 2005. Tracking color and pigment changes in anthocyanin products. Trends Food Sci. Technol. 16, 423–428.

Xu, B., Chang, S.K.C., 2008. Effect of soaking, boiling, and steaming on total phenolic content and antioxidant activities of cool season food legumes. Food Chem. 110, 1–13.

Xu, B.J., Chang, S.K.C., 2007. A comparative study on phenolic profiles and antioxidant activities of legumes as affected by extraction solvents. J. Food Sci. 72, S159–S166.

Xu, B.J., Yuan, S.H., Chang, S.K.C., 2007. Comparative analyses of phenolic composition, antioxidant capacity, and color of cool season legumes and other selected food legumes. J. Food Sci. 72, S167–S177.

Yang, Z., Zhai, W., 2010. Optimization of microwave-assisted extraction of anthocyanins from purple corn (*Zea mays* L.) cob and identification with HPLC–MS. Innovative Food Sci. Emerg. Technol. 11, 470–476.

Yasmin, A., Zeb, A., Khalil, A.W., Mohi-ud-Din Paracha, G., Khattak, A.B., 2008. Effect of processing on anti-nutritional factors of red kidney bean (*Phaseolusvulgaris*) grains. Food Bioprocess Technol. 1, 415–419.

Yilmaz, E.E., Özvural, E.B., Vural, H., 2011. Extraction and identification of proanthocyanidins from grape seed (*Vitis Vinifera*) using supercritical carbon dioxide. J. Supercritical Fluids 55, 924–928.

Yvonne, C.C., Walker, L.T., Verghese, M., Bokanga, M., Ogutu, S., Alphonse, K., 2007. Comparison of extraction methods for the quantification of selected phytochemicals in peanuts (*Arachis hypogaea*). J. Agric. Food. Chem. 55, 285–290.

Zhang, H.-F., Yang, X.-H., Zhao, L.-D., Wang, Y., 2009. Ultrasonic-assisted extraction of epimedin C from fresh leaves of *Epimedium* and extraction mechanism. Innovative Food Sci. Emerg. Technol. 10, 54–60.

Zhang, R.F., Zhang, F.X., Zhang, M.W., Wei, Z.C., Yang, C.Y., Zhang, Y., et al., 2011. Phenolic composition and antioxidant activity in seed coats of 60 chinese black soybean (*Glycine max* L. Merr.) varieties. J. Agri. Food Chem. 59, 5935–5944.

Zougagh, M., Valcárcel, M., Ríos, A., 2004. Supercritical fluid extraction: a critical review of its analytical usefulness. Trends Anal. Chem. 23, 399–405.

Index

Note: Page numbers with "f" denote figures; "t" tables.

Printed and bound by CPI Group (UK) Ltd, Croydon, CR0 4YY

08/05/2025

01864855-0002